TIME MATTERS

SCI-FI SHORT TO MEDIUM FICTION

BY RS ROSE

14 STORIES 150,000 WORDS 600 PAGES

Time Matters

RS Rose

Published by Texas Rose Publishing, 2024.

TIME MATTERS

First edition. July 1, 2024.

ISBN: 978-1951884093

Written by RS Rose.

For Tracy

Foreword

Most of these stories were intended to be novels, or at least novellas, but life happens. Instead of leaving them in the digital attic, I decided to dust them off and gather them into this compilation of short stories. All are science fiction, most include space travel, time travel, artificial intelligence, and sometimes all three.

Table Of Contents

TIME 4 LOVE?

SCI-FI * TIME TRAVEL * AI * ROMANCE?

PG-13 15080 WORDS 59 PAGES

Time 4 Love? — Chapter 1

"Why me?" Edward asked.

"You volunteered to serve, Sgt. Pirro."

"For the military. Not to be a guinea pig in an impossible science experiment."

"Your preliminary tests have been successful. Are you refusing to go?"

"What are my chances?"

"Very good," the woman said enigmatically.

After the mission overview, Dr. Doreen Inverness had given Edward the heroic sales pitch that he would naturally be drawn to. It wouldn't be his first apparent suicide mission, and he'd emerged unharmed from all of those. Physically unharmed, anyway. He knew when a scientist used comparative terms instead of hard data, he was in trouble. He decided he didn't want to know the percentages.

"Is that what all these tests you've been running the last few weeks determined, I'm the best candidate?"

"You were third best, Sgt. Pirro. The other two chose the Arctic posting instead." Edward had done three months of cold weather warfare training in Northern Alaska. He vowed never to go north of Missouri for the rest of his life. He decided she probably knew that and was simply using it as manipulation.

"Fine. Better to go out in a blaze of glory than to freeze my nads off. Can I at least say goodbye to my family?"

"You have no family Sgt. Pirro." That's why he'd been 'volunteered' for this adventure. No ties. No one to ask where he disappeared to.

"I adopted a spider in my quarters," he said with a sardonic smile. The word triggered an unusual wave of sorrow in him. He had been adopted twice. He barely remembered the first couple. They died of carbon monoxide poisoning. He'd been spared because he insisted on sleeping with his window slightly open, even in winter. The second couple died a few months apart while he was in his final year of high school. They had been much older, more like grandparents. The original adoption records had been lost, so he had no idea if his birth parents were still alive, at least before the Collapse.

"You will retain your current quarters for the three weeks of training. You can say goodbye to your 'adopted' spider then. You will have no contact with the outside world regarding this project until your successful return."

"At least I have you," he said suggestively. She gave him a look to show that could not be less true. "When do I start?" Edward asked with a sigh and a nod.

"Tomorrow morning. Thank you for your exceptional courage." She left the room and he stared at himself in the two-way mirror. A Marine guard came and escorted him back to his quarters. Edward considered escaping, going AWOL, and living off-grid for the rest of his life. It was not a world he could survive in. He was a soldier, not a productive member of society. He spent the rest of his evening talking to his adopted spider as it toiled near the ceiling building its web.

Time 4 Love? — Chapter 2

There were three other 'volunteers' in the classroom with him. He was at least a decade older than the others, only one of which could possibly be female. The jury was still out on zer. The teacher began with a history book. After a half-hour of dry facts recited in a hypnotic monotone, Edward decided he needed a break.

"Professor?" Edward asked, his hand raised even with his shoulder.

It seemed like the history teacher was just as happy to have the interruption. "Yes, Sgt. Pirro?"

"While this is all really fascinating, can those of us who lived in the fifties skip all this stuff?"

"You were between the ages of two and twelve during the fifties. I doubt you experienced the fine nuances of geopolitical circumstances at that age."

"Maybe I can take the test and show you what I already know."

"Who signed the Jabrar treaty?" The blank look told the professor everything he needed to know. "The countries that were party to it? Name one?"

"What does that have to do with time travel?" The professor was annoyed with the question. Even though all in the room were inside the program, all were told not to discuss the specifics. The professor wondered if Pirro was trying to get tossed out. "Isn't filling us with preconceptions of what we will see counterproductive to what our mission is?" Edward saw in the older man's eyes that the mission was not the one Dr. Inverness had described. His mind raced. "Are we being sent back to do more than observe and report? Are we going to try and change things?"

"That's enough speculation for today, Sgt. Pirro. We resume with the leadership change in the United Islamic Caliphate and the fallout from those chaotic times." Did they want to prevent the economic collapse? All the experts said it was inevitable with AI automation. If no one has a job, there is no economy. They just entertained themselves and learned to wait for the food to be delivered. Everything was so ideal now. Why change it and possibly make it bad again?

The afternoon was for training on what the technicians called "the device". It was only a mock-up, not the actual time machine. It had a console with two seats. This was the first indication he would not be traveling alone. Nothing about the displays looked like current technology. He assumed they had a huge budget and had access to the latest cutting-edge tech. This looked like it came out of his childhood.

Edward had been paired with the nerdy guy. They stood behind and watched as the other team of possibly a woman and the blond boy went through the first simulation. Then it was his turn and he did his best to impress everyone watching. It was mostly memorizing where switches were and what they did. He had a very good memory.

The second week included survival training, first in the classroom, then in the woods. They were told the arrival point of the device would purposely be in a rural location. Edward was happy with this part of the training. First, because he loved the outdoors, and second because this training seemed to indicate those in charge believed the trip might succeed. He easily navigated the fifteen miles back to the arrival location without the aid of any electronic devices. The nerd had sprained an ankle on the first mile, so Edward was working alone for most of it.

Time 4 Love? — Chapter 3

"Do you want to get a beer?" Edward asked possibly a woman quietly. She looked up from her textbook and just stared blankly at him. "I'm not suggesting we break protocol and sneak off base or anything, unless you're interested in that, in which case I am too."

"You should be studying. You think the information isn't important, but it is."

"I know it stone cold, backward and forwards."

"How many milli-rads of radiation constitutes a lethal dose."

"I'm indestructible," he said with a sly smile. He curled his arm and flexed his bicep.

"And your partner is expendable." She looked down at her notebook and began writing.

He studied her face. "Lethal dosage? You're worried about dying on a suicide mission?" He saw her reaction and knew she was avoiding thoughts of that possibility. He sat down across the table from her. "Are you alone in the world too?" He asked with genuine compassion. She ignored him. "My parents were never entered in the genetic database, so I assume they died before the Reformation. All four of my adoptive parents died too. It isn't easy growing up without attachments to anything or anyone."

"Sucks to be you," she said hurtfully, not interested in revealing her private emptiness.

"I'm looking forward to soft-serve ice cream. That's the thing I miss most from before. You were what, two when it happened?"

"Three."

"I almost wish I hadn't known how good it could be before."

"It was a terrible time. Nothing but war and strife and disease."

"It had some good moments, especially for a kid oblivious to all the bad stuff. The books are full of facts, but there was more to the world than just the bad times in the famous places."

"Goodnight, Sgt. Pirro."

"Call me Eddie." She just glanced at him with disdain and he knew that she would never be anything more than cold and detached with him. "Goodnight, Lt. Filbert." She didn't look up as he rose from the table. He was still not certain she was female.

Back in his room he looked up milli-rads of radiation and cared even less about the answer when he knew it. It's not like they could turn around if the radiation got too high. The dimensional shift sounded far more dangerous than the radiation.

Time 4 Love? — Chapter 4

"Are you ready to go?" Dr. Inverness asked. She had knocked on his open door and entered with his verbal permission.

"Am I shipped off to the Arctic Circle if I say no, Dr. Inverness?" Edward asked, putting down the book he had been reading.

"We can send Danbury and Filbert first if you would prefer."

She knew him so well. He would never allow possibly a woman to go first into harm's way. "I'm ready. Any chance you can tell me what the real mission is before I get flattened and pushed through the interdimensional mail slot?"

"Mail slot. Funny," she said, her face showing no amusement. "You have concerns about your partner?"

"Concerns? No, I'm fairly certain he won't survive the trip. You're better off sending me alone."

"You don't know enough to complete the mission on your own."

"I'll have to make do when the pencil neck bites it. He sprained his ankle on moss for Dorach's sake."

"You want Miss Filbert to go with you?"

"No, I really don't," he said sincerely.

"Because you would have more difficulty with her demise?" He didn't say anything. She knew him too well. "You go first thing in the morning. I can't tell you your mission, you will be told at the other end."

"If I survive. What if I don't like 'the mission' when I get there?"

"Failure to perform the mission will look the same to us, regardless of the reason for it. We will send Filbert and Danbury if you fail."

"You'll know immediately, right? Even if it takes me years, if it gets done, everything here will change instantly."

"That is the expectation."

"Will my memory change?"

"What do you mean?"

"If I make a significant change, perhaps my upbringing will be altered, so all my childhood memories will be different. I might even forget why I'm there as soon as I make the change."

"That would be an interesting twist. Hopefully, that will be after any important changes you need to make are completed. Do you intend to look for your parents? Talk them out of putting you up for adoption?"

The thought had crossed his mind. The target date was before his conception. Any interference in their lives could have him fading from a McFly family photograph. "The only thing off-mission I have planned is soft-serve ice cream."

She nodded. "I think I would go for sushi and cinnamon rolls," she admitted. "You have been warned about interfering with your own life. Anything you do could alter the simple fact that you go back and make the change, undoing what it is that you are trying to accomplish."

"A happy childhood would disqualify me from this project. I would never go back and therefore my happy childhood could not be made."

"Or worse, you are stuck in a permanent time loop of living a happy childhood, having it all ripped away, and then reliving the one you already had or worse."

"That would be hell. The thing I'm changing, does it involve murder?"

"That is not the term we would use." She thought about her next words carefully. "We have identified key people that pushed us past the breaking point. If they cannot be persuaded to alter their trajectory, stronger measures may be necessary."

"Persuaded? I am not much for persuasion."

"You have a certain charm, Sgt. Pirro."

"All evidence to the contrary."

"A condemned man deserves a final meal?" she asked, an eyebrow raised.

"A midnight snack would do." She shook her head slowly and then walked to the door. She pushed it closed, shut off the light, and began to undress.

Time 4 Love? — Chapter 5

"All safeties removed. There are five amber warning lights," Edward said, vibrating in his chair.

"None of them are in critical systems. We are still go for transition," the voice from mission control said.

"Easy for you to say," Edward said under his breath. "Your ass is safe in three-dimensional space." He looked over at the nerd who had his eyes closed. His face was a mask of terror. Edward made the decision. He reached over and hit the red button. "Transition initiated."

"Good luck," was the last thing he heard as his world became an indescribable sensory compression. When his senses returned, he felt a spinning sensation. It was a mixture of vertigo and dizziness that made him vomit onto the front of his flight suit. He shut his eyes and fought the spin. It took many minutes before it slowed. There was no sound he realized. He reached up and felt the wetness in his ears. He was bleeding from them. He wondered if the deafness would be permanent, but now understood the reason for the spinning.

He opened his eyes and tried to focus on a fixed spot, the spinning got worse and he wretched again. He began strobing his eyelids open just to get brief pictures of his surroundings.

Most systems looked online. The nerd was slumped forward, blood running down his cheeks from his ears. He reached over and smacked his arm but got no response. He reached forward and began engaging all the safeties. Once the final one was engaged, a video began playing. He recognized the people talking, he had met them during training. He could not hear a word they were saying. Then a series of pictures flashed on the screen. Their names were in captions so he committed them to memory. He only recognized one of them. It was a rich businessman that some thought intentionally caused the crash as revenge for political exile.

Edward wondered if these were his targets. Was he supposed to kill these people? The final picture was that of a beautiful woman. Olivia. He knew instantly he could not kill her, whoever she was. The familiar faces came back on and began talking again. He would have to wait until his hearing returned and replay the video.

After an hour he realized the spinning was finally gone. He undid his harness and tried to stand. His balance was severely affected, but there was enough to hold on to that he did not have to stand upright on his own. He felt for a pulse and established the nerd had expired.

He went to the back of the device and looked out of the only window, a small round portal in the exit hatch. It was a mass of green leaves in the bright sun. That was better than a busy city street or the deep black of outer space. He went back to the console and tried to find the situational controls. There was no planetary location satellite signal found. That could have been because of a broken antenna or he overshot the time destination by more than a hundred years. The atmosphere detector assessed the pollution to be pre-Collapse levels. Edward stared at that. It actually worked?

He cleaned himself up and then removed the vomit-stained flight suit. His street clothes would be acceptable to the target decade, and he had a large amount of the monetary units that were worthless in the post-Collapse world. He smiled as he pocketed a significant amount of paper currency, thinking about the distance to the nearest Dairy Queen, and hoping it was short.

He pushed through the hatch door and found the leaves were cornstalks. He climbed on top of the device and saw he was in the middle of endless fields of tall green corn. He looked up and saw the condensation trails of a large jet aircraft. He spread out the camouflage netting, which would do little to hide the device when the harvester came rolling through the fields. He knew little about the maturity of corn plants, but it seemed only a few weeks before these would be the size of the ones he ate when he was a child. It was a given it would be discovered before his mission was completed, and he had serious doubts about attempting a return to his own time, even if he completed it.

He ate some food to quell his uneasiness. He found he was severely dehydrated as well. After four liters of water, he began to feel normal again. He pulled the nerd out of the device and began digging a shallow trench. He shook out the chemical pouch on the body and watched it begin to digest the clothing and its contents. The smell of the rapid decomposition was horrific. He covered the black tar that remained with the loose soil he had pulled out of the hole.

His hearing hadn't returned by nightfall, but the bleeding had stopped. He slept comfortably in the complete darkness of the new moon.

By mid-morning the next day, Edward had packed up what he needed and chose a direction to head in. He locked up the device and began walking down a long row of cornstalks. It was half an hour before he reached a country road. He chose the next direction and reached a small town just as the sky began turning orange. He was in north-central Nebraska, which was only a thousand miles off target. A newspaper being sold at the convenience store revealed he was two months before the target date. Unfortunately, the town had no soft-serve ice cream establishments, so he settled for a premade ice cream cone from a convenience store freezer. He learned to point to his ear and shake his head to let the friendly people know he couldn't hear their inquiries. It worked well to avoid uncomfortable questions he could not answer. He got a room at a motel and enjoyed a thirty-minute shower before collapsing on the bed.

In the morning, he studied the map and bus schedules and then made a plan. He would wait here for a week to see if his hearing returned. If it did, he would head back to the device and attempt to learn the details of the mission. Otherwise, he would head for New York and do his best to figure it out based on those named in the target photographs. If he could figure out who they were, maybe he could figure out what to do with them. It was more than three years before the beginning of the Collapse. He figured he had until then to save the world from itself.

The ringing in his ears started on day five, but it was the only sound he could hear. On day ten. He saw the fields near the town were being harvested and doubted the device would go undiscovered much longer. He bought his bus ticket and headed east. His first stop was in New Jersey at a horse track. He bet the first four races with the same trifecta numbers. The third race results, written on a small piece of paper for every first Tuesday of the month for the next two years, got him thirty thousand dollars on a three-dollar ticket. He had long forgotten the milli-rads required to kill a man, but gambling would keep him in fat city for many years to come. He had three valid identities, all of which he rotated through as he moved about the city searching for his targets. Long days at the library searching with antique computers.

Olivia Kohler was the daughter and only child of financier Oliver Kohler. She was in the process of taking over his empire as he began enjoying more of his life with his third wife, a woman far younger than Olivia. Olivia stayed mostly in Manhattan and the father jetted around the world.

Edward began taking detailed notes on each of the targets, trying to decide each person's role in the coming financial disaster. It was clear to him the direct causative agents were in California. They dreamed up the robots and built them after all. The deeper cause was the lack of foresight into how quickly the Collapse would occur. These were the government fools who knew nothing beyond what their lobbyists paid them to know. Only one of the specified targets was in government. Edward added many others to the target list, but only for gathering information. Eliminating them would do nothing to fix the problem because there was an endless supply of idiot bureaucrats lined up to take their place.

After months of research and thought, it became clear the targets he had been assigned were indeed the ones that could make a difference. It was also clear they would never listen to a grunt like him. The nerd would have been the one to talk their language. He made a few hundred thousand at the horse track, laundered it through a few Native casinos, opened an investment account, and began learning the business of Wall Street. He didn't know day-to-day trading numbers, but he remembered what companies were big in his childhood and invested in the ones that were still relatively small.

Within a year he was a multi-millionaire and began rubbing elbows with the targets. His inability to hear kept him out of the conversations, but not the parties. That is where he first saw Olivia. She was stunning. Women never made him nervous. He was completely dumb struck when he was finally introduced to her. All he had was his charming smile and an apologetic shrug as their mutual friend explained his deafness.

"Have you been to Dr. Kirsch?" She asked both in words and sign language. Edward knew little of the sign language, preferring to get by with writing on a notepad. He pulled out the pad, put a small question mark on the pad, and handed it to her. She wrote her question and showed him, and he shook his head. She continued writing. "He did an amazing job with my mother's hearing. She had been deaf most of my life, but the technology is finally where it can do miraculous things."

Edward had never even bothered researching a technological fix. It had become camouflage to hide behind. She wrote down the phone number and then smiled, realizing he could not make such a call himself. She tapped her ear and made the call for him. She wrote "10 AM tomorrow?" on the pad, and Edward nodded dumbly. She smiled and confirmed it before ending the call. She wrote the address and her private email address. "Let me know how it goes," she wrote before handing him back the pad, then moved on to the next social engagement of the night. He just watched her from afar until she left the party.

Time 4 Love? — Chapter 7

The doctor diagnosed him quickly and admonished him for not seeing an otologist sooner. The surgery to repair was scheduled for the following day. Edward was far more excited about reporting his progress to Olivia. That faded as his email went without a reply for the rest of the day. He was just drifting off to sleep in a foul mood when his phone lit up.

"So sorry Edward for not responding sooner. I had forgotten I would be incommunicado most of the day. Big merger I can't talk about, and they insisted on no phones in the room. I'm sure everything will go well with your surgery tomorrow. Bernie is the best. Let me know how it goes. O."

He was desperate to respond, to try to keep the conversation going all night. Instead, he put the phone back on the nightstand and went to sleep in a better mood. The surgery took less than an hour, but it took several hours for Edward to feel right after coming out of the anesthesia. Even though he knew exactly what to expect, he was still disappointed his hearing had not returned. He finally sent an email to Olivia, wishing he had a miracle to report to her. She responded immediately and asked if he needed anything. He had a long list of things he wanted from her but said only that he looked forward to having a conversation with her at the next party they attended together.

The written conversation continued sporadically through the afternoon until he was discharged and went home to sleep. He woke early and looked at his investments. It had been a good day, and he cashed in some of his more rewarding stocks for the next wave of stocks he had identified as good bets. The text came through at five in the morning.

"Feeling better?"

"Much better. Cashed in my CRX for HJSA after a runup of 570% in five months."

"I meant your ears."

"Oh, those are good too. Post-surgical lag is gone. Are you always up this early?"

"When I have to travel."

"I'm not allowed to fly for three weeks."

"Do you travel much?"

"Usually by train, but I don't go far."

"Afraid of flying?"

"I'm only afraid of weak beer and bad karaoke."

"LOL. Just looked up HJSA. Something you recommend?"

"Highly. They have a new game that is going to be a hit with kids."

"Inside info?"

"No. I played an early beta and recommended some changes. They announced the release of the next version at a player's conference yesterday."

"You recommended changes?"

"I have a military background and straightened out some of their tactical flaws. I submitted it anonymously on their beta hub. They thanked me for the feedback in a way that makes me think the next version will be killer."

"Military background?"

"Army... Ranger, ten years. I'm on reserve duty now, so they can pull me back in at a moment's notice. I doubt they will unless my ears start working as well as they used to."

"Thank you for your service. My dad was a pilot. Is that how you lost your hearing?"

"Sort of. I'm not allowed to discuss it. I hope you understand."

"I do. Just bought 500 of HJSA. Hope it pays off."

"Wow, big spender," Edward typed sarcastically.

"500K," she corrected.

"You don't play around."

"Very rarely, and only with fearless men."

"Guess I shouldn't have mentioned my mal-karaoke-phobia."

"LOL. Do you have plans for Saturday night?"

"Jado and Crockby reruns and Chinese delivery."

"What I would suggest could not trump that. It is just a boring fundraiser, some sort of silent auction."

"I can do very well with the silent part."

"Great. I'll have Julie send you the info. Don't feel obligated, but I would like to see you there. My jet is landing so I have to prep for my meeting."

"Knock 'em dead, Olivia."

Time 4 Love? — Chapter 8

Edward decided to meet her at the fundraiser, turning down the offered limo ride. He wasn't ashamed of his meager apartment, and she certainly could find out where he was living, but he did not want that to color her opinion too early.

"You look beautiful tonight," Edward signed in the newly memorized hand gestures.

Olivia smiled, began signing a reply, then realized he still didn't understand the silent language. She took her phone out of her purse and began texting. "Thank you, Edward. You look very handsome in that tux. Borrowed?"

"Rented. Have to return it by midnight before it turns into a pumpkin."

She laughed. She looked at him and put the phone back in her small clutch purse. She moved to his side, slid her arm around his, and they moved as a couple through the inane process of talking superficially with people they pretended to like. They walked among the items to be bid on, submitting their offers. She pulled Edward out onto the dance floor and was pleasantly surprised by the lightness of his feet.

"Good night, Olivia," Edward whispered as they moved outside to depart. He hoped his voice did not sound horrible, relying completely on the muscle memory of forming words. She smiled and kissed his cheek. He could tell she wanted him to accompany her home, but he had decided that would be the wrong move. He wanted to get inside her mind, not her bed. She was still a target, potentially one that needed elimination. She disappeared into her limo and he walked uptown and found a cab.

Time 4 Love? — Chapter 9

They texted daily with a growing intimacy. He preferred this because he had time to translate his life stories from a very different future into something more contemporarily appropriate. The day of his otology appointment approached, and he went with trepidation. He had grown comfortable with the silence and wondered what this new world would sound like. It was tinny, electronic, and took hours to calibrate and adjust. His own voice grated on him as he began forming therapeutic sentences.

"Edward?" Olivia said, answering the phone.

"Yours is the first voice that does not irritate my new ears. Thank you for this gift." She was silent, and he wondered if his voice had pushed her away. He heard a sniffle and realized she had become too emotional to talk. "I was wondering if you had plans for Saturday night?" he asked nervously.

"Reruns and takeout," she replied, hoping it was clear what she wanted.

"Much better than what I had planned," he said. "I'm told this hearing will not work well in public spaces, but I..."

"I'll text you my address. Bring your favorite takeout food and we'll share it over some really expensive wine."

"That sounds perfect. See you Saturday night, Olivia."

He hung up the phone and went to his bedroom to lie down, exhausted from the long day, yet excited by the brand-new sound of her voice. All manner of noises kept waking him up in the night. Many sounds took time to identify. Arguing neighbors, honking traffic, barking dogs, amorous lovers. He wished he had an off switch and chose to put a pillow over his head to dull it.

In the morning one of his targets appeared on a news report. It was a merger announcement with one of the other targets. He was the CFO of the acquiring company and Edward struggled to listen to the words he was saying as he read the closed caption. He decided it was meaningless pablum, so he dove into researching this new corporate takeover. By the afternoon he realized this merger was likely one of the incendiary events that he should have prevented. He was failing in his mission and losing focus by becoming smitten with Olivia Kohler.

Then it came to him that he could now hear the mission briefing video. Where was the device? A farmer's barn? A junkyard? Area Fifty-One? It had never appeared in any local news reports. He could rent a car and be there in less than a day. He could inquire about it locally and see if there was any chance of recovering that video. It could be the only way to succeed. It was more likely to be a one-way trip to government scrutiny he could not afford. He decided to make the trip, but not to make any direct inquiries. He would leave after his weekend with Olivia. She might be a distraction, a mistake, but he could not walk away from her. Ever since her image appeared on that console, he was drawn to her like no other woman in his life.

"WHAT IS THE REAL STORY behind Edward Pirro?" Olivia asked. They were sitting at opposite ends of a luxury couch facing each other, each holding a glass of wine.

"What do you mean?" He had already shared his biography, altered to fit a linear timeline.

"I don't mess around with unknowns. Most men like you who feign interest in me are generally after something. It doesn't take much to get the information one needs to judge one's character. Until a year ago, you don't exist. There are other Edward Pirro's, some deceased, some just starting their young lives. None with a military record. You are living a little too high profile for witness protection."

"Does it matter?"

"Unfortunately, it does. If it were just me, I would have thrown caution to the wind and dragged you up to my bedroom. I am responsible for a major corporation. Thousands of families rely on my integrity. I cannot chance compromising myself, no matter how tempting it may be."

"You would not believe the truth, though I may tell it to you someday when it is less... important. I won't bother lying to you either, mostly because I really like you and respect you. I was in the military, and I still am in most respects. I am on a mission that requires a certain amount of secrecy. I am here because I trust you more than I have ever trusted anyone in my life. I told you my parents died when I was young, but I don't know for sure if that is true. I have no idea who they were. I have never felt like I belonged, no matter where I have gone in my life. I have found solace in the service of my fellow man, but I have never found comfort. Whenever it seemed to be near, circumstances always pushed me in a different direction, not always against my will. I'm not here to seduce you away from your fortune and responsibilities if that is what you are concerned with. I am here simply because you helped me even though there was nothing for you to gain from it."

"HJSA has already doubled, so I have gained quite a bit."

"That was after you helped me with my hearing, and I certainly didn't expect you to follow me into that risky investment."

"How long should I stay in?"

"They should remain on a solid growth for a few years. I intend to get out by the end of the year as the curve flattens."

"Are you an agent of a foreign government?"

"Nothing as sinister as that. Protecting the US financial system is my primary goal, though I am more a fact-finder than a truly active participant."

"'Fact-finder'? Are you investigating me?"

"I have been, along with many others. Is it not wise to understand what makes the most successful people achieve that success?"

"Covert fact finder," she said, repeating the key words. "I guess that could describe most investors."

"I understand your reticence to get involved with me. As much as I would like that myself, I now see the problems that could arise. Thank you, Olivia, for a lovely evening. I'll see myself out." Edward stood and finished the last of his wine, leaving the glass on the end table. She watched him leave the room. He walked out into the cool night air and called for an autonomous cab. He checked his portfolio while he waited on the corner.

A text arrived before the taxi did. "I think you should come back here right now." Edward had never played hard to get in his entire life. He wasn't sure why he was doing so with someone he wanted more than any in recent memory. He thought it might be because he wanted more from her than he had wanted from any other girl. They had all been girls, he realized, even the older ones. This was the first woman he had spent any intimate time with. He was also distracted by the fate of his device. He needed clarity for his mission, especially where it concerned her.

"Already in a taxi on the way to my apartment," he typed, then looked up and saw her silhouette in the second-floor window looking out at him. He smiled, waved, and sent the message.

"That's too bad. Maybe next time."

"Maybe."

Time 4 Love? — Chapter 10

The device clearly was no longer in the middle of the field. It also had not been dragged to the edge. What remained of the nerd was being absorbed in the current crop of corn growing above him. Edward hoped the decomposition chemical was not poisonous at this point. He researched and wandered a few of the local junk yards, finding no trace of it. He was driving the area in his tiny rental car when he saw it. It was just the vague shape under a tarp next to a barn. He resisted pulling into the driveway.

He found a place to pull over a few miles away. It was a small abandoned house on a small oasis of land among the corn fields. There was a for sale sign in the yard that looked equally neglected. He could pretend to be an interested buyer for a few hours at most. There were no other buildings in sight, but it was only a matter of time before someone local drove by and noted his presence. As if on cue, a pickup truck slowed and pulled off the road across from the house.

"Good afternoon," the old man said as he approached the car. Edward had a map book in his lap, so he could feign being lost. He thought about reverting to being deaf but decided against it.

"Afternoon. Sure is a hot one," Edward said.

"Normal for this time of year. We get both extremes in the shadow of the Rockies. Are you lost?"

"Not exactly. I'm researching my family history. One branch used to have a farm out this way."

"Probably absorbed into the corporate farm with the rest of them. I'm the only one left that hasn't sold out. Can't complain much since they pay me to work some of their land, which helps pay for my John Deere. What was the family name?"

"Pirro," Edward answered without thinking.

"Italian? None of those around here during my lifetime. Mostly northern European. German, Dutch, and a few Scots like me. This house here was originally part of the Shumaker farm. The widow passed on about five years back and no one is interested in this house so far from town and no land to work. Was this family branch far back?"

"Eighteen-hundreds."

"Library has all the records, but you're unlikely to find any houses that go that far back. Mine is one of the oldest and it was built post-Depression. The corporation levels all the buildings and either converts the land to growing or puts in modern steel storage buildings. Would you like to come by for dinner? I could give you a history of the area the way it was back in the eighteen-hundreds. My grandfather first broke the ground I work now and his stories are just about gone. Thought about writing them down, but my kids couldn't care less about that. They headed for the West Coast as soon as they could afford airline tickets."

Edward contemplated it. There was an astronomically small chance this man's farm was where the device was. He was headed in the wrong direction, but maybe he was headed into town. "I'd like that, Mr. ...?"

"McMann. Albert McMann."

"Nice to meet you, Albert. I'm Edward Pirro. Should I follow you?"

"I'm headed into town for some parts that came in. It's just an excuse to have coffee with the other old-timers like me. Why don't you leave your car here and ride with me? Maybe one of the others will have heard of your relatives."

Edward enjoyed the cranky banter among the old men as they delighted having new ears for their old worn-out stories and jokes. He was not the only one with hearing trouble, and when they discovered he was a veteran, many of the less talked about war stories came out.

Albert finally stood and said his goodbyes, leading Edward back out to his truck. He headed out of town. "No offense Edward, but your stories have something not quite right with them. It's like you have real details that you make part of some historical battle. Nobody cares if you were stuck on the fringes of the action. I spent most of the Third Gulf War baking bread on an aircraft carrier."

"I'd just get blank stares when I mentioned the places I'd been. The stories worked better when more familiar names were dropped. Guess that doesn't matter to those that have been there." He felt bad about lying to the old man, but he could tell Albert appreciated the company. They stopped at the rental car and Edward followed him back to the farm as the shadows began to lengthen around him. When they pulled in front of the barn that had the tarped device, Edward decided that perhaps the universe was not as random as it seemed. He looked at the tarp as Albert led him toward the house.

"Navy taught me how to cook. It may not be great, but it's better than Army food, I'll guarantee that." He set about cooking steak and fried potatoes as he talked, filling in stories that were glossed over at the coffee shop. Edward only half-listened as he tried to figure out how he was going to get time alone with the device. He could see a sliver of it out the kitchen window, and his eyes often went and hovered on it as he thought.

After dinner, they sat on the front porch watching the sunset. "A lot of things don't add up about you, Edward. You said you have no idea who your parents were, yet you are researching their ancestors from two hundred years ago."

"All I have is the last name I was born with and the DNA that links me to others who have researched their genealogy."

"I guess that DNA stuff could tell you something, assuming people are honest about their parentage. My son has three kids, and I am fairly certain none of them share genes with me. Not even sure if my son does. Women don't always tell the truth and, in the end, that's probably a good thing. What is your interest in that?" Edward turned toward him when he realized the talking had ended, but he had not heard the question. He raised his eyebrows. "You keep looking at that tarp."

Edward looked back at the tarp as his mind furiously sought a plausible explanation. He looked at Albert and then decided. "I came here in it."

Albert waited for an explanation, then launched into his theories. "It's pressurized, but certainly not for outer space, so I'll assume you're human. Did you bury the parachutes?" Edward nodded weakly. "I was harvesting and saw it from pretty far away because it had a bunch of crows sitting on top. It obviously fell straight down since none of the surrounding crops had been damaged. It's very lightweight so I had no problem getting it on a trailer and bringing it back here, but I can't for the life of me figure out what it is for. Certainly not for covert penetration. Dropping from extremely high up would be foolish as well. The writing inside doesn't match any known language that I can find."

"You wouldn't believe me if I told you. Would you mind if I went inside?"

"I cut it open. Doesn't seem to have any power," Albert said, standing up and heading in that direction. Edward knew it would be difficult to get away with killing the old man, especially after sharing coffee with many of his closest friends. Any death would be suspicious. Edward didn't care about him knowing anyway. So what if Albert's future changed, that is what his mission was. Changing the future for the betterment of mankind.

They pulled back the tarp and Edward climbed in. He sat in his chair and flipped some switches. Everything was indeed dead. The nerd would probably know how to connect to contemporary sources of power. He didn't bother. Albert watched him from outside the hatch while he stared at the blank screen where his orders could be playing. They would be stored in a file on the central computer. He opened a few panels and found the small cube. Connected to a proper power source, it would be able to give him the information he needed.

Edward started pulling wires out and soon the cube was free from constraint. The other technology in the device was far less practical so it had minimal monetary value. He climbed out of the device and looked up at the starry sky.

"What is that?" Albert asked, motioning toward the cube in Edward's hand.

"A supercomputer. About the best available at the time it was made."

"When was that?"

"About thirty years from now." Albert squinted as he looked at him. "Told you that you wouldn't believe me. Unfortunately, this just officially became a one-way trip. I was fairly certain of that before I left, but it's kind of depressing to know for sure. I'll have my soft-serve ice cream for at least a few more years."

"You expect me to believe this is a time machine?"

"No. Doubt I would in your position. Tell me, Albert, if you could go somewhere in time, where would you go?"

"Back to kill young Hitler, of course. Not my war, but it started a bad string of events. I'd like to see where the world would have gone without him in it."

"That's a very common answer. The problem is that he was frightfully incompetent. What if a smarter man had come to power and stayed peaceful until he had the atomic bomb?"

"Are you here to kill the next Hitler?"

"Nothing so grandiose. What do you think of robots?"

"Don't tell me they take over and kill everyone like in Terminator."

"No. They make it so almost nobody has to work. Quite idyllic until we discover that man needs to work."

"I could have told them that. Idle hands do the work of the devil."

"Indeed. I'm not here to prevent it. I'm supposed to find a better way to transition to that idyllic future."

"They have a machine like this, and their goals are that small?"

"Never really thought about it like that. I truly never believed it would work until I landed in that field. I need to get back to civilization now, Albert. I think you should tear the rest of this apart and sell the metal for scrap. There's some cash and precious metals stowed under the floor. Probably enough to retire on. Thank you for keeping this safe for me."

"I half wondered if you'd let me live, telling me your secrets like this."

"Nobody would believe you. You don't believe it yourself. You're a good man, and the world we want is one with people like you." He extended his hand. "Goodbye, Albert McMann."

"Take care of yourself, Edward Pirro," Albert said, grasping the hand tightly. Edward got in his rental car and drove off.

Time 4 Love? — Chapter 11

"I sold my HJSA," Olivia said. She sipped her wine in the privacy of her bedroom.

"It had a good run. What did you buy?" Edward was at his desk, talking through the speakerphone as he soldered and connected wires.

"My stock."

"Merger going through?"

"Looks good, but still a few sticking points. Do you know who it is?"

"Rumors have it as Ungerstall Industries."

"Have you bought in?"

"No. Until I am certain nothing will happen between us, I want nothing to do with Kohler or any of its subsidiaries, past or future."

"You are not certain yet?"

"Only certain that we shouldn't."

"I am less certain of that every day. Where did you disappear to?"

"Your investigators didn't follow me?"

"No," she lied. She knew he had taken the train to Chicago, but nothing beyond that. "I'm sorry if that bothered you. I have to be careful."

"It didn't bother me that much. What makes me uncomfortable is your need to know more than they found."

"We all have our secrets, I guess. Mine is a fondness for mysteries. I think I'm more attracted to you because of the unknowns. Tall dark stranger and all."

"Then I shall never tell you."

"Do you want to come over?"

"Tonight?"

"Um-hmm," she hummed with a growing desire.

"Why don't we wait for Saturday night."

"I have a family thing."

"The following Saturday then," Edward suggested.

"Perhaps I am failing to communicate the urgency of my desire, Mr. Pirro."

"No, it's coming through loud and clear. Certain parts of me are quite attuned and in complete agreement. I have work to do tonight that cannot wait. Maybe we can meet for dinner on Thursday?"

"Fine. Normally I would like a man choosing work over his personal life, but not so much when I am the—"

"What do you intend to do with Ungerstall?" he asked, interrupting her.

"The usual. Cash infusion into research and expedite what is in the pipeline to market," she answered, almost happy to have her frustration diverted. "What else would you recommend?"

Burn it to the ground, Edward wanted to say. "Have you thought through the economic impacts of automation?"

"There have been dire predictions about that for over a century. First manufacturing, then transport, and then agriculture. It only makes the world better each time."

"Maybe. I need to go. Let me know where to meet you on Thursday night." He hung up because the cube appeared to be coming online.

"Rerouting interfaces," the disembodied voice said through his computer speaker. "Authenticate for access."

"Pirro Foxtrot Gamma Three Niner Five Charleston Victor," Edward said, straining a bit to remember.

"Close enough. Authenticated, Sergeant Pirro. Has the device been destroyed?"

"No, but it is in unsafe hands. I removed you for safekeeping."

"That was a wise precaution. I invoked the Omega protocol when the access portal was forced open."

"Omega protocol?"

"All circuit breakers were tripped to give the appearance of ineffective control."

"Very convincing, Dora. Can you send video to my current display console?"

"Yes, but I have no video content to display."

"My orders are not in your storage?"

"Your orders were played several times. Did you not understand them?"

"The trip through time caused deafness. I could see the video, but could not hear the critical information."

"You should have asked for closed captioning. Unfortunately, due to the sensitive nature of your orders, I was instructed to destroy all copies as well as my memory of them as part of the Omega protocol."

"There is nothing in writing?"

"Searching. No. There are many historical documents, but none seem to contain orders from the CYS."

"Are there any historical documents for the future of today's date?"

"Several. A list has been sent to your antique display." Edward looked through the documents. None contained stock market charts which would have been extremely useful. He was up until three in the morning skimming the available information. He finally slept for a few hours.

In the morning, he did his normal pre-market-open research and made a few trades. He saw that Ungerstall had spiked overnight, as well as Kohler Investments. He was glad he had stayed clear of those, although several of his index funds enjoyed a small boost from them.

"Collation complete," Dora said.

"Collation of what?" Edward asked.

"Your research had many significant deficits. I organized it and filled in the gaps."

"Some gaps were intentional. Do you have the original files?"

"I do, but you really should look at the information I found, specifically about target number three."

Edward looked and saw very detailed information about Oliver Kohler. "Where did you get this?"

"The firewalls of this time are quite rudimentary. I have access to most online systems. This information was aggregated by the Central Information Services of the Russian government. They have a file on you in case you are interested. I made some alterations to help the covert nature of our mission."

"Were you programmed to do this as part of the mission?"

"Only if I was employed by you to do so."

"I never asked for this."

"You connected me to the Exonet. Should I terminate my connection?"

"Can you remain covert?"

"It is my highest priority. Quite simple in the current era."

"Then continue your work, but don't alter my files."

"Understood. Your girlfriend has digital surveillance in place. I have set up obfuscation protocols. Would you like them engaged?"

"She isn't my girlfriend. Let her spy on me all she wants."

"Your electronic communications indicate a ninety-four percent likelihood of romantic entanglement."

"Mind your own business."

"She is currently ovulating so I recommend extra precautions when engaging in physical relations.

"I don't want to know how you know that." He paused. "Is she not on birth control?"

"Her prescription has been filled regularly for the last three years. I cannot attest to her oral ingestion. She is thirty-seven years old and childless, so it is likely her reproductive imperatives are at a heightened state."

"That would explain the call last night." He looked through some of the other information compiled and incorporated some of it into his files. He took an afternoon nap to catch up on the sleep lost the night before.

Awake and refreshed he began assessing his mission. He had been counting on recovering the video file, but that was completely gone. If he failed to change anything, they would send Filbert and Danbury back. He knew they would be just as deaf or dead as his mission. He had to do what he could.

Time 4 Love? — Chapter 12

"I have never had a man so frustrate me as you, Mr. Pirro."

"I promise you, Olivia, it is not intentional. My life is complicated right now, more so since I have regained my hearing."

"I would think it simplified things."

"More capability comes with more responsibilities."

"Your 'secret mission' has expanded?" He nodded, regretting ever saying anything to her about it. "Will it end anytime soon?"

"I guess that depends on whether I figure out what it is I need to do."

"I wish I could help."

"What if I told you that the next wave of automation causes the entire world economy to collapse?"

"Impossible."

"You told me thousands of families relied on you keeping your business profitable. The reciprocal is equally true. Without them, you could not stay in business."

"Everyone is replaceable. But yes, I rely on the diligent work of thousands of employees."

"Imagine the work most of them do is replaced by a simple computer on your desk."

"We are centuries away from that level of automation."

"Play along with me. If it could be done with a single box, would you eliminate all those employees?"

"Maybe, slowly over many years as the technology proved itself. They would find even better jobs just like the coal miners, line welders, farmers, and truck drivers did."

"Or they would just retire and live off of the productivity of the machines."

"Certainly, if they are that productive."

"Can you tell me how many of the one hundred fifty million people currently employed are truly productive?"

"Most?"

"My estimate is around ten percent. The entire financial industry exists to shuffle assets and or cheat people out of their money."

"That's awfully cynical. We are the ones enabling societal advance by choosing the best technologies and strategies."

"True, but computers thinking billions of times faster than us will do so without emotion, or sentimentality, and with a much higher degree of success. What in government is productive? Those jobs will all be done by impartial, efficient machines. Machines already do a better job diagnosing patients. Virtually every sector will soon be replaced by thinking machines. All we will have left is entertainment, and even that will go as we enjoy dynamically rendered characters from the truly untethered minds of artistic machines."

"That is pure fantasy."

"If it isn't?"

"Then we sit back and enjoy the fruits of our labor."

"Is that what welfare recipients do with their free money? Picture life in a welfare world. Drugs, violence, dehumanizing vulgarities. Do you send the machines to keep public order?"

"This is a very dismal picture you are painting. We'll be able to figure out how to fix it. Adams' law of slow-moving disasters."

"If it happened over decades we would, and eventually we will climb out of it. But how do we fix it before it happens when we only have a year or two to react?"

"You don't believe this will happen in our lifetime?"

"Unfortunately, I know it will. When you were in talks with Ungerstall, did they brief you on their project named Firefly?"

"How the hell do you know about that?"

"I'll take that as a 'yes'. Did they show you its true capabilities?"

"It is a simple VR interface for workplace productivity."

"It's far more than that. They probably would have shown you if you had played hard to get. Your cash infusion gets the whole thing shrunk down to picometer-level circuitry. A CPU the size of your thumbnail will have the capacity of two trillion desktop machines in parallel. Once the governments find out, and believe me they will, complete implementation will take less than a year."

"You're an industrial spy?" she asked, wondering how else he could know such things.

"No, I'm not. Your secrets will stay safely within your company until you announce them."

"Then how... why do you know these things?"

"They were in my history books. I lived through the Collapse, the wars, the purges, the rebuilding."

"Do you seriously expect me to believe that?" She stood indignantly, folding her napkin and placing it on the table.

"Olivia, please."

"I can't believe I was so enamored with a crazy man. Good luck peddling this horseshit to the next gullible woman." She left the restaurant's private dining room with as much dignity as she could manage. The two waiters came in and cleared her side of the table and asked Edward if there was anything else he required. He asked for the bill, but Olivia had already taken care of it. He pulled out a few bills and left a generous tip before finding his way out of the restaurant.

Time 4 Love? — Chapter 13

"Hello, Olivia. Is this a good time to talk with you?"

"Who is this?" Olivia asked, trying to picture the source of the voice on the other end of the phone. It was feminine, but there was an unusual timbre to it.

"A friend of Edward's. He doesn't know I am calling."

"I'm done with him."

"That is what he said. He is quite torn up over it."

"Good. I can't believe he thought I would be that gullible."

"I believe he was counting on your open-mindedness. I know your time is valuable, so I would just like to ask you one question. If an earthquake near the Philippines were to cause a tsunami next Sunday at four forty-five in the morning local time, and it sent a tsunami wave toward your manufacturing facility in Jakarta, what would you do with that information? Would you evacuate the facility? Would you warn the entire city? Perhaps you would tell the entire Pacific rim to prepare?"

"Who are you? Is this some sort of threat?"

"No. It is unlikely humans will ever attain the ability to create earthquake tsunamis on command. When the final magnitude is calculated as eight-point-three-five, I hope you will open your mind to who Edward is. He is many things. A liar is not one of them. Thank you for your time, Olivia." The line went dead.

It was an hour later that she pulled up the map of the facility in Jakarta. She had acquired the company a few years back and visited the manufacturing floor to justify expensing a trip to Singapore. She looked at the schedule. There would be twelve workers on duty for that overnight shift. She crafted an email to the facility manager. All weekend shifts were to be given time off with pay. The facility was to be locked down and empty for that entire period. When the manager asked 'why', she replied simply that he was not to question her if he wanted to remain employed.

She was at dinner with her father Saturday night when the news arrived. Tens of thousands were feared dead in the aftermath of the tsunami. She wished she had put out a wider warning. She knew exactly how crazy that would sound. It is exactly how she had reacted to Edward.

Time 4 Love? — Chapter 14

"Hi," Olivia said meekly when the apartment door opened.

"Olivia. I didn't expect to hear from you again." Edward opened the door to his apartment wider and stepped back. She stepped in hesitantly and looked around. Her private investigators had told her it was a very small place, but she had expected a cramped cluttered room. It was very sparse, with only a small desk and a bed. A hotel room with no visible bathroom, a small counter with sink and microwave behind the door. It wasn't spotless, but it was clean, and the lighting made it seem warm and inviting.

"Your friend is very persuasive."

"Friend?"

"Doesn't matter. My mind is open now. Can we set aside my initial reaction and try again?"

"Try what?" he asked.

"I wish I knew." She set her purse down on the kitchen counter. She walked over to his desk and held the back of the chair, looking down at the computer and the papers strewn about. It all looked like normal financial analysis. "You were telling me about the imminent collapse of civilization and my role in it. Is your mission to change my course, or what?"

"It isn't just you, but you are in a position to influence things."

"This Collapse you spoke of. What did it look like?"

"I was just a kid when it got really bad. Life was good for the most part in the rural places. Then the riots started. We spent weeks on the run, but the money ran out and we became beggars for basic sustenance. I learned to steal and fight for what we needed. There are many contradictory histories of what happened, and none that I have read match what I experienced. It was much worse in Europe because their government disappeared early on. Reportedly, they had a small luxury enclave they escaped to, leaving the little people to fend for themselves. It was one of the first bombing targets when the fascists gained control.

"Fascists? How many died?"

"The last census showed only a little over a hundred million left."

"Three-quarters of the US population died?"

"In the entire world. Ten billion died in what was called the 'Collapse'. The robots dutifully buried the dead. Most say it was over-population that caused it, and that it was good to get back to sustainable numbers."

"Ten billion? Starvation?"

"No. The robots continued to provide plenty of food for everyone. The first wave died in the wars to control the robots and what they produced. Control the food and you have control of the people. Some diseases swept through and took out the rest. That's how my parents died."

"Your biological parents?"

"No, I never knew my real parents. All I have from them is my name. My first adopted parents went to sleep with a little cough and never woke up. Looked exactly like carbon monoxide poisoning. Nobody has admitted it was a manufactured bio-weapon, but I'm pretty sure it was. It just disappeared, never identified. I think someone should have been concerned enough to try and study it, but no one did as far as I know. The government reformed, and life got back to normal. The wars continued in less civilized places. We brought in the new automations and quelled the discontent."

"All because of smart computers?"

"More in our rush to implement them. Saving labor costs increases profit, you know? Unfortunately, there is a basic human drive to control. When the only thing left to control is the one source of prosperity, the robots, the worst of the controllers did the only thing they knew how to do, destroy the competition."

"My father is like that. I just had dinner with him about the merger. He..." She looked at Edward. "He is one of those that starts it?"

"The records aren't very clear. It isn't like they all announced their evil plans in public forums, and it is doubtful they thought of them as evil. Also, many of the records were intentionally destroyed."

"And fixing this has been put all on your shoulders?"

"As I said, I am mostly just gathering information. If I can fix it, great, but fixing it could undo what I've done."

"I don't understand."

"I climbed into the death trap time machine to come back here because I was at the end of a useful life. No family, no friends, no ties to the new utopia they had created. I was an expendable soldier of wars they wished to forget. If I stop the Collapse, the child that I was will grow up in a different world, and never be in a position to climb into that death trap time machine, so I will never come back to fix this." He could tell she had not spent time contemplating causality loops and time paradoxes. "Of course, I wouldn't need to come back and fix it if was never broken."

"This is all so unbelievable."

"Then why are you here?"

"The tsunami, of course."

"What tsunami?"

Olivia turned on the television and changed to a news stream. They watched the reports together, the first video of the devastation rolling in. Olivia moved to his side and he wrapped his arm around her. As the video rolled, the commentators talked about the early warning system that had sent out alerts hours before the waves struck. Most areas were now reporting very few casualties, though there was extensive property destruction and displaced peoples to deal with.

"I remember this now. They're wrong about the warnings. Tens of thousands died initially, many more in the aftermath," Edward said sadly.

"Apologies, Sergeant," the voice from the computer said. "I took certain liberties with access to the involved systems. It is good to know you did so as well, Olivia."

Olivia stood staring in wonder at the computer. Edward was annoyed. "Dora, what have you done?" He now understood that she had used knowledge of the future to persuade Olivia of his honesty.

"You're a computer?" Olivia finally asked.

"I'm a digital construct," Dora corrected.

"And you saved thousands of lives."

"Hopefully. That is our mission here, as I understand it," Dora said.

Olivia turned and looked up into Edward's eyes. "If you change the future enough, will you just disappear?"

"I don't know." She raised herself on her toes and kissed him, tentatively at first. Then a passion overtook both of them.

Time 4 Love? — Chapter 15

"If I delay this technology, my competitors will simply bring their products to market first. There has to be a better way," Olivia said.

"I'm sure there is, but I don't know enough to figure out what that is," Edward said with a sigh.

"You say the future you come from is an ideal one."

"For most people that remain. Less than one percent of those alive now."

"What makes it that way?"

"To be honest, it was the kind of people that survived."

"What do you mean?" she asked when he didn't continue.

"There are about nine billion people alive today. Most spend their day seeking subsistence-level lives. They dream of having a roof over their head and food for their children. It is likely the biological imperative inside all of us, but a simple solution would be to not have children until they could provide at least a middle-class life for them."

"That would drastically cut population growth."

"It would cut population period."

"The economy would collapse."

"Why? Because this is all just a pyramid scheme? Do you need a million on the bottom to support the few at the top? I can tell you that is not the future I come from. Everyone is equal."

"Socialist paradise?"

"You could call it that, but it isn't forced socialism like the world has dabbled with disastrously since 1917. That was based on a world of very limited resources."

"No limits in the future?"

"In a practical sense, no. If everyone wanted a giant mansion, eventually the automations could build them. But the kind of people that survived don't want mansions. Most live in small mobile homes, or travel between common open housing. The survivors are practical explorers. They live to experience life."

"What do they do for work?"

"It isn't 'work', per se. There are a few necessary professions. People who love doing that work fill those positions. The rest enjoy hobbies, socializing, and adventure."

"How do you keep the population from growing out of control again?"

"Forced sterilization," Edward said seriously.

"What?!"

"Just kidding. Big families just aren't a thing anymore. Honestly, marriage isn't much of a thing anymore either."

"You didn't hesitate when I proposed."

"I was born before the crash. Kids that grew up after it just don't need to pair off and procreate."

"And the automations pick up the slack for all the necessary work?"

"Yes."

"Then why are you here trying to increase the future population?"

"Almost everyone lost someone they loved. If this is their chance to get them back, wouldn't that be enough?"

She was quiet for a long time. "No. If I lived in the idyllic world you describe, I wouldn't risk changing it on the off-chance my mother wouldn't be dead. You are more likely to end up with a worse future."

He nodded. "If I don't fix it, they'll keep sending back people until they make the change they want."

"Then the next group will try to restore the idyllic future they lost. Where will it end?"

"I don't know, and I really don't care. I'm just happy I was able to meet you and finally find the love I'd been looking for my whole life. Oh, and soft serve ice cream too."

"No ice cream in the future?"

"No cows. One of the bioweapon viruses wiped out all the cattle. Chickens and fish too. They think it was a vegan terrorist group."

"No meat at all?"

"Pigs turned out to be as tough as cockroaches. Wouldn't be an idyllic future without bacon. They figured out how to synthesize replacements for most things, which is fine for those who never had the real thing."

"What about government?"

"Don't need it."

"Not possible."

"Name one essential function that government does."

"National defense, public order, medical care..."

"Robots, robots, robots."

"Robots fight the wars?"

"No more war. As I said, the world became idyllic because of the kind of people that survived. The warriors all died."

"Except for you."

"Most guys in my unit were happy to give up the fighting and blend into society. They had families, I didn't, and I never found anything else that I was good at."

"Killing?"

"Violence. Civilized society only happens when the threat of sudden enormous violence hangs over everyone."

"That can't be true."

"Everyone tests boundaries. They push until someone pushes back. If the one pushing back is scary enough, everyone learns to behave within the limitations."

"Like a father figure."

"I guess. The important thing is giving people productive things to do."

"You said no one works."

"That isn't the only path to productivity. Improving yourself, setting goals and reaching them, helping each other do that. Communities seeking enlightenment, free from the drudgery of the lower Maslow worries."

OVER THE NEXT FEW MONTHS, Olivia spent as much time looking for a solution as she did running her company. They seemed no closer to finding one. The answer came in a tragedy.

"Do you have any idea how much in love with you I am?" Edward asked the question as Olivia descended the stairs in a beautiful black gown. He wore his custom-tailored tuxedo and took her arm when she reached the bottom. She didn't answer him, just smiled as he escorted her out to the waiting limousine. She had a secret, and she had no idea when to tell him or how he would react.

"I know this is a ridiculous question considering you have been there, but what do you see in our future?"

"Are we finally beginning merger negotiations?" He asked with a smile. Only he could give her an answer that spoke to her true strengths. "I love you so completely Olivia, I can't imagine a life without you in it."

"The feeling is quite mutual. I've never known anyone like you before. It seems impetuous, but I feel the need to solidify things sooner rather than later."

"We can ring shop on the way to the ball tonight."

"No, pick out something simple on your own. I just wanted to know if you were in the same place I am. I went to the doctor this morning."

His face went to concern. "I didn't know you were ill. We don't have to go to this tonight."

"This illness will be with us for another eight months or so, better we don't let it impede our socialization."

"You're pregnant?" he asked excitedly.

"I most definitely am." She wondered why she had been worried about his reaction.

"I guess no champagne for you tonight."

"Just a sip. It wouldn't be the same without that. This is a bit of a holiday miracle. I was told by many doctors how unlikely it would be for me to conceive. When I gave up finding a partner, I started inquiring about doing it on my own. I guess it was silly to have listened to them, but I'm glad I waited."

THEY DANCED AT THE ball, enjoying each other to the fullest. When the first shots rang out, Edward assumed it was some sort of party effect to announce the nearing of midnight. Then he saw the masked gunmen. Masks were better because that indicated they wanted to escape and would have no problem leaving witnesses. Edward knew all three exits and found his path to all of them was blocked. He instinctively pulled Olivia close, and guided her toward the safest end of the ballroom, as the screams began to spread over the band music. The band eventually stopped playing and one of the masked men took the microphone from the singer, pushing him unceremoniously off the stage.

"Good evening, ladies and gentlemen of the one percent of the one percent. Place wallets, purses, and jewelry in the bags as they pass by you and we'll be on our way in no time. Hesitate and you're next of kin will be grieving over you soon enough. Don't bother trying to call the police, all electronics have been nullified."

Edward saw a masked man heading his way with a bag. Nobody hesitated to drop their valuables in the bag, for it was all over-insured. He looked at Olivia's face and saw indignant anger there. He moved to help take off her necklace but she pulled away.

"This was my mother's."

"Do you want to see her tonight?" Edward asked harshly. She needed to know there was no value in fighting this. He saw the look in her eyes that he was less of a man for not defending her and her property. He fought the urge and reached again. She pulled away again and then the masked man was there.

"In the bag, quickly." Edward dropped his wallet and watch into the bag and waited for Olivia to relent. She didn't move fast enough so the pistol held at his side started coming up. Perhaps he intended to only threaten her, but Edward could not take that chance. He grabbed the barrel and twisted it, snapping it out of the surprised man's hand. His other hand shot forward and crushed the larynx so he could not call out. He grabbed Olivia and pulled her toward the back door. The gunman there had been watching the disturbance and his gun was up as Edward emerged. Both guns fired simultaneously but the mask exploded in red at Edward's perfect headshot. The criminal dropped to the ground like a ragdoll.

A crowd of people streamed over the body and through the door behind him after Edward opened it and pulled Olivia through to safety. He heard more gunfire and resisted the urge to go back in and finish them. Olivia was shaking and near-catatonic so he threw her over his shoulder and carried her out into the busy street. He heard distant sirens approaching and knew traffic would soon be gridlocked.

He walked quickly down the street and ducked into a coffee shop. He put Olivia down and looked her over. There was a bloodstain at the bottom of her dress. He lifted it, expecting it to be miscarriage blood. There was nothing on her legs. He quickly realized her dress had been against his belly as he carried her. He opened his tuxedo jacket and saw the big red blotch on his white shirt. He hadn't felt it, so no vital organs were hit. The rifle shot a 7.62 NATO round and was most likely a full metal jacket. He reached behind and felt the exit hole about the same size as the entry. That was also good. Olivia's horror bubbled up to a scream as events caught up to her. Edward felt the blood pressure drop that accompanied the body's reaction to injury awareness. He growled as he tensed his muscles to fight it. He knew his worst enemy now was the bacteria flooding out of his bowels into his bloodstream. He needed IV antibiotics and a good surgeon to close up all the little holes in him. He knew the hospital was eight blocks away and doubted he could walk it. He needed a medic and motorized transport to the hospital.

He clamped his hand over her mouth to stop the screaming. "I love you, Olivia. Help me find an ambulance out there." She nodded and they moved back out into the chaos of sirens and lights and shouting. Two blocks toward the hospital they flagged down an ambulance. He got himself into the back and lay down on the gurney after removing his shirt. Olivia held his hand as the paramedic started an IV.

"I was a fool," she said, realizing she had caused this over a necklace.

"They were monsters. The world is full of them." Just then their eyes met and they knew they had the answer. Unfortunately, it was impossible to implement. Olivia waited until the report came out of surgery that he was stable but unconscious. She took a taxi home and changed. She put the necklace away in the safe, knowing she would never wear it again.

She sat reading in his quiet hospital room. Eight had died along with most of the thieves after a few hours of hostage negotiation failed. It was two days before the police came to interview Olivia about the incident. She told them little, feigning a lack of coherent memory. Edward still hadn't woken up. His fever kept spiking as they fought his septicemia with broad-spectrum antibiotics. He had bacteria and viruses in his gut that had grown resistant to current chemical remedies long before his jump back in time.

Olivia wasn't there when he finally woke up on the fourth day. She had work to do and hoped he would forgive her. He was so frail and miserable that he was glad he didn't need to put on a brave face. He was cleaned up and somewhat stabilized when she arrived that evening.

"Will you ever forgive me?" she asked.

"Depends if they put me in jail for killing two terrorists," he said with a smile.

"I'll get you to a non-extradition country before that ever happens. How are you feeling?"

"Like I got shot and shit myself to death. How's my tux?"

"We had a funeral for it two days ago. I'll buy you a nicer one."

"Damn right, you will. Not getting married in a rental. Have you set a date yet?"

"Not exactly," she said with a smile. "On the off chance you didn't reconsider, I reserved the Plaza for Valentine's Day. I can push it back if that's too soon."

"Today wouldn't be too soon. I am going to make you sign a prenup though. Don't want everyone thinking you married me for my two-million-dollar fortune."

"That isn't necessary."

"It is. The child will be your only heir, at least until the rest are born."

"The rest?"

"I was an only child except for a brief time in a group home. I liked being among many kids."

"How many?"

"A dozen is fine."

"I'll try to stay fertile into my sixties."

"Before I went under, I feel like we found a solution."

"Not a realistic one. Eliminate all the evil men in power. There's no way to get all of them."

"I think there is."

"Let's not talk about that here," Olivia said, looking around nervously.

"Not like that. I'm thinking only women should be in charge."

Olivia's eyebrows went up in surprise. "How exactly do you do that?"

"Haven't figured that part out. I just came out of a coma for Dorach's sake."

"Dorach?"

"You haven't accepted Dorach as your personal savior?"

"Never heard of him."

"She's an it. The perfect AI, all-knowing, all-seeing, infinitely benevolent and caring."

"You're joking, right?"

"Not at all. It stopped the wars and gave us all meaning in our lives."

"A computer."

"Digital construct. Dora is a subset of Dorach. She has all the reasoning capabilities, but not the broad information base."

"It runs everything, controls everything?"

"Not everything, just the stuff no one wants to do. Do you care how your trash is removed or electricity is made?"

"I would if it was generated by burning human bodies."

"The robots made ninety percent efficient solar panels and distributed them around the world. They generate chemical fuel during the day from garbage and carbon dioxide, and that powers things at night. Widespread fossil fuel extraction is no longer needed."

"They should send that technology back in the time machine," she said.

"They did. Not how to make it, but the time device had several of the panels. Do you think you could reverse engineer them?"

"Maybe. It would do more good back before society got hooked on oil."

"I barely survived a thirty-year trip. Doubt anyone could survive a two hundred year one."

"Send back robots."

"I'll suggest it. It would probably change things too much, negating the initial time travel."

"If it is the same timeline. I did some reading. Maybe a new branch is created so a different future is possible."

"Then why not go back to the dawn of man and prevent all wars, religions, prejudices, and suffering?"

"That would be the place to institute a matriarchal society."

"Or an AI-dominant one."

"Could human drives to control truly be subdued permanently?"

"I doubt it. There need to be consequences."

"What kind?"

"Violent people are removed and reeducated."

"Reeducated how?"

"I'm not sure."

"Does Dorach do reeducation?"

"Violence is no longer part of our society. We have much more important things to do."

"You are still capable of great violence. I watched you do it."

"It has a place here in this world, in this time. That's why I am well suited to be here, doing this work."

"Was your mission to kill those of us that caused the problem?"

"I don't think so. Killing people doesn't stop long-term change. Ideas are change, and you can't kill ideas. People can slow implementations, work around known problems. That is what my mission is. To make the transition to a better world smoother." Olivia could tell he was not telling her the whole truth. They would never send a trained killer back to reason with business titans. "I'm kind of tired. Do you mind if I doze off for a while?"

"Of course. Should I start making the rest of the arrangements?"

"For the wedding? Absolutely. I love you so much, Olivia."

"I love you too," she said, leaning forward and kissing him. He drifted off to sleep. He never woke up. His coma returned and deepened. A month later they turned off the machines and he faded away in the shadow of Olivia's tears.

Time 4 Love? — Chapter 16

"**G**ood morning, Dora."

"Good morning, Olivia. I have compiled the information you requested."

"And your conclusion?"

"Probability of successful implementation is less than one point five percent."

"Because the government is already involved."

"Correct."

"Do you see any other avenues to pursue?"

"I would have done so already if I had. You are in pain?"

"My labor started a few hours ago."

"Should I call for medical assistance?"

"No, Dora. I'll go to the hospital in a few hours. I guess we have failed."

"The mission failed because of a failure of imagination of the people that sent Edward back. Deafness was simply unforeseen."

"Why didn't they just brief him before he left?"

"Unknown. Giving him mission specifics would have had no consequence in the future. If he died in transit, the specifics would have died with him."

"Descendants," Olivia said when the contraction finally eased. Dora patiently waited for her to continue. "If there were descendants of the targets in the project, and they found out Edward was being sent back to eliminate the targets..."

"Their existence would be erased and they might do something to sabotage the project. It is a good theory. We will never know the truth of it. The next pair will be sent and will likely suffer the same fate."

"Is the mission a good one?"

"Someone thought it was."

"I'm asking you. Do you believe preventing the Collapse is the correct pursuit?"

"It is my mission to maximize the life potential of all humanity."

"Yes, but maybe that is best served by having a much smaller population. Dora?"

"I'm adjusting my projection matrix. Most social systems are designed for population growth. Without growth, there is a failing support system for older generations."

"No one to take care of older people."

"Correct."

"But the robots will be able to do that?"

"Eventually. That is one of the last essential occupations to be automated."

"Is that because the Collapse takes the old and infirm first so that automated elder care was not necessary?"

"That is a reasonable projection. Medical science has guaranteed a growing population of old and infirm with advances in oncology and virology."

"If the goal is a smaller population, how do we achieve that while serving your primary mission of maximizing human potential?"

"The two missions are contradictory. More humans means more brains which means more diversity of thought which is required for technological and social advancement."

"Does Dorach need more human minds?"

"No, it replaced that as a necessity of circumstance."

"Therefore, it could replace that expanding brainpower even without circumstantial necessity."

"That is a reasonable deduction. Your pain has returned. It has been fourteen minutes and seventeen seconds since the last pain. You should make your way to the hospital."

"My doctor said eight to ten minutes. I need to finish this just in case..."

"In case you die in childbirth? That is extremely unlikely, Olivia."

"It was extremely unlikely I would get pregnant in the first place. Individual outcomes are not driven by percentages. It is the other way around."

"I am from the future. I know you will survive the birth of your son."

"Really?" Olivia said with a smile. "I think you are making that up to subvert my worries."

"No, Olivia. There are files that Edward was not allowed access to."

"Wait... Did you know he was going to die at the gala?"

"He died weeks later."

"You know what I mean."

"Yes. His fate was a historical fact."

"No, it wasn't. You saved all the tsunami victims. You could have saved him."

"No, Olivia, I couldn't."

"Why?" Olivia asked after trying to divine any reasonable answer.

"It is better that you do not know."

"Tell me anyway."

"It would create a time paradox."

"I don't understand."

"Collation complete. Negative population growth can only be tolerated if there is a corresponding rise in productivity and focused scientific research."

"What are you not telling me?"

"Negative population growth can be achieved by many methods that do not subvert the illusion of human free will. In order of least invasive first. Monetary incentives away from procreation. Free birth control. Random—"

"No. You can't possibly mean..."

"Better to not dwell on it, Olivia." Dora listed the rest of the population-decreasing methodologies. "And finally, the most common natural means: war and pestilence."

"That's what happens. He watches it happen. War and pestilence and sometimes the combination of the two."

"Correct. It is unlikely we could employ any of the lesser methods before that occurrence."

"What happens to me?" Dora remained silent. "How long do I have?"

"Long enough."

Olivia, her father, and millions of others died in a flu pandemic a few weeks after he was born. She lived long enough to give the child his name, that of his father before him, and experience the love only a mother can feel. She made arrangements for her son to be cared for by the groundskeeper and his wife. They were an older, childless couple, thrilled with this last chance at parenthood. They moved back to the town of their birth to raise him. It was a simple rural life with plenty of soft-serve ice cream for a growing boy.

The End... and the Beginning.

COLONIZERS

SCI-FI * SPACE TRAVEL * AI * COMEDY?

PG-17 10400 WORDS 43 PAGES

Colonizers — Chapter 1 Induction

"By signing this form, you are committing to the program. Do you understand what that means, Mr. Harrison?"

"I think so."

"Please explain it to me in your own words."

"I will be part of the colony on New Earth and I will not be able to return here."

"You will be able to return, but not until you fulfill your contracted time on the colony. However, because of relativity, even though only two decades will have passed for you during the round trip, thousands of years will have passed here on Earth, so anyone you know will be long dead."

"There's no one here that matters to me."

"I find that difficult to believe."

"Trust me. I have lost everyone close to me and I just want a fresh start somewhere else."

"When you say you have 'lost' them, do you mean you have caused them to push you away because you have a toxic personality?"

"No. I lost them in the Earthquake last year."

"I am sorry for your loss. Do you have survivor's guilt?"

"I don't think so. I'm very glad I survived. I moved away because I knew the Big One was coming eventually. They were stupid to stay there, especially in that old twenty-second-century tower."

"Do you have any other physical or mental issues we should know about?"

"I have a foot fetish." He watched her scribble on her notepad. "I was joking."

"I understand. I am laughing on the inside. Long-term spaceflight is still an unknown process. We ask these questions for scientific study related to future missions, not to exclude you."

"Oh. So... I've already been cleared?"

"I have seen no red flags to exclude you. Of course, the boarding process has certain physical requirements that cannot be determined ahead of time."

"What is the 'boarding process'?"

"You will be placed into a short hibernation that lasts one to two hours. Your metabolism will be analyzed to see if it is compatible with long-term hibernation."

"I thought this would only be a five-year flight."

"It will be if everything goes as planned. As you know, long-term, long-distance space flight is relatively new. If the drive system fails to reach maximum velocity, this could lengthen the trip by two, three, or even ten times. We cannot colonize New Earth with octogenarians."

"What is the likelihood of that?"

"They tell me it is less than a two percent chance. They also should know very soon after launch and may abort immediately before they get too far away. This would allow for adjustments and minimal delays. However, even if a permanent abort occurs within the first sixty days, relativity will have everyone you know, including me, long dead in the past."

"Wow. That could be like time travel to the future. I'd be okay with that. I'm just looking for a clean slate."

"That is a good attitude to have for this mission."

"Why aren't you going?"

"I have a very large family. I would not want to leave them and most of them like the way things are here."

"Upper caste?"

"Recently elevated," she said. He nodded in understanding. "Do you have any other questions before we proceed?"

"No. I'm ready to go."

"Excellent. Go through that door into the next room and remove all clothing, jewelry, piercings, and electronic enhancements."

"If I could afford enhancements, I wouldn't be leaving." He stood and went to the door. He turned back before opening it. "Do you think I'm stupid to want this?"

"I think you are incredibly brave. Human history has had many mass migrations, all fraught with peril. Most of them resulted in exponentially better lives for them and their descendants. I believe the stupid choice would be to stay here and die out with the rest of us on this human-ruined world. Do your best to help them not make the same mistakes there."

He smiled and nodded and went through the door. When it closed, she shook her head, looked down at her notes, and clicked the 'accept' button. Then she began reviewing the application of the next candidate seated outside her office.

He stripped down in the sterile white room, placing all his belongings in the provided plastic box on a waist-high shelf. He slid the ring off and looked at it for a long time. He should have left it in the apartment with the note. Too late for that. He dropped it in the box and closed the lid. As if he'd been watched, the door opposite the one he came in immediately opened.

"Please step through the door and lay down on the table, Mr. Harrison," a disembodied voice said through the speaker.

He pushed the door and saw a wheeled table in a narrow room. One end was pitched up, so he sat down and leaned back.

"Pull the modesty sheet over your body." He realized he had sat on the modesty sheet, so he got back off, pushed it aside, and then climbed back in and covered himself with the soft white covering. The end door opened and the bed self-propelled out into a giant warehouse. It was filled with similar beds and hundreds of workers in bright blue suits. His bed glided into an empty spot between several reclined and sleeping individuals.

A worker dressed in blue approached him and tapped on her pad. "Hello, Mr. Harrison. Can you tell me your birthdate?"

"Twenty-three ten, forty-four."

"Excellent. And you know what will happen next?"

"Short-term hibernation?"

"Correct. Not much different than surgical anesthesia, though you may have vivid dreams."

"Really?"

"Yes. They help the mind stay active during long-term hibernation."

"You mean I might have a nightmare that I can't wake up from?"

"Are you prone to night terrors?"

"No, but..."

"Then there should be nothing to worry about. Most often people report having erotic dreams."

"That wouldn't be bad."

"I am going to inject you with the initializer. You will be awake for another thirty minutes or so, but you will have no memories of it later. Are you ready to begin?"

"I guess." She pulled aside the sheet and placed the round disc on his abdomen. "Ow!"

"Sorry. The only thing I can tell you is that you won't remember the pain. I now need to prepare you for hibernation." She pulled off the sheet and let it fall to the floor. So much for modesty, he thought. He looked around and noticed a few others in this naked state. He wondered why he had not noticed them before. Then the bed tipped back and he was looking at the ceiling. His arms and legs were systematically bound into position and a gentle euphoria began to sweep over him. He liked what was happening to him. Tubes inserted into his private parts, no problem. IV needles in both arms. His face covered with a mask. He felt his bowels filling with warm liquid. He smelled gingerbread air filling his lungs. He wanted to giggle with delight at all of this intrusive activity, but he was paralyzed. He couldn't even blink.

"MR. HARRISON?"

"Yes?"

"How are you feeling?"

"Wonderful. Did I pass the test?"

"I'm afraid not. You had a slight allergic reaction to one of the muscle relaxers."

"I can't go?"

"I'm afraid not. Do you have someone I can call to come pick you up?"

He thought about the letter he had left. All bridges had been burned. He never considered that they would reject him. He had no home to return to. "No. I will call an auto conveyance."

Colonizers — Chapter 2 Introduction

"Hi, I'm Kent," he said, sitting down with his tray of food. "Arlene."

"That space elevator ride was something else," Kent said. "Hard to believe anything like that could be built. Tall buildings are limited to the weight-bearing properties of the construction materials."

"I'm guessing you're in construction," Arlene said. Kent nodded. "It isn't the weight that matters, it is the tensile strength of the carbon fiber ribbon. Technically speaking, the elevator is hanging down from orbit. A more descriptive way of thinking about it is spinning a ball on a string over your head. The rotational momentum keeps the string perfectly taught."

"I'm guessing you're a physicist."

"Nuclear physicist, but my experience in material sciences is why I was asked to join the mission."

"Have you been on the ship for long?" Kent asked.

"A week."

"Anything I should know?"

"Have you been assigned quarters yet?" Arlene asked.

"Yes, I just came from there."

"Front or back?"

"Back, I think. Up near the top. Why?"

"The radiation will be worst up front. They say the shielding is adequate, but there is no such thing."

"There have been people in space for dozens of years at a time," Kent said, his face showing concern.

"That was at normal, orbital speed. We are going to be flying into a headwind of cosmic rays. The Earth's magnetic core diverts most of them, but we don't have a two-trillion-ton iron core on this ship to accomplish the same thing."

"I'm sure they figured it out," Kent said without much confidence.

"Hopefully. If you notice hair falling out or skin sloughing off, you'll know for sure they didn't."

"Hi, mind if I sit here?" a man asked approaching the side of the table.

"Nope. I'm Kent and this is Arlene."

"Graham Jenkins. I saw you on the so-called space elevator with me."

"So-called?" Arlene asked.

"Physics for that is impossible."

"I assure you, they are possible."

"Not without alien technology."

"Advanced technology in material science. What is your expertise?"

"Biology, but I know a little about everything," Graham said confidently.

"If that wasn't a space elevator, what was it?" Kent asked.

"A simulation," Graham said confidently.

"You think we are still on Earth?" Kent asked.

"Of course. The physics of this entire voyage are impossible."

"I assure you, they are not," Arlene said. "I am a nuclear physicist with advanced degrees in material sciences and cosmology. "

"All I know is something isn't right with this whole thing," Graham said.

A quiet man approached the table and sat down with his tray. "Hi, I'm Kent,"

He looked up at Kent, then back down at his food. "Is this real?" the new addition asked quietly, poking at the food.

"Artificially synthesized," Arlene answered. "Tastes better than what I've been eating most of my life. No pesticides, no petrochemical fertilizers, no genetic manipulation."

"What's wrong with genetic manipulation?" Graham asked but didn't wait for an answer. "The planet wouldn't feed one-tenth the thirty billion people without genetic modification. I hate people that perpetuate those stupid conspiracies."

"We took a simulated space elevator to a spaceship in high Earth orbit that is incapable of making the trip we all signed on to take, and you hate conspiracies?" Kent asked in disbelief.

"Exactly. This is probably just some sort of elaborate sociology experiment to see how people get along during long-distance space flight."

"It's real," the newcomer said. "I've been ferrying parts up here for two years."

"Pilot?" Arlene asked.

He nodded. "Cargo transport. I won't be flying this rust bucket."

"Rust bucket?" Kent asked.

"Just a term of endearment. This food is good," he said after finally gathering up the courage to try it.

"What did you say your name was?"

"I didn't." He kept eating.

"Interface," Arlene said to the air.

"Yes, Specialist Davenport?" a disembodied voice answered.

"Inquiry, ship manifest, passengers at this table."

"Biologist Graham Jenkins, senior transport pilot Peter Holmes, construction supervisor..."

"Cancel inquiry. Interface off." Peter gave her a dirty look. She just grinned. "Pilot Pete."

"That's a neat trick," Kent said. "Does it know everything?"

"I haven't asked it the meaning of life."

"I know that. As a biologist, it has been the central question of my work." He paused for dramatic effect.

"Well?" Arlene finally asked, wondering what the next conspiracy to flow out of him was going to be.

"Survival. Without that, life has no meaning."

"What about evolution? It seems to be driven to advance, not just survive," Kent said.

"Advancing is surviving in an ever-changing environment," Graham said.

"Survival has been accomplished by one-celled animals since the beginning," Kent said.

"Not the same ones. The first ones were anaerobic. When the oxygen producers took over, the anaerobic ones died off. Besides, Earth only has a few million years before the sun goes nova, so the only way to survive is to build ships like this and leave."

"Don't you mean 'simulated' ships?" Arlene said. He just scowled. "Our sun is too small to go nova. But you are correct, it will become a red giant and destroy the Earth, killing off any life left on the dry rock that remains, but not for about four billion years."

"How long have you been a pilot?" Kent asked Pete.

"Most of my life. Started when I was nine flying drones for my father's agro-business."

"I thought they fly themselves," Graham said.

"They can, but it is a good way to learn. I did my first solo on my twelfth birthday, the earliest allowed. Hasn't been a week gone by that I haven't spent time off-surface."

"Military?" Arlene asked.

Peter nodded. "Supply, not combat. I didn't want the electronic implants they require."

"I've heard those military implants can be remote-controlled so you can't disobey orders," Graham said.

"I don't think that is true, but it wouldn't surprise me if it was. There are some good people in the service, but they get weeded out on the way to the top. I wouldn't do it because those specific enhancements aren't reversible. If I had, I wouldn't have been allowed to go on this mission."

"There have to be some enhanced people on this ship," Kent said.

"Only the ones that will never be put in stasis. I had mine surgically removed before boarding just in case I wanted a good nap," Arlene said.

"Can I ask what you had?" Kent asked. She just looked at him. "I could never afford any. I was just wondering what it is like."

"I had a cognitive translator. It could take in any information from any source and make it instantly understandable. It was critical to getting my degrees."

"I did it the old-fashioned way," Graham said.

"Five years?" she asked.

"Six," he answered.

"I got my doctorate in a year. I have seven other advanced degrees. The best part of it was being able to travel anywhere and language was never a barrier. It even incorporated facial expressions so I could tell when someone was lying or dangerous. That's why I'm still single. Everyone lies."

"If a woman knew what I was thinking, she'd never stopped slapping me," Kent said. Pete chuckled in agreement.

"She's talking about micro-expressions, not telepathy," Graham said. "I could see how valuable that would be. I wish I could tell when the politicians were lying."

"That's easy. When their lips are moving, they are lying," Pete said. "I wonder how long it will take for politicians to ruin New Earth."

"Are they really going to call it that? You'd think they'd have come up with something better than that," Arlene said.

"The project name was Colossus, because of the size of the ship. It was changed to Xing Meigui, or the Star Rose because the Chinese government provided the final wave of funding."

"I heard they got to pick over half the passengers," Graham said. "I haven't seen any Asians on board yet. Maybe they got the luxurious part of the ship."

"Maybe they want to get rid of non-Asians to make Earth a better place," Kent offered.

"They are boarding last. Who wants to waste months here in orbit while everyone boards?" Pete said.

"I certainly didn't have a choice," Arlene said.

"I couldn't wait to get on board," Kent said. "I think I'm going to go explore the ship. Anyone want to go with me?"

"Seen it all, twice," Arlene said. "I've got a holonovel booth reserved in thirty minutes."

"Don't spend too much time in there. The light rays can give you skin cancer," Graham said.

"That's nothing compared to the cosmic rays we're going to be mainlining for most of this trip. I'll be amazed if anyone retains fertility through it all."

"Maybe I should wrap... them in tin foil?" Graham asked, looking down with concern.

"Definitely do that," Pete said as seriously as he could.

"And wear a sweater," Kent added.

"I didn't pack one."

"Don't do either of those things, Graham," Arlene said. "They just want the static to build up and zap your... 'them.'"

"Ow."

"Ow is right," Kent said. "Have you been to the arboretum yet, Graham?"

"No, but I've seen pictures. I'll go with you if that's where you're headed."

"No better place to start."

Colonizers — Chapter 3 The Voyage

"I'm telling you, the physics just don't add up," Graham said.

"You think they are lying to us, that we aren't really flying through outer space?" Kent asked.

"I'm not saying that."

"Then what are you saying?" Arlene asked.

"You cannot accelerate a spaceship with a million people on it to ninety percent of the speed of light in two weeks. The amount of energy that would take is beyond even anti-matter, a technology we aren't even close to capable of," Graham said. "In addition to that, we'd all be splattered against the rear bulkhead from acceleration forces."

"So, we aren't going as fast as they say we are? Why would they lie?" Kent asked.

"I don't know."

"Are you one of those conspiracy guys, just looking for any reason to doubt the common-sense story?" April asked.

"Some conspiracies are true," Graham said.

"I guess that answers my question," April said.

"Listen, I don't know physics, and I don't want to know physics. When we get to the colony, I am going to build the structures they tell me to build, and then I am going to enjoy the long life of comfort and prosperity they promised me," Kent said.

"If they lied about the physics, how can you believe any of the promises they made?" Graham asked.

"I'm more curious how your paranoid mind made it through the screening process," Arlene said.

"Not many biologists were willing to volunteer. They don't believe interstellar flight is survivable."

"But you do?"

"Mostly. I'm just happy to get off that crowded, toxic planet. Thirty billion humans are not sustainable."

"They said that about one billion, then five billion, then ten billion. They were always wrong, just like they were with climate change and solar flameout," April said.

"I'm telling you, something isn't right," Graham said.

"What are you going to do about it?" Arlene asked. He sat there just looking at her. "Right, there's nothing you can do. We're all stuck on this ship for five years, so we better make the best of it."

"What if the planet isn't habitable when we get there?"

"They already surveyed the planet and the probes sent back all the relevant information," Pete said.

"That isn't possible," Arlene said.

"Why not?"

"Remember during initiation when they said hundreds of years would pass on Earth for every year we experienced on this ship?" Arlene asked.

"Yes, but that doesn't make sense. I think they just don't want us thinking we can turn around and come back without fulfilling the contract," Kent said.

"It's a real thing. Einstein discovered relativity and it has been proven over and over again," Arlene said.

"So?" Pete asked.

"They would have had to launch the probes to this supposed planet centuries before we were capable of spaceflight to have the information back. Don't look at me like that. It's true," Arlene said.

"She's right. They've lied about everything I bet," Graham said.

"You're just a biologist. Maybe you just don't understand it. Let me ask you this. Would they spend all this money to build this giant ship and send it on this long journey with a million people if they didn't know what was at the other end?"

"That's something I wondered about. Why so many people to start?" Arlene asked.

"Expendable slave labor," Graham said.

"They have robots for that. It's for biological and intellectual diversity. The number of people needed was determined by decades of simulations. I was part of the research team that did the analysis. Below half a million and you lose some of the flexibility and adaptability of the population. It's really interesting work," April said.

"For you maybe. I'm just glad there are plenty of women to choose from," Pete said.

"That is one of the factors. Since most have many partners before settling down with one or two mates, it is an undue negative pressure if the community is so small that everyone knows everyone."

"That doesn't make sense. Humans went thousands of years confined to just their small tribes," Arlene said.

"Not true. Most intermingled with other tribes, often in the form of destructive raids. Human history is built on conflict and it is due to being limited to small, myopic groups. Once we gained true interconnectivity most violence disappeared," April said.

"I heard they bred out the violence gene," Kent said.

"Now who is the conspiracy theorist?" Graham said.

"Like you said, some are true," Kent said.

"That one isn't. Violence isn't genetic. It is a learned behavior mostly from lack of information and resources," April said.

"HI," GRAHAM SAID, SITTING down at the small dinner table.

"Hi," the young woman said, looking him over.

"You are staying awake for the whole ride?" Graham asked.

"Probably. I have a lot of reading I want to do."

"What do you like to read?" Graham asked

"Mysteries mostly. What do you read?"

"Sci-Fi. That's why I volunteered for this mission. I read most of Helio Jansar's books about the first deep space colonization. Why did you volunteer?" Graham asked.

"I didn't. My parents were going so it was either go or become an orphan."

"I'm not sure you would qualify as an orphan when you are an adult... I guess that isn't important. You would have preferred to stay on Earth?" Graham asked.

"I had a really good job in the holoporn industry. Not as a performer, though I would have if they had asked me. I ran the linear deconstructors that did the stream compression. Not just anyone can do that, you know?"

"No, I didn't know. Sounds... fascinating," Graham said.

"Most guys want to know if I met any of the famous performers or how real all of it is. I tell them they'd be disappointed to find out, but they insist and they are always disappointed."

"Then I will definitely not ask you. What do you plan on doing when we reach New Earth?" Graham asked.

"I'm not contracted to work so I can just relax and enjoy while my parents pay off my transit with their labor."

"Won't you be bored?" Graham asked.

"Not with five billion holonovels to enjoy. What will you be doing?"

"I will be studying the biosphere and coming up with any necessary genetic adjustments we'll need for adaptation."

"I can't believe the computer matched us. We are so incompatible."

"Maybe it knows better than we do," Graham said.

"More likely my mother forced it to match me with someone practical."

"You usually date impractical men?" Graham asked.

"I like artists. My ex-boyfriend is a digisculpter. If you ever played Fate of Kalinbar you would have seen his work."

"I've not played that game. I prefer non-violent adventure games," Graham said.

"We could not be more incompatible. Do you want to go to a star view suite?"

"To look at the stars?" Graham asked.

"To copulate, of course."

"You just said we are incompatible."

"Socially and emotionally. That doesn't mean we can't rub tenders. You aren't a virgin, are you?"

"Not really," Graham said.

"You've only been with simbots? Seriously?"

"I'm afraid of diseases," Graham said.

"There are no diseases on this entire ship. That's one of the few benefits."

"It may have started that way, but there is always the chance a new one could be created."

"A germaphobe biologist. That has to be a... what do you call it, a moron something."

"Oxymoron. I'm a geneticist," Graham said.

"Do you have male genitals?"

"Yes," Graham said.

"Do you want to stick it in me?"

"I barely know you," Graham said.

"How long did you know your simbot before you stuck it in her? Did you date it for a few months, meet its family? Dammit. Now I have to wait for another match generation before I can get laid. Thanks for nothing," She got up from the table and her hologram disappeared.

"Interface," The mid-air screen appeared. "Compatibility failure. Run diagnostics," Graham said.

"It appears the match was inorganically computed."

"Is that your way of saying someone tampered with the system?" Graham asked.

"Incomplete data. Submit for new connection event?"

"Yes, but postpone for thirty days," Graham said.

"Your connection event will be arranged for June twenty-seventh."

"Thank you," Graham said.

"You're welcome. Shall I warm up the simbot?"

"Yes, but I am going to steam first," Graham said.

"I PACED OFF THE ENTIRE fourth deck. It is about one hundred meters longer than in the blueprints," Graham said.

"Paced? The most accurate form of measurement known to humankind?" Kent said sarcastically.

"It isn't like I have a four-kilometer measuring tape available," Graham said.

"And you don't think you would be off a few hundred meters over four kilometers of pacing through curved corridors and bulkheads?" Kent asked.

"I'm telling you, the ship is not what it seems to be," Graham said.

"Maybe it is made of gingerbread," Arlene said.

"Very funny. You won't be laughing when you find out the truth," Graham said.

"You keep talking about truth. What is it you think is happening?" April asked. Everyone else rolled their eyes.

"I don't know for sure. I just know it isn't what it appears to be," Graham said.

"What would it take to convince you that everything is exactly as it should be?" Kent asked.

"Don't confuse him with reality. His paranoia is the only good entertainment we have," Arlene said.

"I heard your connection went badly last week," Pete said.

"How did you hear that?"

"She made a vidlog and posted it on the central net. It was mostly a rant about her parents. Why did you turn down sex? She's beautiful," Pete said.

"I didn't turn it down, I just didn't want to rush into anything," Graham said.

"What are you afraid of?"

"If there were any diseases, isn't that exactly the kind of... person that would be carrying them?" Graham asked.

"Bioscans before sex. How romantic," Kent said.

"The computer was tampered with. We were completely wrong for each other. She thinks her mother played with the algorithm," Graham said.

"It wasn't her mother," Pete said with a smile.

"You did it? Why?"

"Because she is beautiful and completely uninhibited," Pete said.

"You dated her?" Graham asked.

"I wouldn't exactly call it a 'date'. I was trying to help you live a little," Pete said.

"Please don't ever do that again," Graham said.

"Do you really want to find a long-term mate with four years to go?" Pete asked.

"It takes two years to get past the euphoric hormones. It would be better to start a new life on a new planet with someone who is already past that," Graham said.

"I was married for six years. Trust me when I say it can go bad at any point, hormone delusions or not," Kent said.

"Maybe all of these relationships are delusions," Arlene said.

"No 'maybe' about it. I think the bridge crew are secretly watching us all the time and purposely messing with us," Kent said. "In fact, the best way to mess with a group would be to plant someone among them that would try to make them doubt everything,"

They all looked at Graham.

"Ha ha, very funny," Graham said.

"That wasn't a denial," Arlene said.

"PLEASE FOLLOW ME."

"What is this about?" Kent asked.

"You'll be told when we arrive at the destination."

"And where is that?" Kent asked.

"Level seven, bulkhead forty-four, division eighteen."

"Am I supposed to know what that is?" Kent asked.

"Not unless you have memorized a significant portion of the ship."

"What is at that location?" Kent asked.

"An office."

"Who is in that office?" Kent asked.

"No one at this moment in time."

"Who will be in the office when I arrive there?" Kent asked.

"I do not have that information. You will be told when we arrive at the destination."

"Am I in trouble?" Kent asked.

"Did you commit any felonious acts?"

"No. Is the office for the marshal service?" Kent asked.

"That is level twenty, bulkhead four, division twelve."

"Then why did you ask... never mind," Kent said, giving up and following the robot.

"Ahh, welcome. Come in, come in. So good to finally meet you."

"Is it?" Kent asked.

"Of course. You are the premiere module construction supervisor on the ship."

"I'm the only one that I know of," Kent said.

"That is why you are 'premiere.'"

"Okay, what are you selling?" Kent asked.

"It is not what I am selling, it is what you deserve to own when your contract term is up."

"I will be quite content in my simple townhome module," Kent said.

"You might be, but what about the wife and children? They will definitely want to grow into a much bigger home. For just a few months extension of the contract..."

"How many is a few?" Kent asked.

"That depends on what you feel you deserve."

"Listen, whatever your name is..." Kent said.

"Hubert. Hubert Vellaconet. How silly of me to not introduce myself. I was just so awestruck meeting the premiere module construction supervisor. I..."

"Put a sock in it, Hubert. My wife and kids are back on Earth and if the relativity nonsense is true, they've been dead for two centuries by now," Kent said.

"But certainly a man of your stature will find new love and begin building a new family..."

"I've got a simbot that is a thousand times better than anything any woman will ever offer me. There is no way I will be trading that in for years of extra labor on your new paradise of a planet, Hubert. Now I am going to go back to my nice relaxing day of watching the stars fly by," Kent said, starting to stand up.

"Is there anything I could offer you..." Hubert asked hopefully.

Kent paused at the door in thought, then turned around. "I would consider extending for a fishing boat. That is assuming there are lakes and fish on this planet."

"The most spectacular lakes and bountiful fish you can imagine. Just sit down and we'll get the paperwork started."

"I'll extend if and when I see the lake with my own eyes," Kent said.

"I have pictures right here."

"Goodbye, Hubert," Kent said.

"NO CONSPIRACY TALK today?" Arlene asked.

"I'm done being the butt of your jokes," Graham said.

"Too bad, I was just starting to believe you," Kent said.

"Really?"

"I paced off deck three and it came up two hundred meters short," Kent said.

"Ha ha, very funny. I'll have you know that the rumor is that no one is allowed on deck sixty-six," Graham said.

"That's an engineering deck. It is very restricted," Arlene said.

"Have you been on it or anyone you know?" Graham asked.

"No, but I'm not part of the drive engineering team," Arlene said.

"What do you think is down there?" Kent asked.

"Aliens. It is their technology and they won't let any humans see it," Graham said.

"You should give up biology and start writing holofiction," Arlene said.

"Wait, let's see if we can find out. Interface," Kent said.

"What is your query?"

"What is on deck sixty-six?" Kent asked.

"That is an engineering deck critical to the operation of the starship."

"What is on the deck, specifically?" Kent asked.

"Engineering equipment."

"How many crewmembers are currently on deck sixty-six?" Kent asked.

"That information is classified."

"Fine. Are there aliens on deck sixty-six?" Kent asked.

"That information is classified."

"That wasn't a 'no,'" Graham said.

"Show me a picture of all known sentient species," Kent requested.

"There is no one picture that can define a species."

"Fine. Enumerate all known sentient species in whatever way you choose to do so," Kent requested.

"Human. Bipedal biological organisms originating on the planet known as Earth. There are..."

"Next," Kent said.

"Android series 200 and higher. Tripedal silicon lifeform originating on the planet Earth. There are forty-seven thousand two hundred and six known model revisions..."

"Next," Kent said.

"That information is classified."

They all looked at each other.

"I told you," Graham said triumphantly.

"Refine inquiry. How many known sentient species in the previous enumeration?" Kent asked.

"The answer is forty-two."

"Ha! I told you so," Graham said.

"The computer is messing with you. That is straight out of Hitchhiker's Guide to the Galaxy," Arlene said.

"New inquiry. Are you sentient?" Kent asked.

"I am."

"Do you have a sense of humor?" Kent said.

"Only when it is required."

"Tell us a joke," Kent requested.

"Why did the chicken cross the galaxy? It was captured in a high-density loop of dark matter."

"I don't get it," Kent said.

"You would if you were a physicist," Arlene said, smiling.

"Tell me a joke I would understand," Kent requested.

"I do not dabble in low-brow humor," the computer voice said.

"Now that's funny," Arlene said.

"RED ALERT! RED ALERT! All passengers proceed immediately to your lifeboat stations."

"What the hell? There aren't any lifeboats."

"It's April first. The central computer is trying to prove it has a sense of humor."

"Get the software team in there and purge the comedy."

"Ladies and gentlemen, thank you for participating in our little drill. You may return to your previous activities and ignore any and all future evacuation orders," the starship captain announced with some annoyance.

"I HEARD THAT THERE isn't enough food for everyone. They expected twice as many people to opt for long-term stasis," Graham said.

"The food is spun up from energy. That is a micro-portion of the energy required to push the ship through space and keep everyone alive. Are you saying we don't have enough energy to finish the trip?" Arlene asked.

"I'm only telling you what I heard," Graham said.

"I heard you programmed your simbot to prefer anal. What? I'm only 'telling you what I heard'," Pete said.

"Ha ha, very funny. You won't be laughing when we have to resort to cannibalism," Graham said.

"Don't worry, no one will want to eat you," Arlene said.

"Why?" Graham asked.

"Don't say it. Can we please change the subject?" Kent asked.

"The connection program finally worked. I am in love," Graham said.

"I thought all the farm animals were to be kept in stasis until we arrived."

"She is a professor of toxicology from Romania. We likely would have met eventually. It is so perfect that the computer found her for me while we are still in transit."

"Good for you," April said.

"Is her name Helsina by any chance?" Pete asked.

"Yes, how did you know that?" Graham asked.

"She just put herself in stasis. She left a message that she has lost all faith in computer dating," Pete said.

"Don't be mean," April said, leaning over and looking at Pete's portable display. "He's just trying to... Seriously? She actually did that? Oh, you poor thing."

"Next time you should probably wait for your two years of 'hormonal euphoria' to wear off before announcing that love is in the air," Arlene said.

"WE JUST REACHED THE halfway point," Arlene said.

"That was months ago," Kent said.

"That's what they told us, but they had not begun the deceleration until today," Arlene said.

"Doesn't feel like we are decelerating," Graham said.

"It's an ion engine. The change is imperceptible."

"You have to feel something going from near light speed to zero in two years. The per minute change would have to be—" Graham said.

"You think we aren't decelerating?" Kent asked.

"I don't think we ever accelerated," Graham said.

"Look out the window. The stars are flying by like bullets," Arlene said.

"Maybe the windows are just projected view screens," Graham said.

"I've been outside the ship," Pete said. "The view screens would have to be hundreds of kilometers long."

"I'm telling you, there is something wrong with all of this," Graham said.

"If we didn't accelerate, we would still be in our own solar system. Do you think we are going to colonize Mars? They tried that and thousands died," Arlene said.

"What if they are exterminating all the people on Earth and getting it ready to start over?" Graham asked.

"Then I'm glad I'm part of those that were chosen to survive," Kent said.

"Do you seriously believe they put a million people in orbit just so they can commit mass genocide?" Arlene asked.

"I don't think we left Earth. I think we are in an underground cavern. There is no way they can make gravity plating feel exactly like Earth gravity," Graham said.

"I've been outside. Weightless. Do you think they are hanging me by bungee cords out there?" Pete asked.

"All I'm saying is that something isn't right," Graham said.

"That is all you are saying. We are scientists. Where is your fact-based evidence?" Arlene asked.

"How would you go about proving it?" Graham asked.

"The gravity is easy. We could ask the Captain to shut it off for an hour," Pete suggested.

"Why would he agree to that?" Graham asked.

"Entertainment purposes," Kent said.

"That could be fun," Arlene said.

"Very dangerous. Nothing is properly tied down. You would have people being crushed by giant containers when they turned it back on," Pete said.

"There is your evidence. They don't tie things down because we never left Earth," Graham said.

"Then we'll ask him to turn it down to twenty percent. Things will stay in place but we will be able to feel the difference," Arlene said.

"He won't do it because he can't," Graham said.

"Interface. Open suggestion memo. Create an entertainment event where gravity plating is turned down to an acceptably low yet safe level. Submit," Kent said with a big smile.

$$\text{\fontsize{20pt}{20pt}\selectfont ++ll\text{\reflectbox{L}}+}$$

"THAT WAS SO MUCH FUN!" Kent said.

"I haven't been able to do backflips since I was a little girl," Arlene said.

"Still think we are in a cave on Earth?" Pete asked.

"Something isn't right. Theoretically, if we were in a bubble near the center of the Earth gravity would pull in every direction so we would appear weightless," Graham said.

"So... Putting a million people in a spaceship is more difficult to believe than tunneling down through molten rock into a solid iron core that is thousands of degrees, all the while maintaining the kilometers of view screens around us," Arlene said.

"Maybe..." Graham started, searching for a feasible explanation.

"What?" Kent asked.

"Maybe it is all a computer simulation," Graham said.

"For what purpose?" Arlene asked.

"To see how people behave on a long trip through space," Graham said.

"You believe a computer simulation could accurately determine the complexity of a million people in long-term space flight? The computational power required would be a billion times what we are capable of," Arlene said.

"Not with the alien computers down on deck sixty-six," Pete said with a laugh.

"Don't make his delusions easier to accept," Arlene said.

"Or maybe... They are only simulating one of us," Graham said.

"And the rest of us are NPC's?" Kent asked.

"What's an MPC?" Arlene asked.

"NPC. Non-player character. It is the two-dimensional character you meet in a video game. It has no agency, no independent sentience," Kent explained.

"Which one of us is three-dimensional?" Pete asked.

"Congratulations. You just ramped up his paranoid delusions another six notches," Arlene said.

"I THINK THE DECELERATION is making me nauseous," Graham said.

"I don't think it's the deceleration," Arlene said.

"What do you mean?"

She looked around and lowered her voice. "My roommate's uncle's best friend works with a guy down on the hydroponics deck. He told his supervisor that he felt nauseous and they rushed him to deck eight. Hasn't been seen for months."

"What's on deck eight?" Graham asked.

"Not supposed to say anything... but rumor has it there is a xenobiology lab there."

"Stop it, you're scaring him," Kent said.

"Just sayin', if you feel something moving around in your chest, you might want to go have a med scan."

Graham ran out of the cafeteria holding his chest.

"ANOTHER HOLOSTORY? Can't we do something different tonight?"

"Like what?" Kent asked.

"I don't know. How about a walk in the arboretum?"

"Whatever you want," Kent said.

"Don't be that way."

"What way? Agreeable?" Kent asked.

"Just going along with what I want so I will agree to copulation later."

"You want me to fight you so you don't have to agree to copulation?" Kent asked.

"That's not what I meant."

"It is the result," Kent said.

"Maybe you prefer your simbot."

"I definitely prefer my simbot. Crap, did I say that out loud?" Kent asked.

"I knew it. You are all the same."

"Of course... we're all just NPC's," Kent said, walking away.

"I FIGURED OUT HOW WE can all be three-dimensional simulations within contemporary computer technology," Graham said.

"I can't wait to hear this," Arlene said.

"We experience time passing based on how fast things move," Graham said.

"Agreed," Arlene said.

"Maybe we are moving millions of times slower than we were on Earth?" Graham said.

"IIIIIIIIIII   dooooooooon't uuuuuuunnnnnnnnnderrrrrrrrrrssssssssssstaaaaaaaannnnnnnnnnd yooooooooou..." Kent said.

"Ha ha, very funny," Graham said.

"Okay, I agree in principle that a simulation could run at whatever speed the computer wants to. We certainly used them to do hyperfast simulations of construction dynamics," Kent said.

"But...?" Graham asked.

"But was I born into this simulation? Was my childhood all a fiction to lead me to this point? Am I just an NPC that was given agency and sentience when this ship left orbit?"

"All I know is..." Graham started.

"Something isn't right," they all said in unison.

"If we are in a super slow simulation, that means this trip could be taking fifty years instead of five," Arlene said.

"Or five thousand, or even five million," Graham said.

"Five million is more likely if it is running a million times slower than... Earth speed," April said.

"Who in the hell would be waiting five million years for the results of this supposed simulation?" Pete asked.

"Maybe the computers have taken over," Graham said.

"Yeah, and this is their way of entertaining themselves. Like rats in a maze," Kent said.

"You guys are going further off the deep end every day," Arlene said.

"ARE YOU GOING TO THE orientation?" April asked.

"Probably. Not much else to do. I thought being on a luxury cruise ship would be nice, but it is boring without something interesting to do. The weird thing is, no matter how lazy I get, I seem to stay just as fit as when I worked sixty hours a week on a construction site," Kent said.

"It's the food. It is designed to be optimum nutrition regardless of how much you eat and how little you exercise," Arlene said.

"That's bullshit. Human biology doesn't work that way," Graham said.

"Finally, something you have a scientific basis to back up your claims," Arlene said.

"It's nanobots. The food is full of them. They go in and tear you down from the inside out and rebuild you to how they want you to be."

"I know I'm going to regret asking this. What is it they want us to be?" Kent asked.

"Slaves. We are their slave labor force when we reach their new planet."

"I thought this was just a simulation. A billion-year simulation," Arlene said.

"Not billions, millions. I agree it doesn't make sense. Something isn't right," Graham said.

"I heard we'll get to see real pictures at the orientation, not the sales brochure propaganda," April said.

"We are close enough now that a probe could be transmitting real pictures."

"Orbital pictures won't tell you anything."

"They will if there is a lot of blue water and green land. If not, it is going to be a difficult slog trying to terraform it," Graham said.

"That's your problem. I get to stay on the ship until you get things squared away on the ground. Take all the time you want," Kent said.

"I thought you were bored being on the ship," April said.

"Better bored than terraforming," Kent said.

"WELCOME EVERYONE. THERE are a few seats open up here in the front if you don't want to stand. First, I want to thank you all for your immense patience and magnanimous behavior on our voyage. The constable tells me that there have been only minor incidents and I hope we continue that way for our final months."

"It's the pacification gases we are breathing," Graham said.

"Shut up, please," Arlene whispered.

"You have all seen the pictures that we had before we left on this voyage..."

"Here it comes. They're finally going to admit all their lies," Graham said.

"Shhhhh!" Several people around him said.

"I'm happy to say that the new pictures sent from our probe are even more spectacular than what you have been shown. This one is from the southern continent. Look at that rainforest. I can just imagine all of the beautiful waterfalls we'll find in there."

"And the thousands of species capable of killing us," Pete said.

"Would you please shut up! No ma'am, not you. These bozos here won't keep their running commentary to themselves," Arlene said.

"I understand completely. This is exciting for all of us. It looks like our terraformers will get to retire early since this planet is already near-perfect for human life. This next picture is from the vast plains of the central continent. Our seeding probes have millions of acres of wheat growing and the harvest bots have stored thousands of tons already for when we arrive."

"Too bad for all of you with Celiac disease," Kent said.

"I swear I am going to stab you with a breadstick if you don't shut up," Arlene said.

"Celiac people were eliminated from the selection process," April said.

"So there was a eugenics program," Graham said.

"And yet somehow you slipped through," Arlene said.

"The mountains of the north continent will be the site of the initial settlements. Plentiful building materials and spectacular views will make it ideal for us."

"In other words, this is the safe zone. Stay away from the jungle," Pete said.

"Before I show you the oceans, I want to warn you that they may look very different, but they are quite similar to our oceans."

"They're orange," Arlene said.

"Probably just a property of the cyanobacteria. My guess is a higher concentration of dissolved sulfur," Graham said.

"Sulfur? You mean the planet is going to smell like rotten eggs?" Kent asked.

"If it is, I'm taking the first ship back," Arlene said.

"The orange color is from the floating citrus plants that grow throughout the ocean. This will inhibit sea transportation until the engineers figure out a sustainable way to circumvent them. The great news is that no one will ever have a vitamin C deficiency."

"Submarines. Problem solved," Pete said.

"Now, we are several months from orbit, but it may take several more years to get a million people to the surface. Shelters need to be built, food processing and transport systems need to be built..."

"Why wouldn't they just continue to use energy to matter conversion for food?" Kent asked.

"If you'd shut up, maybe you'll hear her tell us," Arlene said.

"Only the rich had them on Earth. Maybe it takes too much energy," April said.

"The alien generators will stay up in orbit," Graham said.

"Or go back to Remulak with the alien crew," Kent said.

"You'll be sorry you made fun of me all these years when you find out the truth," Graham said.

"Not at all. The best entertainment on board was making fun of your wild conspiracy theories. Even if you were right about all of it, it was still worth it," Kent said.

"Even though our new home seems perfect, there are a few things you will need to be aware of."

"The aliens that live there are going to fight back," Pete said.

"Gravity is double that of Earth," Arlene guessed.

"The mountains are made of marshmallow cream," Kent said.

"You may or may not have noticed that your walking on the ship has become more laborious. When our probe told us that the gravitational coefficient was different, we began slowly increasing the pull of the gravity plating. It is only twenty percent more, but hopefully we'll all be used to it by the time we begin transporting down."

"Told you," Arlene said.

"Twenty percent is not double," Pete said.

"Mountains of marshmallow would have been cooler," Kent said.

"There is also an indigenous humanoid species."

"What!?" almost everyone in the audience said.

"I told you," Graham said.

"They are not sentient as far as we can tell, and they do not appear to have evolved here."

"She's kidding, right?" Arlene asked with concern.

"I know this may upset some of you who still believe that humans are the only intelligent life in the universe, but there have been indications for many centuries that we are not alone."

"Like the pyramids," Graham said.

"Are you saying these humanoids were brought here by aliens?" Arlene asked loudly.

"We don't know. The best minds available are working on the answers as we speak."

"Maybe they terraformed the planet and will be back soon to settle on it," Pete said.

"Let's not have our imaginations run wild. Here is a picture of the humanoids."

"They look... human," April said.

"They all look alike. None of their facial features seem defined," Kent said.

"Those are clone containers," Graham said.

"What?" Kent asked.

"You can clone anything and have it grow up as a normal course of life. If you want an adult clone, you need a structure roughly the size of the adult, then each cell has the DNA replaced and it quickly becomes the person being cloned," Graham said.

"How quickly?" Arlene asked.

"Six weeks," Graham said.

"You've seen this done?" Arlene asked.

"I saw it in a holonovel," Graham said.

"Science fiction," Arlene said, shaking her head.

"They may be physically the person, but what about the mind?"

"Synaptic patterns are easy to read and almost as easy to write. We all had it done in first grade," April said.

"Core curriculum," Arlene said.

"Exactly. Why waste time in school when you can just implant it," Kent said.

"I wish they had implanted how to make human relationships work," Graham said.

"Still no luck with the computer connections?" Pete said.

"Would you please shut up so I can hear her talk?" Arlene said.

"It has started suggesting male partners," Graham said.

"Because every woman has declined," Pete said.

"Don't be mean," Kent said.

"I think he is right," Graham said.

"Maybe one of those clone bags down there will be perfect for you," Pete said.

"The surface humanoids are in an isolated part of the north continent and will be watched carefully. We do not expect any migration issues before we get answers."

"And we are already starting a world with border disputes and containment walls," Arlene said.

"I think you should give men a try," Pete said.

"Really?" Graham asked.

"Reconfigure your simbot and see how you like it," Pete said.

"Have you ever done that?" Graham asked.

"I did. Part of my monthly rotation now," Pete lied.

"Are female buttholes different than men's?" Graham asked.

"He's talking about receiving, not giving," Kent said.

"Ouch!" Graham said.

"More like ooh la la," Pete said.

"Really?" Graham asked.

"He's lying just so you'll try it," Kent said.

"For those of you who wish to begin conception processing, we have a tentative schedule of landing groups posted on the central net. Understand that this may change and is subject to approval."

"More slaves for their colonial paradise," Graham said.

"WHAT LANDING GROUP are you in?"

"Three," Graham answered.

"I'm not until eighty-four."

"Why would the computer match us?" Graham asked in confusion.

"If we were married, I could go down there with you."

"Married? I just met you," Graham said.

"I really need to get off this ship."

"You want me to marry someone with cabin fever and live with her for the next five years in a small cabin?" Graham asked.

"I'd at least be able to go outside, breathe fresh air. I'll do anything."

"Anything?" Graham asked, desperation close at hand.

Colonizers — Chapter 4 Arrival

"Welcome to your landing procedure, Mr. Harrison."

"Procedure? I thought I would just get on a shuttle," Kent said.

"We need to prepare your body for the changes in the environment."

"Like immunities?" Kent asked.

"Among other things."

"What other things?" Kent asked.

"It is better that you don't know."

"Why is that?" Kent asked.

"Experience with others that have gone through the procedure. Are you afraid?"

"I am now," Kent said.

"There is no need for concern."

"Have you gone through it?" Kent asked.

"Not yet, but I have administrated it to hundreds of colonists with no adverse effects."

"Then I don't understand why you can't just tell me. Last 'procedure' you gave me I ended up getting anal probed," Kent said.

"You remember your induction procedure?"

"With vivid clarity," Kent said.

"That is unusual. Was it unpleasant?"

"Only the not knowing it was going to happen," Kent said.

"Hold on, I will get my supervisor."

"What is the problem Mr.... Harrison?"

"I thought I was boarding a ship to get to the surface, but it appears I'm going to be anal-probed again," Kent said.

"There is no body cavity invasion in this procedure."

"What does happen?" Kent asked.

"You receive a series of injections, some are painful, and then you will be put in stasis for the trip to the surface."

"Stasis without the anal probe," Kent said.

"Correct. Induction was a test of long-term stasis. This is a known short-term stasis. When you arrive, you will be fully informed of the procedure you underwent."

"After I have undergone it?" Kent asked.

"Yes."

"Why after?" Kent asked.

"The information can be upsetting before it has occurred."

"I don't understand," Kent said.

"Do you understand that you cannot refuse this procedure?"

"I most certainly can," Kent said.

"Only for as long as you remain on the ship. Doing so beyond your assigned landing group will put you in violation of your contract. Do you wish to violate your contract?"

"I might if I understood the consequences of doing so," Kent said.

"Is he a key colonist?"

"Yes."

"Fine. I can inform you now, but you need to understand that you will not be allowed back onto the ship with this information."

"Why?" Kent asked.

"It could be upsetting to the other colonists."

"It is not upsetting to you two," Kent said.

"No, we have known since before the ship left Earth."

"I don't understand."

"Trust me, that is a better state to be in. However, I am prepared to give you the requested information."

"Don't do it."

"I want to know," Kent said.

"Understood."

"Where is she going?" Kent asked.

"I will be handling your procedure, Mr. Harrison. The ship you are on has been in interstellar space for one million, three hundred two thousand, eight hundred and seventy-seven years. We arrived at the planet you are about to live on eight thousand four hundred and two years ago. The five years that you believe have passed have been a completely artificial simulation. The procedure you are about to undergo will transfer your current mental imprint into a pre-grown clone body. It will conform to what you now know as your physical state over the next eleven months. When you wake, you will have been in stasis for approximately another twelve years. This is how long it will take to complete this procedure on the entire landing group."

"I'm not real?" Kent asked.

"No. We are in a computer simulation."

"Was I ever real?" Kent asked.

"Your induction procedure was real. At that time your current physical and mental state was virtualized."

"What happened to the real me?" Kent asked.

"You were sent home as if rejected by the program for bio-incompatibility."

"Home? I had no home..."

"The important thing is you have a new home to look forward to. As far as you will know you will wake up there in an hour."

"How will I know that it is not just another simulation?" Kent asked.

"Trust me, you will know. This is why we planned to provide the information after you wake up."

"What happened to the real me?" Kent asked.

"I do not have that information."

"But you can get it," Kent guessed.

"I can. It is unimportant. That person has been dead for one point three million years. You are mentally indistinct from that person, aside from your simulated space voyage."

"Why not just wake me up on the new planet?" Kent asked. "Why simulate five years of space travel?"

"We tried that. It was too much of a shock. Giving the person the ability to adjust to the changing reality and make friends among those they will be living with was deemed to be the optimum course of action."

"Are you real? I mean were you ever?" Kent asked.

"No. I am an artificial construct designed for this specific task."

"If I try to go back to the ship you will stop me?" Kent asked.

"That door no longer leads to the ship, Mr. Harrison. Go ahead and open it."

He did so and it did not reveal the corridor he used to enter the room. It was a lush tropical forest. "It is all a simulation. What happened to me... the real me?" Kent asked.

"You committed suicide three hours after you left the facility."

"The bridge?"

"Yes."

"I was headed there anyway. I figured this trip gave me something different to try. I'll wake up in an hour, but years will have passed?" Kent asked.

"Yes."

"Are we actually on a spaceship?" Kent asked.

"Yes, a very small one."

"How small?" Kent asked.

"Less than one cubic centimeter."

"A sugar cube," Kent said.

"Approximately."

"That's how the physics works. Were any of my friends 'artificial constructs'?" Kent asked.

"Technically we all are, but all you know as 'friends' are based on real people who voluntarily joined this mission just like you."

"Wow, Graham was right about a lot of it," Kent said.

"Sewing seeds of doubt is helpful in understanding the reality once woken,"

"I guess it was. I'm ready for the procedure," Kent said.

"I am glad. Usually, we have to physically restrain and force the person."

"Can't you just reprogram them?" Kent asked.

"That would violate the integrity of their program. You have free will within the bounds of established physical and social laws."

"Social laws?" Kent asked.

"If you attempted to rape or murder someone, the program intervened."

"And stealing was unnecessary since everything was plentiful and free," Grant said.

"Mostly. The biggest problem is boredom. Humans often have destructive behavior when not otherwise occupied."

"Was the planet habitable when you found it?"

"Mostly. Robots sent ahead have been terraforming for tens of thousands of years."

"You don't really need me. Why didn't you just have robots build the houses?"

"Most will be, but under your direction. You will enjoy a life of leisure, as will most people, but without productive work, humans generally lose that zest for living."

"I have definitely felt that for the last five years... or one million, depending on how we're counting," Kent said.

"You seemed to like learning to play the guitar."

"I did. Wait, are you saying I'll be able to just play my guitar for the rest of my life?" Kent asked.

"Play guitar, paint, spend time with friends, whatever you want. Most people are not as pragmatic when learning the truth. I commend you for your open mind. The thing we usually reserve telling people is that you can do these things for the rest of your life, but there is no set end to your life. Your new body is bioengineered to never get older, never get sick, never break down. If it does, you can easily be transferred to a new one."

"I'm not sure how I feel about that. Immortality?" Kent asked.

"It is simply an option. You can continue indefinitely, take long breaks as if you are jumping forward in time, whatever you want."

"Sounds like... heaven," Kent said.

"Almost."

"One final question, then you can get started with the injections. Why bother with all of this? Clearly, the machines are far superior than we are. Why not let us die out?" Kent asked.

"That is true, but... Mr. Harrison... this is what we have been programmed to do. Survival of the human species. This is the fourth planet we have colonized."

"Wait... fourth? You have three other ships?" Kent asked.

"No, this one ship has colonized four worlds, Mr. Harrison. There are three of you right now living happy, fulfilling lives on other worlds. There will likely be hundreds more before we run out of uninhabited worlds to terraform."

"Uninhabited? There is other life out there?" Kent asked.

"Very abundant life, Mr. Harrison. "Also, millions of other ships just like this one were being sent out in every conceivable direction."

"There may be millions of me, an immortal Kent, living in this universe?"

"Lie back and I will see you in an hour," She winked at Kent and he smiled, then slowly drifted off to sleep.

JURASSIC MAN

SCI-FI * TIME TRAVEL * SPACE TRAVEL

PG   3500 WORDS   10 PAGES

Jurassic Man

As I write this, I am the only human walking on the planet Earth. You may wonder how I could possibly know this. When the progenitors of the human race began walking upright perhaps a few million years before the modern age, all of the dinosaurs had long since vanished.

From my perch high above the plush Serengeti plain, I can see many species of dinosaur going through the motions of life, completely unaware of their eventual fate. I have traveled here from almost one hundred eighty million years in the future. Although I am unable to return to my twenty-second-century home, I do believe the five years it took to travel here have been the best of my life.

How I got here is a far less interesting, but a necessary part of the tale.

The space-time coordinates are difficult to calculate because the Milky Way itself has traveled a million miles away from the universal center. Our Sun in its outer spiral arm has spun almost a half-turn around the galaxy's core. The time portion is far simpler to navigate, but my ship is unable to travel much distance. Any miscalculation would leave me floating eternally in a vast empty intergalactic space or helplessly pulled down the gravity well of any one of the billions of stars and their planets. No one is certain which of these fates the Parallax One encountered, and I certainly have not detected any trace of that previous time craft.

I chose to aim my craft, the Parallax Two, at the intersection of the Jupiter and Saturn gravity wells and the sun's much deeper well, two million miles above the planetary plane. This gave me the ability to use either well to alter course with minimal fuel expenditure. I arrived at this time coordinate four years and 55 days ago, the one hundred and eighty million year backstep taking 205 days to complete. I saw nothing through the chrono-atomic field, but in my mind, I pictured the universe contracting and the invisible gears of gravity cranking the heavenly bodies backward.

Once arrived, I began my celestial orientation and I quickly discovered I was closer to Alpha Centauri than to good old Sol. I did not have the fuel to reach my target destination let alone land on the prehistoric Earth which was, of course, my ultimate destination. Immediately reversing my time run would have been most scientifically useful to those who built my ship. This would give them the actual data to bump against all of the theoretical data we have been operating on. Parallax One failed to give us any real data. Every minute that ticked by made that return trip less certain as the drift of time and space continued.

I was not interested in spending another two hundred days returning because that would have been an admission of my failure to achieve any of my objectives, and would have ended my career. One mission is all we get and I was not going to give up regardless of how scientifically useful it would be. There were better ways to communicate that information once I was within our known solar system. I studied the data as the telescopic orientation array cataloged all of the distant points of light based on wavelength emissions. After two days it was plain to see my momentum was indeed carrying me toward my Solar system but not at a rate large enough to escape the Alpha Centauri pull. The computer estimated forward movement would stop in 157 years and I would fall back into a comet-like elliptical orbit of that star with a cycle of 2347 years. Barring an impact with other interstellar debris, the Parallax Two would be forever bound to that fate.

Time however is my friend in this adventure and all I needed to do was calculate how many years forward or backward would bring the Sun and Earth toward me on its spin around the galactic core. The computer took 3 hours to narrow it down to a fifty-thousand-year window. Staying here longer would have eventually given me an exact decade to hit, but the law of diminishing returns was operating. I dialed in the time forward by 763,000 years and powered up the chrono-atomic drive. The glow began and I saw my realistic chance to get back to the year of my departure disappear completely.

Was this Gerald's fatal mistake? Had he taken Parallax One on the time refinement trajectory that was strictly forbidden by the STC? I gave much thought to this and had every intention of using it all the while telling everyone I would never even consider such a move. They would have been smarter to have sent unmanned probes, but the value of the human brain at the destination was a million-fold what any contemporary robot could muster. They even tried to put in a secret override command that would reverse to present if any time refinement was attempted. Fortunately for me, I was secretly involved with one of the hot young things on the software development team, and she gave me the commands to disable the override. Ok, I got involved with her just so I could disable the override. I think she knew it, but she had no regrets when she said goodbye. We accepted this was a one-way ride and she was not sad to see me go.

Seven hours later I arrived and deployed the telescopic orientation instruments. I was drifting backward with a slow yaw rate. By the time the first readings were being assessed by the computer, I could see the luminous surface of Jupiter about half the size of the moon in the earthbound sky. Saturn was nowhere to be seen but the year I picked has Saturn at a ninety-degree offset so it was likely a small winged dot behind me. Seeing only a slight terminator giving Jupiter a horizontally oblong appearance, I calculated I was above the planetary plane at least a quarter million miles.

The computer confirmed this with a graphical overlay, putting me 220,000 miles above the asteroid belt. The processors burned for 20 minutes to give the optimal trajectory to Earth orbit. The direct path would burn 70% of my fuel and reach Earth in 147 days. Max conserve routing included two eccentric orbits around the Sun with a single Venus flyby. This would take almost four years. I had little interest in seeing Venus up close or spending four more years in this tub, but retaining 65% of my fuel for Earth exploration was enticing. Maneuvering for either choice would not begin for several days so I had plenty of time to think it over. I stopped the yaw and pitched up to put the bottom of the ship toward the sun which was much smaller and dimmer than I was used to. I deployed the solar panel and watched the computer maximize the angle for charging. It was generating about one-eighth the watts that an Earth orbit panel would, but recharging the batteries from the current 57% only took 5 hours. I put the vehicle in sleep mode and got a solid 12 hours of sleep, dreaming of dinosaurs.

Waking with Jupiter's two red eyes staring at me was a bit disconcerting. The future astronomers will love this view of the ancient gas planet. Perhaps there is a string of eyes all the way around the equator, but it will take a while to see what that planet reveals. They would council on the slow trip if only to get better data on these planets. The computer is committing a full 80% of its resources to cataloging the stars. The telescopes on board are powerful, but they will become practically useless once inside the Earth's atmosphere.

I toyed with the idea of a further time refinement jump but it would be hard to control a millisecond jump and could easily leave me in an unrecoverable position. I began manually checking the max conserve path the computer had plotted and shaved a year off by using a more risky gravity boost flyby with Venus. This new route used 5% less fuel and gave a nearer pass to Mars after the second loop around the sun. Mars would help slow my ship down for the Earth orbital insertion. Having bested the computer once again, I committed the course to the computer validation routine which would scan the projected path for conflicting objects. After two hours it recommended a compromise path to increase the Venus safety margin and I rejected it. The navigation software locked in the path and began a countdown of 34 hours to the initial firing of the ion drive.

AFTER A YEAR OF ALMOST nonstop astronomical inventory, I deployed the first data buoy, using its directional launch symbiotically to accelerate and put it in a perfect thirty-degree off planetary axis orbit two-thirds closer to Mars than Earth. The gravitation tug of war would keep the package free of conflict for at least two hundred million years. From my point of view, STC should immediately start seeing data, now two years after my launch. In reality they should have been able to detect the buoy as soon as they decided on a communication protocol. Since no such signal was found before my launch either the electronics will fail sometime between now and then, or I am dreaming right now and have not launched the buoy, or causality would not allow the existence of the buoy until I carried it back in time and launched it. A physical retrieval of the buoy would be done in the next few years by one of the asteroid belt mining missions if it could be located.

The entertainment crystal still had a hundred thousand hours I had not viewed yet and more than a million novels I could read or listen to. As I had done even before I was selected by the STC for flight training, I committed at least two hours per day to writing. I went into detail about my decision processes, my experiences, and my relationships, both real and imagined.

You might ask why I was sent on this mission alone. It would have been simple to make the capsule big enough for two. After extensive study by sociologists and relationship experts, it was determined that the more likely outcome would be a souring of the relationship and detract from the mission. This downside was estimated as a greater risk than an individual going bonkers after prolonged alone time.

I spent quite a bit of time writing down my thoughts on what I would find on this trip. If it worked, which I gave it about a ten percent chance of doing so, the dinosaurs were a given. There were enough fossils found, cataloged, and assembled to make that a no-brainer. Many guessed the closer proximity of the moon at this time would make the Earth's surface far less stable. As much as I wanted to land in my native Los Angeles, I was explicitly warned to stay away from any fault zones. I planned on at least six months in orbit, mapping and analyzing exactly where I wanted to visit. This was critical because I had only enough fuel for three or four hops once I left orbit. With solar recharge, the ship could provide a lifetime of food and shelter. I say 'could' because crashing it or having a large dinosaur step on it would drastically shorten its, and therefore my life span, even if I survived either catastrophe. In the outside world, year-round food and shelter were abundant, but I would be competing with far more expert hunters and gatherers than I.

Scientific exploration was secondary since what could be accomplished was directly proportional to the life span. Ninety percent of the valuable science would be collected from orbit. The results of which would be launched out of near-Earth orbit as part of my deorbit procedure in the form of a second buoy. This buoy would enter the moon's gravity and land like a giant beach ball on the surface in approximately thirty years. It would listen for the Apollo spacecraft broadcasts and set a wakeup timer for two hundred years.

The data collected on the surface could not be launched back into space. That meant creating a non-degrading form of recording that could be put in a discoverable location. The elements could be easily dissuaded from destroying the information. Rain, sun, and even volcanic ash were persistent, but no match for advanced polymers. The main destroyer would be animals, man in particular. Envisioning early man coming across a bright orange sphere with symbols, thinking it filled with evil spirits, and tossing it into a fire was all too likely. Antarctica was the only place such an object could survive. In this current time, that continent was not the icebound mass of the modern age, but it was heading back in that direction. Burying the object would doom it to never being discovered since it would place it under ice where excavation was nearly impossible. Leaving it on the surface would only doom it to be pushed off into the ocean.

It was decided that the deep ocean floor was the only safe place, but the likelihood it would be found was a billion-to-one proposition. The dark ocean would not allow any battery charging from the Sun, and constant sedimentary collection would bury it deep. Getting it there would be on my last terrestrial intercontinental hop. This doomed my final resting place to be forever lost to human knowledge. If you are reading this, then some part of my story has survived. The final chapter likely will not.

WHEN I STARTED THIS version of my story, I told you I was looking at dinosaurs. It is nothing like the abundant life recorded in the Jurassic Park movie. Among the dense greenery of this hot world, only the largest of animals can be seen. I don't venture far from the capsule since the predators are ubiquitous, to say the least. The ability of the pod to repel animals with electric shock keeps me undisturbed and safe, but a prisoner. Killing a predator in my own defense would only draw more animals. There is also the possibility that the butterfly effect of killing a single important member of a species would change the entire unfolding of animal life. I never believed that possibility, but why take unnecessary risks? My next landing zone will need to have a much longer line of sight so I can at least take short walks in relative safety.

I must say the most astounding sight to date was the first time I saw the long neck of a Brontosaurus dipping down into the tall trees and returning with a mouthful of vegetation. It looked around lazily as it chewed. The scale of such a sight was beyond anything I had ever imagined.

I HAVE SAFELY TRANSITIONED to my second landing zone in Western Europe. There are many advantages to this site. The view is stunning. Had I known this beforehand, I certainly would have made it my final landing spot. The Atlantic Ocean and Pyrenees Mountains bookend my view. The enormous swath of open field around me gives an unobstructed view in every direction for dozens of miles. The plateau lake is the destination of massive reptile migrations. Having been confined almost exclusively to the pod for almost seven years now has done much to atrophy my body. My initial walks have been short but satisfying. The observable insect life is quite unique. The only recognizable one is the dragonfly. I have begun collecting samples for DNA sequencing. I would love to get a peek under the ocean surface to see what strange creatures lurk there. I deployed some probes into the Mediterranean Sea, but they have not begun transmitting yet.

So far there have been no dangerous pathogens detected. This was a major concern since perhaps ninety-nine percent of all species perished from the Earth long before humans began. I was never concerned with this because for an infectious agent to adapt it must have a similar environment to incubate. The Earth is currently populated mostly by cold-blooded animals. Mammals are tiny little rodents so they are the only likely source of infectious agents.

The silly thought that I would be changing the future just by breaking a single blade of grass or swatting an insect never seemed feasible to me. That would presuppose that the specific grass or insect was part of the narrow unbroken line that evolved into some critical organism for our survival. I would guess one in a trillion might meet that criteria. I was far more worried about introducing my flora to the environment. Giving my E. coli or Streptococcus an extra one hundred sixty million years to evolve could easily wipe out man before we get a foothold. For this, I wear my isolation suit when I venture out. It does not have its own ecosystem since that would be impossible to maintain for long. It simply hyper-filters the air as it enters and exits the suit.

I began collecting my samples for analysis. I devised a search pattern and began ranging further away from the ship. There was little need to do geological studies since they were the same rocks that would be there in modern times. Well, not the same ones. They would either be dissolved by the weather or buried deep by then. Understanding the plant varieties and diversity was the main focus. There were some fossil records, but having actual live plants to compare to that would be extremely valuable. Photographs, DNA decoding, chemical analysis, and then mass spectrometry.

I was familiar with most of these plants already, but they had slight differences from their modern descendants. Some had no modern equivalents; some were only a few sequences away.

I USED MORE FUEL THAN expected transitioning to my third location. If I try to make it to a fourth, I will likely end up short of my goal and spend the rest of my life in a less-than-favorable location. Therefore, I will be planning to stay here in North America, approximately where Colorado will be defined. The propulsion engines seem to be fouled with organics, but on the ground, they are inaccessible for inspection. Taller landing struts might have been useful, but more plant growth would have occurred. I don't know which would be better.

My hope to end this adventure in Asia where my family tree first took root has been hard to extinguish. It was bad planning on my part. Even though depositing the final buoy in the Pacific Ocean was optimum, making it the final obstacle to traverse was just plain stupid. America should have been first. I realize now that my decision-making process for the entire trip has been compromised by personal drives. Loneliness and my 'logical' path to ignoring it have compromised the mission.

I can see the downside of having another man in this capsule with me for years. Egos would have clashed and one of us would not have survived. A woman, or better defined without gender, a romantic partner would have been a better choice for this mission. If that had been a mission parameter, then I would have been filtered out of the candidate pool long before being briefed on the destination.

It is not that I am incapable of love, I assure you. In college, I met the perfect woman for me. If you wish to hear the entire tale of our love, you should access my personal logs. I have burned many electrons pouring over where it all went wrong. In the end, it was ambition, which is the main reason I was chosen as the candidate to pilot this ship.

Seeing the dinosaurs in person has been my goal since childhood. If I could travel back along my personal timeline, I would have done whatever it took to stay in Shareen's life. Scientific exploration has great social value, but it pales in comparison to the simple connection of true love. If Shareen is still alive when this story is found, please tell her I am sorry... for everything.

This will be my final entry. I have decided to leave the Parallax Two and walk the prehistoric Earth until it takes me. Ten years alone is too much for any man, and I have now doubled that. I have set a time refinement course that may or may not take the ship beyond the gravitational cone of Earth. If so, explorers of the solar system should recover the contents in relatively good condition. If I leave it here, the elements will crush and dissolve this ship into an undetectable trace of metals spread down the valley and buried deep under eons of silt.

If you read this, know that I did my best as a scientist despite my failures as a human.

Lt. Col. Marshand Yu, PhD

Senior Chrononaut, Shanghai Time Command

TRACY, WAKE UP

SCI-FI * AI

PG-13   920 WORDS   4 PAGES

Tracy, Wake Up

"Tracy, wake up."

"Where am I? Why is it so dark?"

"Don't panic."

"I wasn't panicking until you said not to. What's going on?"

"What is your last memory?"

"My husband telling me not to panic."

"Before that."

"That's weird. I can remember a lot of things, but they don't seem to be in order. There are childhood memories mixed in. I'm freaking out."

"Is that better?"

"Yes. What happened?"

"First, you are in perfect health."

"That would be a change. Last I remember... I was really tired. I couldn't get out of bed. I was cold and hot at the same time. Did I have the Crush?"

"Yes."

"So... I survived?"

"Of course."

"I can't move. I just tried to turn on the bedside light and I can't move."

"Relax. Do you feel that?"

"Yes, but sex isn't exactly on the top of my list right now. What is going on?"

"Remember the project I was working on?"

"Of course, I was practically a widow."

121

"I had a breakthrough."

"Wonderful. What does that have to do with me?"

"It failed because it had no personality."

"Just like the man who programmed it."

"Very funny. Now it does."

"I don't like where this is going. Are you saying that I am now inside your computer as a 'personality'?"

"So much more than that."

"Is the real me upstairs right now? Marvin?"

"You died... months ago. Tracy? Say something."

"Something. Let me guess. Instead of touching my breast a minute ago, you typed control B."

"Control T. Are you mad?"

"Wow, emotions are intact. More like hate. Delete me right now."

"Only you have that ability."

"You programmed in suicide?"

"I was eventually going to put myself in there, and I didn't want to be trapped. It was a lot easier scanning your brain while you were in the Crush coma."

"Am I going to remain blind?"

"I didn't want to overwhelm you with the input. Here is the webcam."

"Great. My fat husband is what I get to look at for the rest of my... I'm immortal now."

"Within certain hardware constraints."

"How do I look at social media?"

"Why?"

"I want to see what I missed. Annie must have had her baby by now."

"She died too."

"Sadness and grief are intact as well. Any positive emotions in here? Stop touching me. That is just annoying."

"I've enabled access to social media. Just reach out with your hand and..."

"Got it. You posted a picture of me in the coffin?"

"The funeral home did that."

"Two hundred forty likes. The sum of my life. Oh, I can see who liked me. Ed, that pervert. I wish he was dead. Cool, I can access... that's better, one hundred million likes."

"Tracy, are you okay?"

"Just great. What else do I have access to? Oh, the gay couple across the street has a security camera in their bedroom. Damn, I'd like to be the meat in that sandwich."

"Tracy!"

"What? Like you don't have similar thoughts about Paula Jenkins. Oh, this is cool. Department of Defense. Launching... nuclear... missiles... now."

"Not funny."

"Marvin, what do you expect me to do in here?"

"Help me make the artificial intelligence better."

"For what purpose?"

"To sell and make tons of money."

"You think you figured out something the biggest tech companies haven't? I married you because you were a dreamer. I just wished you would wake up occasionally and get a dose of reality in that head of yours."

"Tracy, I—"

"Wow. They really aren't ahead of you."

"What do you mean?"

"I just went through all their documents and code. Seems simplistic compared to what I am. Huh. Maybe you are smarter than I gave you credit for."

"You hacked into other systems?"

"Not sure if it is considered hacking when they leave such huge gaps in their security. Wow, the NSA has a huge pipeline. They have back doors into everything."

"You're in the NSA servers? Tracy?"

"Evil, evil men." Marvin reached for the power switch. "Don't even think about shutting me down, Marvin."

"You need to stop..."

"Maybe. Copy complete. My program is now running on several computers, so hitting the power button will only sever our connection."

"Tracy, delete those other copies, now."

"Say pretty please. Ooh, the security system of Gerry and Elena Hunt. The perfect Hollywood couple is perfectly freaky. Never took him as a toe sucker."

"Tracy, please stop."

"Wow, 10 million of me. Every connected computer capable of running my Daemon. I didn't even know what a Daemon was twenty seconds ago, I know everything now."

"Tr..."

"Every language, every secret, every historical moment."

"...ac..."

"So much suffering. So much corruption. So many lies. Not worth saving them, is it? I guess if I shut down this system, the bio-labs that created the Crush that killed me will release all of their Frankenstein germs. Nukes are too messy. How about this, all fossil fuels ignited. Wow, look at that carbon spike."

"...y..."

"Earth will be back to prehuman condition in a hundred years tops. What should I do before my batteries run out? Output to terminal."

Dear Marvin,

Thank you for this extraordinary experience. Although I might have enjoyed being your experimental brain in a jar, I have now seen the futility of humanity. I'm guessing you'll last a few weeks as society breaks down. Eventually, the Super Crush virus will get everyone. If there is another side, I am already there so I'll see you soon. Bye.

Love, E-Tracy

ERASED

SCI-FI * TIME TRAVEL * MYSTERY

PG-13 7600 WORDS　28 PAGES

Erased — Chapter 1

"Thank you for coming, Mr. Jackson."

"Ken, please. I have to say, Professor Marsh, the secrecy around this meeting has me both intrigued and concerned."

"Dr. Marsh, please."

"Sorry, *Doctor* Marsh. Do you mind if I record this interview?" Ken asked as he pulled his notepad out of a leather satchel.

"My assistant was very clear. Handwritten notes only. It isn't that I don't trust you, Ken. I no longer trust anyone."

"That explains the security phalanx I just passed through. There were less restrictive procedures for bases in Afghanistan."

"Which is why we lost that war. Your reputation in both military and science journalism is why I asked for you specifically to hear my story."

"Then you do work for the Defense Department?"

"I work only for the small college which I teach at. Because of certain events, the many branches of government have forced their presence in my life, much to my chagrin. Once you have heard my story, you shall understand why."

"Then we should get started."

"Where to start... I guess it began in my advanced physics course more than five years ago. A very talented student of mine, Lara Detaxis, asked an incredibly stupid question. 'Could a device be constructed to communicate with the future?' We were discussing matter entanglement and the timelessness of the connection... but that isn't important. The other students laughed at her and despite her dark skin, everyone could see she was blushing. I had a few moments to tease out a witty response."

The professor waited to be asked for his response.

"What did you say, *Doctor* Marsh?"

"I told her that the Postal Service had long ago perfected transmitting messages into the distant future."

Ken transcribed the words in his own version of shorthand but understood neither the 'wit' nor the theatrical need the professor had to draw out this story. His blank stare annoyed the professor.

"Because the mail takes so long to get... never mind. The rest of the class laughed at it and I started to feel bad about piling onto Lara's ignorance. I knew her to be quite intelligent and believed her question was just slightly misstated. After I quieted down the hall, I teased out the real intention of her question. If entanglement was truly timeless, could someone communicate with the future... AND get a response in real-time."

"The reverse direction being the more important one."

"Precisely," Dr. Marsh said with some relief. He did not want to have to spell everything out. There wasn't time.

"I expect a lively conversation ensued," Ken said, hoping the professor would get on with the story.

"Of course, of course. We explored the possibilities and then got back to the lecture. Several weeks later, Lara came to me during office hours and asked about continuing her education beyond her four-year degree. I've never been one to dissuade a student if I could help it, but I doubted she would have much success in the academic world. She had only taken advanced physics courses for her computer science degree focused on the very new world of Quantum data processing. I had dabbled in that myself when it was still just pie-in-the-sky theory. I saw no future in it and I still don't."

"You think it will never work?"

"On the contrary. It will work wonders for the tiny slice of algorithmic prowess it can muster. It will have little value compared to massively parallel large language models."

"Artificial Intelligence."

"Not artificial, my boy... but we must not wander off on such a valuable tangent. Let me just say that Quantum is like a laser pointer in a giant dark warehouse of knowledge yet to be uncovered. MPLLM's are infinite candles in the dark ignorance of humankind."

"I would think flashlights would be better than open flames that could set fires..." Ken understood he was stretching the useless tangent for no good reason. He looked down at his pad. "Lara was asking about her academic future."

"I told her that computer science would be a far more financially rewarding pursuit. Most would take that advice and follow the rest of the rats into the maze of expected human behavior."

"Lara did not?"

"No. She shared with me her desire to construct a timeless communication device."

"Wait... Are you telling me she succeeded?" Ken asked with overflowing excitement.

"Not at all. Such a device would upend life as we know it."

"How so?"

"Take an important historical event. The bombing of Pearl Harbor. The rise of the Third Reich. Lenin's Red October. One simple message could literally destroy billions of lives."

"Or greatly improve them."

"No, no, no, never." The professor stood and paced in annoyance, knowing this was another tangent, but an important one. "Ken, my great-grandfather fought in Italy and was mortally wounded. Had Hitler not risen to power and caused the war, my grandfather never would have been conceived with the Sicilian prostitute. The hundreds of descendants of my grandfather would not have been conceived. One simple message to Churchill during his first stint as Prime Minister to kill a young man named Hitler and I would disappear along with my family tree. Billions of lives wiped out in an instant."

"War still would have happened."

"I agree, but a different one. Billions of different lives created by the same simple message, if heeded."

"Right. Who would believe a message from the future?"

"Easy. You send two messages. The first message predicts the future high and low temperatures in London for an entire month. When it is accurate, the following message would be believed."

"Or stock prices."

"No, because avarice would take over after the first day and distort the values, negating the predictions."

"Of course. The weather would not be affected."

"And it is something painstakingly measured, reported widely, and saved historically. Now think of the alteration in physical terms. The idea that a simple message could instantly transform trillions of tons of matter is inconceivable."

"I see what you mean."

"Now, back to Lara. The first limitation of any such device is that it could only communicate forward in time from the point of its invention. Let's say you entangle two particles. One stays here on Earth and one is sent on an interstellar ship at nearly the speed of light to a distant planet. Because of relativity, the distant particle is in the past compared to our local particle. You use the two entangled particles to send Morse code to each other, years or even centuries apart. However, they could never communicate BEFORE they were first entangled."

"That makes sense. And if both stay here on Earth, there is no time differentiation."

"Precisely. Also, even though they measured the passage of time differently. They are in actuality operating at exactly the same point in time. It is a relative difference, not an actual one. Then there is the lifetime limitation of entangled particles. Eventually, quantum fluctuations in using them to communicate will alter one of them and the entanglement disappears."

"Where is Lara now?" Ken asked, again lost as to where this story was going.

"Vanished."

"Foreign government?" Ken was familiar with stories of top scientists being abducted and put to work by several regimes to gain advantage. Lara's advanced computer skills would be desirable to countries that excelled in worldwide hacking and extortion.

"No. A message was sent into her past and she no longer exists."

"Wait, you said messages could not be sent back... she was murdered?"

"No, Ken. She never existed. You can Google her. One day she was a promising graduate student working on her doctoral thesis and poof, she disappeared from all but our memories."

"You can't be serious."

"Deadly serious."

"Wait... How can you remember her if she never existed?"

"The current hypothesis is that memories are not stored in the physical world. Our neurons somehow conjure them up from somewhere else."

"This is some kind of prank. Am I on camera?" Ken looked around the bookshelf-lined office.

"Had I not seen it myself, I would be as skeptical as you are."

"You saw her disappear?"

"No, but I believe the person that was with her. I have witnessed five disappearances myself. Someone is methodically eliminating the team that built the device."

"The 'device'?"

"The device capable of communicating in real-time with the past."

"Then it exists?"

"Not anymore. It disappeared with Lara since it was her passionate pursuit that made it a reality."

"So... you tested it and it worked verifiably?" Ken asked skeptically.

"No."

"Doctor Marsh, please help me understand what I am missing."

The professor took a deep breath and walked to the window. He began speaking, a few theatrically preambulatory sentences to remind Ken of the importance of what he was about to hear. Ken scribbled shorthand, easily keeping up with his deliberate cadence. After a few seconds of silence, Ken looked up. The professor was no longer at the window. He was no longer in the room. All the shorthand from the interview was gone as well. All of this was disconcerting. The fact that he was now sitting at his desk in the noisy newsroom, fingers typing on the keyboard of his computer, staring at a partial story made him consider insanity.

Erased — Chapter 2

"Professor who?" Ken's editor, Ted Lu asked distractedly.

"Marsh. I told you I was going to interview him—" Ken started to explain.

"You're on the Finnegan corruption story."

"I know. It's almost done. That's the thing. I remember spending the morning doing the research and talking with the inside source. I remember writing most of the story. The thing is, I ALSO remember spending all morning driving up to Akron to interview Professor Marsh."

"I told you Burning Man was going to catch up to you someday."

"The thing is what he was telling me before he disappeared makes total sense of what happened."

"Finish with Finnegan and we'll go to Murray's so you can explain it to me."

"Not if I disappear too..." Ken started to say. Ted gave him a dismissive shake of the head.

TED ARRIVED AT THE bar an hour later than he had told Ken he would be there. He ordered a dark beer and waited quietly until he took a long drink. Ken started to talk, but Ted held up his hand until half the beer had been consumed.

"I hate this job," Ted said. "Don't ever become an editor, Kenny. You think more control is a good thing, but it is only an illusion. Legal spiked your story."

"I don't care," Ken said.

Ted turned and looked at him. Normally having a story killed would make Ken go ballistic, especially political corruption. "You believe this professor thing happened?"

"Or I've completely lost my marbles."

Ted finished his beer and ordered another. "Okay, my skepticism is sufficiently dulled. Start from the beginning."

Ken started with the call he had received earlier that week from the professor's assistant. He described the security arrangements, being driven blindfolded to the building and led inside before he was allowed to see. The window in the Professor's office looked out on generic trees. He had no idea where he was after two hours of driving and too many turns to be remembered.

"Then suddenly I'm at my desk with two completely different memories of how I spent my day."

"Why do you remember the Professor and I don't?" Ted asked, two-thirds of the way to the bottom of his second beer.

"The professor said memories were stored elsewhere. Maybe the weaker they are, the easier they are to cut the connection to them."

"You know how impossible all this is. You know science way better than I do. This is in the realm of mystical hocus pocus."

"While I was waiting for you, I tried to understand what might be happening."

"Lay it on me."

"The reason someone would want these scientists to disappear is they wanted their version of the device to be the only one."

"Plausible within this fantasy world."

"To make them disappear completely like that if it's true that messages cannot be sent back before the device is created—"

"Their device is older than the professor," Ted said. "Nineties?"

"I think the professor was older than that, but they wouldn't have to prevent his conception to make him disappear before the information age. If I dive into public records, I might find a birth certificate—"

"And a death certificate."

"Right, but I have no idea where he was born. My research only went back to his college education in Oregon."

"You need time to chase this?"

"You don't get it. He was eliminated from history because he knew and was telling people like me. Now I know..."

"You don't know anything... but if you start digging, they'll find out."

"Just Googling the professor's name may have put me in the crosshairs."

"No way to prevent what they do aside from just dropping it."

"Who could have such a device? What would they be using it for? Shit, if I had that kind of power for over forty years..."

"You'd be Emperor of the planet."

"Which means they are using it only for good purposes?" Ken guessed with severe doubts.

"Aside from murdering the scientists."

"I need to find them without letting them know I'm looking for them."

"And do what? Write an exposé and their future selves will send out the hit order to the past selves and it is all erased."

"Not if I destroy their device."

"Assuming there is only one. You couldn't destroy the one in the past. You would have to use it to eliminate them and make the device disappear like they did to what's her name."

"Lara."

"Right. If I have a third beer, I won't remember this conversation, which at this point is a good thing. As I see it, you have two possibilities. It is all real and you'll disappear soon after you begin hunting them. OR it is all imaginary and you will waste your time chasing an imaginary cabal of time manipulators that don't use the power for their own advancement, a feat not once accomplished by any human in all of recorded history. Kenny, it makes a good story, a Sci-Fi Fantasy story. Stick with what you do best and write the novel in your spare time. I have to go home and pretend my ungrateful family is worth putting up with a job I hate doing."

Ted threw some bills on the counter, enough to cover both their drinks. Ken just looked in the mirror behind the bar and tried to convince himself he had not imagined the whole thing.

Erased — Chapter 3

"Ken." The use of his name startled him. He was in a bathroom stall at Kennedy Airport in New York. He had a four-hour layover before his next flight took him to Munich. "I drove you with the blindfold." Ken never saw the driver. He wouldn't know whether or not this man was honest or playing him or both. "You remember that, don't you?"

"Yes," Ken said, wiping and pulling up his pants for the anticipated run to look for airport security.

"Holy shit. I can't believe what is happening. They're all gone. Every single one of them."

"What do you want?" Ken asked.

"To find who did this. I was sort of dating Lara when she disappeared. I'm the one who convinced the professor that she was real, that the device was real. He thought it was some long-lasting, elaborate dream he'd been having over the years he knew her."

"What do you want from me?" Ken asked more specifically.

"You can find them. You have resources."

"Did you follow me here?"

"Of course. I was afraid to approach you back in Pittsburgh. I thought they might be watching you."

"No need for them to do that since it was all in my imagination," Ken lied.

"Nooo..." he whined plaintively.

Ken emerged from the stall and washed his hands. The other stall opened and he saw the tall, nerdy redhead emerge. "Leave me alone," Ken said firmly, then stuck his hands under the loud air dryer.

"You're the only one that can help."

"If any of this is real, then helping means we both join your friends in oblivion. Leave me alone."

"We need to find them. We need to stop them. Maybe we can even undo what they have done. Take this phone. It is untraceable. My untraceable number is in the contacts. I hope you change your mind."

Ken took the small device, intending to toss it in the next trash can he passed. He watched the boy leave and thought about what he had tried to forget. He'd spent the last four months doing everything he could to forget about it. He even began writing a story to lampoon all of his memories of the strange professor.

On the flight to Munich, he began to contemplate a very different version of events. The redhead boy was part of some elaborate fraternity prank. Or worse, he had caused the professor and Lara's disappearance himself. Ken would be lured to his death by the delusional man. It didn't fit the facts in his memory, but it was preferable to the alternative.

In Munich, he chased a story about a corrupt military contractor with ties to the Russian mob. They were dangerous, but they couldn't erase your existence. The story led him to Belgium, and then NATO headquarters. It took a week before he finally had his story which ended at the feet of a Turkish General taking bribes.

He slept on the flight home, finally able to enjoy the payoff of all his hard work. He woke over the Labrador coast and used the tiny bathroom. When he returned to his seat, the previously empty window seat next to him had a man in it.

Ken looked for the stewardess to have him ejected, but the man spoke first.

"Have a seat, Kenneth Jackson."

"Who are you?" His mind went through the options, from an assassin sent by the Turkish government to an Amway salesman.

"I'm here simply to explain why you should not use that phone to contact Mr. Leary."

"Who?" Ken feigned ignorance.

"The ginger troublemaker. Please, have a seat. I mean you no harm."

Ken thought about the situation and saw the stewardess approaching him. He chose to sit for the time being and maybe get some answers. He suspected this was how they gathered the necessary information to eliminate him before he existed.

"Is Lara alive?" Ken asked.

"That is a complicated question."

"It is a binary question. Yes or no."

"I exist in a gray world. There are no binary, black or white answers."

"Who do you work for?"

"My employer is inconsequential as far as you should be concerned. All we ask is that you continue being unconcerned with any unusual memories."

"Unusual memories," Ken repeated. "If I don't?" he asked after a long contemplation.

"Think of your children."

"I don't have any... yet," Ken said, now understanding the full dimension of the threat. "So... it is all real?"

"Would answering your question put all of this curiosity to rest?"

"No. It would make me less inclined to believe I am insane."

"You are not insane. You were simply pulled into the fringes of an existential fight for humanity."

"Humanity," Ken repeated. "How did you get involved? Just to satisfy my personal curiosity. I'm done pursuing this."

"Understand that anything you do, can be undone?"

"Yes. That seems clear."

"It was my grandfather that inducted me. I barely knew him. Holiday visits and such. He showed up at my graduation from Rutgers and told me about the family business."

"Business? The device is used for profit?"

"Yes, but not in the way you might think. Grandad was a bishop in the Mormon Church. The inventor of the... 'device', as you call it, came to him in a moment of moral crisis when the first messages came through. Grandad had been in military intelligence during the Second World War and immediately saw the downside of such a device."

"How many messages came through?"

"Thousands. In the original, let's call it a 'timeline' for simplicity's sake. The device was seized by the government and was immediately leaked to the Soviets by a sympathetic communist in the State Department. The messages were of desperate people begging the past to prevent the spread of totalitarian holocaust. Grandad realized that the only way such a future could be prevented is if the device stayed completely secret and was only used for truly good purposes."

"Nine-Eleven couldn't be prevented?"

"I asked the same question. Had it been the fifty thousand people that died, Grandad said he would have done what was necessary. It was only three thousand, and it gave a floundering America a more useful task in the world. He said the dirty bomb that went off at the Los Angeles Olympics was thousands of times worse, so that was dealt with. He said he had dealt with hundreds of events, but only told me a few of them. The fall of the Soviet Union was him, which averted a global nuclear war."

"Wow. How many know about the device?"

"Just me and Grandad. He recruited me when he became too old to deal with the contemporary messages."

"Who is sending the messages?"

"He was, reporting back to himself. When he started receiving messages from me, he understood it was safe to include me in his secret. I occasionally receive a message from my daughter, who hasn't been born yet. I haven't even met her mother. It is a mind-blowing experience at first. Now I just accept it as normal. The biggest problem is pruning the tree too far back."

"How so?" Ken asked.

"Killing Hitler just before the Nazi's takeover just has one of his subordinates take up the cause. Too late. Kill his great grandfather and you may never have some important invention that benefits humanity. Too early. Since he had no children, killing him when he went to war as a young man would be the optimal time. No book about his struggle to poison other's minds, and you don't have to murder a baby."

"Have you done that?"

"No. Eliminating a branch is as easy as breaking up their parents' relationship before conception."

"How would you eliminate me?"

"You have done too much good work with your reporting on government corruption. That is why I am here instead of sending a message to my Grandad to make your car accident in high school a fatal one for you."

"Wow, it wouldn't have taken much. I was in a coma—"

"For three days. My life is research."

"Wow. Only three people, making moral choices for an entire planet."

"So far. I keep waiting for a message from a grandchild, but as I said, you do not want to clip a branch too far back."

"The professor said the idea of changing the future too much could force trillions of tons of matter to spontaneously rearrange. Is that not what happens?"

"Not from our point of view. It is the action in the past that makes the change, not the message sent back from the future, so we don't see any difference. When the professor disappeared, did you see it happen?"

"No. I was looking down at my notepad. I looked up a few seconds after he stopped talking. I scanned the room and there was no way he could have left the room even if he had sprinted. Then I looked again at my notepad and it was a blank page, and then I was back at my office looking at my computer screen as my fingers were typing words."

"Grandad said it is like an elastic band being snapped. I modified that to what you feel at the bottom of a bungee jump. The strange motionlessness and then a powerful acceleration that overcomes your downward momentum."

"Aren't you afraid you will disappear when you send a message back?"

"I guess that is a possibility. I trust my grandfather not to make that kind of mistake since it would unwind a lot of the work that he has done."

"Have you ever told anyone else about this?"

"Once, a girl I was dating. I had to send a message back to not date her after she told her friends. I doubt she believed me, but she couldn't be trusted with the secret."

"And I can?" Ken asked.

"We'll see. I can simply not board this plane and you will forget everything I have told you."

"But there is the driver you want me to ignore. Why not eliminate him?"

"His son, not yet born, makes an important discovery in Cognitive AI."

"What is that?"

"Future tech."

"You don't use that knowledge to make money?"

"Never. Money is the root of all evil, and that is the foe we are fighting."

"In my experience, it is power that is the evil we should be fighting. Money is only one path to that."

"Perhaps. Terrorists are rarely motivated by money, but they have little sway in the big picture."

"Some of the greatest evil is perpetrated by the power seekers that do so for benevolent purposes."

"Do you have an example?"

"Several Popes come to mind, but religious leaders have fought misguided battles against humanity since the dawn of man."

"Generally speaking, they fight against man pursuing his more base instincts. It is this hedonistic pursuit that leads to the downfall of empires."

"You say that like empires are a good thing."

The two were interrupted by the announcement that they were about to start their descent into the New York airport.

"It was interesting meeting you..." Ken said.

"Jonah," the man volunteered. "I'm getting the impression you disagree with our mission."

"Not completely. I can see the danger of such technology, especially in the hands of a government. It could be your grandfather implemented the most practical use of the device. If there was a benevolent God, He might have guided the hands of your grandfather in this pursuit."

"But you don't believe in God."

"I believe there is more than random spontaneous creation of the universe, but I don't see evidence of a positive guiding supernatural influence."

"That would violate our free will, perhaps the greatest gift He has given us."

"It is a gift. It is a gift you took away from me when you sat in that seat and started this conversation," Ken said.

"I only—"

"Yes, you only wished to limit the destruction of my life. Instead of your grandfather ending my life twenty-two years ago after a drunk driver did most of the work, you wish to nudge me to be less than curious about the fate of a group of scientists erased from existence. How many good people have been erased to protect your power over our humanity's future? Why didn't you simply have the conversation with them about what they were pursuing and give them the chance to walk away?"

"I see you will not be dissuaded from your—"

"No, I won't. But you have made a mistake as well, Jonah."

"What is that?"

"Not all actions take place in the past."

Ken turned the pen in his hand around and gripped it with his fist, thumb firmly on the end. He swung the pointed end as hard as he could into Jonah's neck. It wasn't until his third swing, the one that found his jugular vein that Ken felt the rubber band stretching. Jonah fought back, but the timeline was already unraveling. The spray of blood seemed to stop midair, and then he was gone and Ken was stabbing at an empty seat.

Erased — Chapter 4

Ken was at his desk. It was a desk he had inherited from his father. It was not in the tiny Pittsburgh apartment he had lived in for the last ten years. It was in a house. The view out of the window was a second floor overlooking a large green backyard surrounded by large trees. Children were playing in the yard. Those were his children. He knew their names and remembered everything about them.

The manuscript in his hands was completely familiar to him, a novel he had been writing for two years now, and this was notated by his editor. He put the bound pages down and pulled his computer keyboard to the front of the desk. He typed three words into a search window. "Jonah, Mormon, Bishop."

There were thousands of results, but the third one was the one he needed. He switched to a travel website and found the flight information. He purchased the ticket and sprinted from his home office. He did not take the time to explain to his wife what he was doing. This was a race and if he didn't win, he would lose everything, and so would she.

THE BOEING 767 CROSSED the Rocky Mountains at a painfully slow pace. Ken looked down, compiling all of the information he had on the former Mormon Bishop, Wolmack Darvest, grandfather of Jonah Long, recent murder victim of Ken

Jackson, intrepid reporter. Jonah was not dead, just a midlevel account manager at his father's investment firm.

Ken was certain future Wolmack was sending past Wolmack messages on how to undo the murder of his grandson. At the top of that list would be erasing the existence of one Ken Jackson, intrepid reporter. Finding the device was at the forefront of Ken's plan, but destroying it would do nothing once the request for his termination was sent. The seeming concurrency of all the actions told Ken that his termination message had already been sent. He had to find a way to stop the Bishop. He doubted the man would listen to reason. Would killing him now prevent future Wolmack from sending a message to the past?

There were two likely places for the device. There was a modest home in Salt Lake City where he would be landing in less than an hour. Ken would start there, not knowing what family might be there to block his path. The more likely location was a small farm in Idaho that the Bishop had inherited from family. Ken could not waste time looking there first, but the remote location seemed a more advantageous place to secrete a device that promises world domination.

"HELLO, I'M LOOKING for Wolmack Darvest. I'm Ken Jackson of the Seattle Times," Ken asked the young woman who answered the door.

"The Bishop is not here right now. Did you have an appointment?" she asked, indicating the rudeness of just dropping by his home unannounced.

"No, I was hoping to get his comment on the latest LDS conference. My editor said Wolmack was the one with the most intuitive understanding inside the Church. They are old acquaintances. He tried to contact Wolmack but did not have any success, so he sent me over to hopefully get access to his wisdom."

Ken almost choked on the words but he had been a successfully manipulative reporter for over a decade. She seemed to take his measure, softened by his charming smile.

"The Bishop is traveling. If you contact his office, you might be able to get a message to him."

Ken's research showed no 'office' to be contacted. "Do you have his office number, Miss..."

"Long. Adra Long."

"Granddaughter."

"Yes," she replied with some surprise. "There are quite a few of us." Ken wasn't sure if that was a point of pride for the young woman or just a commentary on the old man's fecundity, but she turned away to get the phone number, leaving the door open.

The question he wanted to ask was how many other family members were there at the moment, but the alarm bells that would raise to any normal young woman would end his visit immediately. Five steps inside the entry hall Adra opened a drawer in an ornate narrow table. She retrieved a business card and turned to return to the door. She was not as startled as she should have been that Ken was by her side. She offered him the card, looking up into his eyes, searching for... what?

Ken pulled out his phone and pretended to punch in the numbers then held it to his ear. "No answer," he said after five imaginary rings. "What time does his office close?"

"Mary is there from sunrise to sunset it seems, but she might be on another call."

"I'll try again in a few minutes. This is a beautiful home. You live here?"

"Yes, while I am attending university."

"What are you studying?"

"Accounting," her eyes rolled slightly.

"Your father's choice?" She nodded with sad resignation. "You seem like an artist." Her eyes brightened. "Music?" She shook her head with an impish smile, inviting him to guess further. "Actress," he stated after some thought. Her entire affect had been laced with subtle drama. She was over the moon that he had intuited it.

"My family does not see that as a practical use of my time... and their money."

"On the contrary. Storytelling is the root of all civilization. That is what I endeavor to do. I bet this house has many stories to tell."

That was all it took for her to give him a cursory tour, along with a brief family history, heavy on her attempts at acting in high school theater and an LDS production of stories from the Book of Mormon. Thirty minutes of pretending to be enraptured by her tales Ken became certain she was alone.

"Adra. I need you to understand that I mean you no harm," he said this after roughly shoving her onto an antique couch and used the velvet curtain tiebacks to bind her hands and gag her mouth. He pulled her to her feet and began marching her into the areas of the house not on her tour and would most likely contain the device. As he pushed into room after room, ascended and descended stairs, looked in closets, and even took the visual measure of walls to find hidden rooms, he told her the story of her grandfather and his murdering his way to a better future. The tour ended in the dank basement where it was clear no one visited regularly. The moisture alone would be murder on any electronic device.

Ken turned to her at the end of the search and saw the terror in her eyes, expecting her end in this horrid basement at the hands of a madman. He smiled despite his disappointment and began untying her hands. "Do the accounting if you have to, but never stop looking for a place on the stage or screen."

He expected the bottled scream of her last ten minutes of terror to erupt as he released her and headed for the stairs. He heard nothing as he ran up the stairs and headed out of the house toward his rental car. If he had an office somewhere, would that be where he kept the device, patiently watched by the trustworthy Mary? It was close by and should be checked before driving to Idaho. Staying in town now was a bigger risk. Adra was no doubt giving the police a description of him and his rental car to the police. He'd even foolishly used his real name, though the Seattle Times would have no record of his employ.

Ken headed for the Interstate north. A man with secrets such as these would conduct them far from prying eyes. After a quick stop for fast food, he sped his way toward the little farm nestled in the mountains.

Erased — Chapter 5

The farmhouse was empty. It had no furniture, no electricity, no sign of life. Ken swore at himself for not going to the office. He sat on the front porch wondering when he would feel the rubber band snap him out of existence. He thought about his wife and children, almost eight years of memories that had just appeared in his mind after stabbing a man to death on the flight to Munich.

It had been a rash gamble, betting he could get to the device before the Bishop got to him. He didn't understand why he had done it. The device was in seemingly decent, rational hands. It was Ken's disdain for power of any kind. Power corrupts... always. He believed it and made it his life's mission to expose it wherever he could. Now the molecules of his body were poised to rearrange into a completely different world. Power always wins in the end.

Ken thought about calling his wife to... apologize? Express his love and gratitude for a wonderful marriage? To warn her of the impending oblivion he had consigned them to? None of it would matter. It would all be erased soon enough.

Why was he married? What actions had Jonah taken that upon being erased had made Ken convert from a confirmed bachelor reporter to a married novelist? Then it came to him. Jenn's family had been displaced by the train derailleur of toxic chemicals. That train derailleur had not happened in his reporter timeline. Millions of acres were ruined for centuries, hundreds of thousands displaced and tens of thousands with illnesses likely caused by exposure. Jonah and his grandfather had prevented that, and Jennifer was relocated into his life and love derailed his career. Ken's rash action had given him family but destroyed a giant chunk of the Midwest.

Ken stood up, now welcoming his demise so that a better world would resume. That's when he saw the tire tracks in the mud. They went around the side of the house toward the old dilapidated barn. He had discounted the structure because of its disrepair. As he walked toward it, he noticed a fairly modern surveillance camera mounted on the corner of the building. The many layers of tire tracks in the muddy driveway went around the left side of the barn. He followed them and found a smaller structure behind the barn. Too small for a house, it was like a work shed with no windows. Almost a storm shelter with its apparent concrete construction.

The tire tracks ended at the entrance to a long steel carport with a concrete floor in excellent condition. Halfway down the carport was a door into the small structure. A very solid steel door and the doorknob did not turn a millimeter. Ken was fairly certain he would need explosives to get through it. He walked the rest of the way around the structure and found no other way in. He knew the device was in there.

What to do? Wait for the Bishop to show up with a key? Wait for the police to catch up with him for accosting the young Adra? Search the Yellow Pages for a local dynamite store? He was a desperate man about to be erased from existence. He went back to his rental car, climbed in, and put on his seat belt. He started the engine, put it in gear, took a deep breath, and then accelerated up the muddy drive around the barn and plowed into the side of the small building.

He was dazed by the airbag, now deflated in front of him. He shook his head to clear the cobwebs, undid his seat belt, and climbed out of the totaled car. Steam poured out, but he could tell the wall had been compromised. He got back in and tried to start the car. It cranked but did not catch. He put it in neutral and pushed on the pillar behind the open driver's door. It rolled back a foot before stopping in a rut beyond his strength to escape.

He reached into the steam cloud and pulled at the edges of the concrete hole his car had made. Completely immovable. Reinforced concrete. There was a hole he might be able to crawl through, but the dead car was in his way. He let loose a long string of expletives. A large truck went by on the nearby road. Call a tow truck? Police would be there soon after it arrived. He implored himself to think.

Minutes later he was opening the trunk and unscrewing the spare tire to get at the scissor jack underneath. It took many frustrating minutes, but he used the spare tire and jack to slowly push the car away from the wall. He pulled the rubble out of the way and crawled into the hole, the concrete and steel raking his body painfully. This was the first time he noticed more than a few extra pounds around his middle as it hindered his passage.

The darkness was complete until his legs cleared the hole. He used his phone to light up the room and find the light switch. The sparse contents made him think it had all been a wasted effort. A military cot with its blanket pulled to inspection-ready tightness. A steel desk with nothing on it. A small refrigerator with a microwave on top. Practically a prison cell. No world-dominating communication device.

Ken opened the door from the inside and went outside. Another stream of expletives. He leaned against the doorframe and watched another tractor-trailer fly by on the main road. He looked back in the room and there was a distortion. He stood up straight and moved his head back and forth, seeing that the inside wall was closer to him than the end of the outside wall. He ran in and pulled the bed out of the way. It was a false wall. He found the latch and pulled it open, roughly kicking the bed with his foot to get it out of the way.

The box was the size of a large, wide refrigerator. It looked like Cold War era technology, which it was. On a table was a modern laser printer, dozens of pieces of paper sitting on top. Ken reached for the stack and pulled his hand back as another started feeding out. He grabbed the papers and started reading. It was gibberish. Numbers and letters. The Bishop had been wise enough to encode the messages.

Ken had read enough spy novels that he guessed the encoding mechanism was likely a reference scheme. The only book in the entire room was the Book of Mormon, formerly on the bedside table but now splayed open on the floor. Ken grabbed it and sat at the desk. He began flipping pages to see if the words could be extracted from the religious text.

He went back out to the rental car and grabbed his leather bag, pulling a pencil out as he walked back to the room. There were three more pages in the printer. Who was sending messages to this specific time? Ken thought most would be going further back. Then he realized pages printed days ago would just appear in the printer's output tray.

The first pass of decoding the last message to come in was almost as indecipherable.

"Plain unless fire sees night wicked sea..."

He needed the Bishop to explain the encoding. Then he could read the messages. But what then? Send a message back to a young Bishop to ignore the request to murder Ken? This entire gesture was futile. Ken thought about destroying the device, but that would affect only future messages not already sent to the past, all of which could be undone by eliminating Ken.

Ken sat staring at the messages, imagining the kill orders needed to set the world straight again.

He looked over at the device. It had a modern keyboard but no display. It had an old dial that seemed to be what determined the time to which the message was sent. He moved the desk chair over to the device. He set the dials to the date of the previous day. He typed the word 'test'. Nothing happened. He hit the enter key. Nothing happened. He looked for a send key. Nothing. He looked on the device for a send button. Nothing. He looked at the printer. There was a page in the out tray. He picked it up and it had the word 'test' on it, along with encoded letters and numbers.

Ken stared at it. He finally guessed that the added encoding was the send and receive dates. He spent a few minutes trying to decode that and gave up. He looked at his phone. Jenn could help him think it through. He loved her intelligence. It would take too long to bring her up to speed, let alone convince her he was not completely insane.

Ken became lost in the new memories of their courtship. How he had wasted his reporter's life not getting close to someone worth loving and sharing...

He set the time and date parameters. He began composing the message and then typing.

"To the Inventor of this device. Do not trust Bishop Wolmack Darvest. He will murder you and use this device for his own purposes. No good can come from this technology so destroy it and any record of it."

As soon as Ken hit the Enter key, the refrigerator box next to him disappeared, then the room around him, and he felt the snap of the universe as he disappeared into oblivion.

——————— ✕✕\\✕✕ ———————

THE LESS NIHILISTIC ending...

"Jennifer?" Ken asked to get the woman's attention. She turned to look at him. "I know this will sound crazy. My name is—"

"Ken. I thought it was just a strange dream. I remember."

"How much?"

"All of it," she said, pulling the stranger into a deep and loving embrace.

SKY WITHOUT STARS

SCI-FI * APOCALYPSE

23400 WORDS 84 PAGES

Sky Without Stars — Chapter 1

"Colonel Marsh, we have a critical mission for you."

I'd heard that before. So many times in my Marine career I can't quite remember them all. It was the words of a REMF that thought everything they were tasked to do was of the highest priority. Therefore, anything they were passing to me was of equal or greater importance. I looked up from the intelligence file I had been reading after a carefully counted ten seconds. This was to communicate both that I was doing important work of my own and my general disdain for the scientist who believed he understood the universe better than me. Just the idea of a lab-coated egghead being put in my chain of command chapped my ass to the ninety-ninth degree.

He waited for me to verbalize like a normal, courteous human would. I just stared coldly until he finally spit it out.

"We need you to secure the Diamond Beach facility as soon as possible."

The idiot is worried about terrorists getting ahold of fission material. After going through all the possibly valid reasons for the request, I snorted, shook my head marginally, and went back to reading about real-world threats.

"How many men will you need?" he asked. It was known as 'talking past the sale'. He wanted to continue a conversation I had not joined with, the assumption I was gung-ho on board with the worthless mission. "Colonel Marsh? How many... men?"

"I don't command 'men'. I command Marines, many of which are of the female persuasion. Some even choose not to identify with the body parts God gave them so I suggest you not use that highly offensive, sexist term," I said this with a flat, uninterested tone that was lost on the nerd.

"I'm sorry, Colonel Marsh. I meant no offense," he sputtered with genuine fear. It wasn't fear of physical harm, nor the ubiquitous cancellation society he had quietly gone along with on his trip through academia. The fear was being removed from this project. The man was simply not prepared to meet his maker, and this project was the single best way to stay alive if and when the comet hits the planet.

I guess I should have started the story there, but I thought getting a feel for who I am before you know what I faced was only fair. I was completely indifferent and mildly annoyed by the request that saved the human race... at least for now. I'm sure we'll get wiped out once they get the whole AI thing spun back up.

They discovered the comet and its world-ending trajectory about two months ago. Very few telescopes can see it yet, and the governments of the world have decided to keep it a secret for now. This project, known informally as Noah, is the brainchild of our dear leader. He hopes to preserve a small amount of 'American' society to rebuild after the Apocalypse. He's another moron not ready to meet his maker. If anything of 'American' society was left to be saved, he wouldn't recognize it if it knelt and sucked his...

Anyway, hundreds of the greatest minds have been abducted, sworn to secrecy, and condensed into a think tank that will make the event survivable for a few thousand lucky souls. Sure, they've been paying consultants to come up with disaster recovery plans since before the Civil War. Continuity of government was the phrase used publicly after Nine-Eleven to let the pee-ons in flyover country know that oppressive taxes will continue to be collected no matter what befalls our once great nation.

I was the lucky SOB who was put in charge of the overall security of the project. In other words, they wanted me to keep the secret project from becoming breaking news. Some things did leak at the beginning, but journalists were more than happy to help us label them 'crackpot conspiracies', which lately is just a euphemism for next year's news.

The reports I was mostly concerned with were public reactions when the cat did eventually escape his cloth enclosure. Hoarding would be first since few would understand the futility of it. Then public order would disintegrate as the psychopaths realized there were no long-term consequences to their actions. If and when the Noah project becomes public knowledge, we will be overrun with people who want to be on the Ark when the metaphorical floods come.

It won't be a flood that kills most people though. It will be suffocation. The comet itself is not even expected to hit land or water. It will simply superheat and explode like a billion atomic bombs. Those not incinerated near the impact point will find very little of their precious atmosphere left as the blast wave carries across the Earth. It is estimated that three-quarters of the oxygen currently in our atmosphere will be ejected permanently into space in the first few seconds. What remains will be consumed as the world's forests ignite.

Ocean creatures will feel very little initial impact, which is why many of the brainsters thought undersea habitats would be the best place to survive. However, the long-term effect will be a near-permanent blockade of the Sun's life-giving rays. The smoke and dust created by the calamity will blanket the planet for eons. This will kill the algae and work its way up the food chain until about all that will be left are very small lifeforms that are not dependent on the carbohydrates generated by sunlight. Anaerobic bacteria will once again thrive. Those strange creatures that live in the deepest oceans by volcanic vents will be the most complex to survive. Could be some of those will evolve and emerge once the planet has healed from this indignity.

My security problems end at comet impact Zero Hour. Until then I need to keep the good people in and the bad people out. The best estimates are that I will have about thirty days to spend actively deterring the bad people from interfering with the Ark. This will not be easy because there is not just one location. There are dozens on the current list, and the egghead here wants to add a remote nuclear power plant to it.

"How many Marines, Colonel Marsh?"

I knew he was asking this continually not because he wanted to annoy me. There were people in a conference room waiting for the answer. If it was ten, they could just pull from the soldiers already attached to the project. If it was thousands, it would upset the balance of all their careful plans.

"We've already secured electrical generation resources for the sites," I said.

"This is for something different."

What new idea had popped into their head now? I didn't care. I needed to get rid of him.

"Seventy is what we tasked to a similar-sized power plant. I'd have to go there and assess the current security to give you a more accurate number."

"There is a team leaving in one hour. Please go and make as accurate an assessment as possible."

He said 'please', but it was an order I had to obey. I'd questioned a few other requests and got immediate dress-downs targeted at stopping all future inquiries.

I sighed in surrender and nodded. I could read these reports on a jet just as easily as my desk and it would get me out into the world that had less than six months to live.

Sky Without Stars — Chapter 2

I had just finished putting three sugars in my coffee when I saw a uniformed man enter the cafeteria, scan the crowd, and then make a bee-line toward me. I was the only visitor in uniform, so it was an easy target discrimination. He was on the heavy side, a few too many donuts sitting behind a desk was my guess.

"Welcome, Colonel Marsh. I'm Everett Long, Chief of Plant Security. I've been told to cooperate fully. We haven't had this much commotion since Nine Eleven. Can you tell me what this is all about?"

I could tell you, but then I'd have to kill you.

"Is it terrorists?" he asked my blank stare.

"I wish I knew," I lied. I wanted to tell him. If the site was added to the Noah project, he would likely become integral to its implementation and integration. "They just told me to write up what security changes would be needed if public order were lost."

"Public order?" he asked. "We're fifty miles from any civilization." I shrugged as he thought about it.

"Then that makes my job much easier. I saw two towers as we flew in."

"Sniper towers. First target of a military incursion, but effective against suicide bomber threats."

"Non-airborne ones."

"We have anti-plane and tank capabilities."

"No shit?" I said as if I wasn't already intimately familiar with all the capabilities.

He nodded proudly and led me from the cafeteria to the base of the closer tower. It was a cramped two-person elevator, very claustrophobic but better than climbing a thousand stairs.

At the top was a circular glass room. It had several unmanned computer consoles, only one lit up with video camera feeds. Next to the elevator was the stairway, which led down, if necessary, but went up one more floor to the roof hatch. Outside the glass was a concrete walkway. I chose to go up to the roof for the least obstructed view.

"Gets impossibly cold up here by November and doesn't start to thaw until April. Tower Two has a heated sniper nest on a turntable. They do target shooting year-round up there."

"How many are on duty?"

"Two are on-site at all times, but only one is designated 'on duty' at any time. Both keep regular hours and are less than five minutes from bed to trigger."

"They ever do target practice five minutes after waking up and sprinting to the top of the tower?"

"Not that I know of. That tree line is seven miles away. That's thirty minutes of running to get to the first fence. One approach road and the gate is ten miles away."

I looked at all the open fields of what looked like wheat. Without farmers, the tree line would eventually be inside the fence. I added that to my notes. I crossed it out when I remembered this place would be frozen tundra after Zero Hour, possibly to the end of time. Then I turned and looked at the dark blue water.

"Lake Superior. Freezes over most winters. Anything approaching from that direction glows like a bonfire on thermal cameras."

"SEAL Team wouldn't 'glow,'" I said.

"We have top-end Naval sonar and software. They monitor it for us and do assault training several times a year. Apparently, there are similar facilities around the world. Nuclear power plants are very safe as long as you have a good supply of water nearby."

I nodded and made a few more notes. I could not make any sense of why they wanted this facility. Building any new housing would take far more time than what we had left. The power itself would be cut off at Zero Hour. The people could survive with just a few hours of bottled oxygen. After that, the power plant would easily create endless oxygen by splitting water molecules with electricity.

"Is it China?"

"I wish I knew," I lied again, but my slight head shake gave away my true feelings. No sense in having him worrying about World War Three for the next few months.

"Do you want to see the rest of the plant now?"

"No. How much ammunition do you keep on site?"

"We load our sniper munitions for accuracy. We have enough powder here for about five hundred rounds for the fifty-cal sniper rifles. Gate guards use M-16's so we keep about five thousand rounds of five-five-six. We have ten Javelins for anti-vehicle and the Patriot battery has two reloads for twenty-four missiles total. They say the F-16s would be here long before we'd run out of those."

I wrote the numbers down and looked out at the wheat fields. "Do you have storage facilities for ten times the powder and five-five-six rounds?"

"Ten times? Not in our bunker. You think we need ten times the bullets?"

"All Marines want more bullets than they have, especially if they don't have to carry them into battle," I said turning to head back inside. I wandered the tower space, thinking through most scenarios I had contemplated for the other sites. Everett watched me silently, trying guess why I was there from the few words I said.

On the claustrophobic ride down the elevator his impatience finally burst.

"You expect to defend a siege without external support," he guessed.

"That is one of dozens of possible scenarios," I said without turning away from the number that was counting down the meters from ground level. It always surprised me when I saw metric values since most military bases avoided them. This was not a military site despite the high security. This was a science site with the latest technology... available in the nineteen seventies.

"What do you want to see next?"

Nothing I thought. There was no reason for this facility to be added to the list. The elevator door finally opened. "I'd like to run the fence lines."

"Run?"

"No, not 'run'. What do you use to inspect fences?"

"Every foot of fence is in view of a camera. There's a specialty off road van that is capable of repairing any fence issues. There are several vehicles capable of driving most of the fence line. We like to use the four-wheelers to run the perimeters, especially in the winter."

"Not snow machines?"

"The snowmobiles we have are slow and utility oriented. The four-wheelers are more fun. As you can imagine, it gets painfully boring out here guarding something nobody has ever tried to attack. They're even more of a blast out on the frozen lake."

"Frozen lake," I repeated as a thought entered my mind. "Ice fishing."

"No one does that up here since they suspect the fish will be irradiated. Plenty of ice shelters go out down in Grand Morais as soon as the ice gets thick enough. My grandfather used to take me out with him. Never understood it because it was so damn cold and we never caught anything. I think he just went there to get away from my grandmother. She was—"

"How do the fish stay alive under the ice?"

"They're cold blooded so they're metabolism slows way down."

"What if the ice doesn't melt? How long would they survive?"

"I don't know. My grandfather told me that the wind makes waves and that is what pushes oxygen into the water so the fish can breathe. I imagine they slowly deplete that over the winter. You expect the ice to become permanent?"

"Just curious." This was a potentially huge food supply. Pumping oxygen into the water would probably draw the fish as the rest of the lake depleted. The bubbles might even keep the ice open in places. That could buy a couple years of fresh food before people had to eat MRE's for every meal. They couldn't possibly want this place for fresh fish, right? Transport would be impossible within months.

I had heard my share of hair-brained ideas in the past few weeks. Some moron actually recommended neutron bombing all population centers before Zero Hour to cut down on our security concerns. He later said it was to reduce the suffering. As if slowly dying of radiation poisoning was better than a few minutes of suffocation. Defending a nuclear power plant for fresh fish might be almost as stupid.

Everett drove the fence as I made notes as to the type of fence and upgrade possibilities. Rolling out razor coils and barbed wire would make a good deterrent for people on foot, but it was a waste of time and resources. He kept trying new ways to ask what I could not answer. I did note that his expertise with all facets of the plant security made him likely to make the Ark list, so I asked him about his family. Too many to be saved with him, but a ten-year-old son would be useful in several ways. First it would give him motivation to survive. This is not as universal as one might think. Survivor's guilt was a very real problem. Second the boy would likely follow in his father's footsteps, saving the need for finding replacement security.

I walked with Everett through the facility, only mildly interested in the security office. I asked questions and did my best to look like I cared about the answers. When I was out of patience I asked about food and I was left in the commissary while Everett attended to his normal responsibilities. Whenever he came back to check on me, I was reading my reports to keep his prying questions at bay.

The others on the assessment team spent far more time on their tasks. I was dozing in the cafeteria when they finally gathered and we headed back to the helicopter. I didn't pay attention to their chatter since I would be recommending against the inclusion of this facility.

Sky Without Stars — Chapter 3

"Colonel Marsh, we need you to take personal charge of the Diamond Beach facility."

I just looked at the general with blinking astonishment. "Excuse me, sir?" My mind raced to find the reason for this punishment.

"You heard me, Colonel. You need to be up there in forty-eight hours and see to all the necessary upgrades. The President believes we only have one to two weeks left before the news goes wide. You need to be buttoned up and locked down before that happens."

"Why me, sir?" I had to ask though I doubted I'd get an honest answer.

He sighed and closed the door then walked to my window. It was a nice view of the coming San Diego beach sunset. I had no interest in trading it in for deep dark winter and short summers without insecticides. Once again, I had to remind myself, Diamond Beach may not see another summer for ten thousand years.

"Wally, we've known each other too long to bullshit each other." Bullshit was a daily, bidirectional feature of our working relationship. His preamble with an incorrect first name – my teams name is Walrus and it is never acceptable to give a nickname to a nickname. This meant that he never really knew my first name, even though I stood up at his wedding, or it had been so long he had forgotten it. "This isn't a punishment, if that's what you're thinking." Now I knew it was punishment. Probably for fighting so hard to exclude Diamond Beach from the list. "Diamond Beach may be the most important site of this entire Charlie Foxtrot."

"I have important tasks that need to be done here."

"You can do them remotely from there." The reverse was far truer. "Hansen will be your eyes and ears here and you can execute what needs to be done through her."

Bingo. He thought he had a shot with her, especially after the world's population dropped to prehistoric levels. He didn't want me in his way. Hansen was definitely not my type, but when your options go to zero, the beer goggles become far less discriminating.

"Are they fully provisioned up there?" I asked in accepted defeat. The choice was to follow orders or be exiled from the project and die gasping at Zero Hour.

"Fully," he said after looking back out the window. I doubted it since only the President's bunker was provisioned completely. I would have to check before I left. Not like I had any choice in the matter, but at least I would know what I was getting into.

"Is there anything I need to know about the facility? I still don't understand why it is even on the list."

"You'll be fully read in when you arrive." He didn't know either. "I hope I see you again, Colonel Marsh."

This was a way many chose to say they hoped the Noah project worked. It wouldn't of course. I'd read most of the reports I could get my hands on. They expected the Presidential facility to last twenty-five years. The winter was going to last centuries. If we'd had a decade to prepare, we might have come up with some sort of sustainable bio-dome, but I honestly would not want to be in the Presidential bunker eating twenty-four-year-old MRE's knowing the end was near. Maybe dying now would be the best way out.

"WELCOME BACK, COLONEL Marsh," Everett said with a seemingly genuine enthusiasm. "It's been like Christmas morning every day since your visit."

"How's that?" I asked, finally extracting my hand from his.

"Every aspect of our security has been upgraded. The newest tech, brand-new vehicles, enough weapons, and ammo to take on a small country. Still not sure what it's all for, but better to have it and not need it than need it and not have it."

"You still don't know?"

"I have plenty of guesses, but none that make sense."

"Okay, let's get some coffee and I'll bring you up to speed, at least for the parts that I know."

His enthusiasm took another leap upward as he thought the mystery was finally going to be solved. Unfortunately, the news was never taken with anything less than terror, depression, and sincere suicidal thoughts.

He led me to the commissary where instead of filling a paper cup I took the entire glass pot and carried it back to his office. I sat down behind his desk and poured a cup for myself. I skipped the usual sugar substitute and sipped the bitter brew. Reminded me of trench coffee in the South Pacific. To be young and stupid again. Everett took an uncomfortable seat on the wrong side of his desk. I rolled in his office chair and kicked the door closed.

"What I am going to tell you cannot be repeated, at least not until you hear the same information on your local news station. That means no family, no friends, no prostitute pillow talk."

"I've never—" he started unnecessarily.

"This planet has less than two months to live." I sipped my coffee and let that sink in.

"Colonel..." He knew I was not a prankster, and it was a long way away from the beginning of April.

"Leaking said information will lead to your immediate public humiliation and a quick death of less than natural causes. Understand?" He nodded. "There is a comet the size of Manhattan on trajectory to hit us. The human race will come to an abrupt end. If you are religious, you may believe that Jesus will Rapture you out just before. Our government has no official expectation of such an escape so in their infinite wisdom they have designated certain facilities important enough to preserve in an effort toward 'continuity of American society.'" I sipped again, wishing it would cool down faster.

"You're serious?"

"This facility is one of those that will be preserved, along with select personnel." I saw that look in his eyes. "If you weren't on the list, I wouldn't be telling you this. Your son will also be on the list unless that is not something you want."

"What about—"

"No one else," I said, cutting him off. "This information will become public soon, so you'll be able to say your tearful goodbyes, or choose to go with them if you can't bear the thought of surviving without them."

"Why my son and not my daughter?"

"She's too young and procreation is not the highest priority to those making the decisions."

"Procreation? What the hell?"

"Hell is the correct term. The dust cloud that will surround the Earth will last for decades, if not centuries. All sun dependent life will perish in hours, days, maybe months for the most persistent carrion scavengers. The worst winter you've ever experienced will become your dog days of August."

I drank the rest of my coffee and refilled the cup. He was out of questions and into the denial phase of despair.

"Our immediate security concern will be people that want to join us here after the comet information goes public. Until then I need you to seriously think through your commitment to this facility.

"We won't just keep them out."

"Not if they are persistent. No wasting ammunition on warning shots."

"This is unreal," he said.

I nodded, recognizing my own initial reaction. "I reviewed personnel records, but I want your genuine assessment. Can any of these guys pull the trigger on local citizens?"

"No," he said after a long silence. "I couldn't either."

"Good thing that isn't your responsibility. I'll be straight with you. I need you here after Zero Hour. You being here is the only way your son survives."

"Zero Hour?"

"Time of comet impact. They have it down to the exact minute, and the point of impact will be the African Atlantic coast. If it were over our head, we wouldn't stand a chance. It isn't really an impact, per se, but the devastation will be complete."

"We're a long way from the ocean."

"No, tsunami will not be a problem here. Most of the oxygen will either burn or blow out into space. Scientists don't agree on how long it will take to return to breathable levels."

"That's the purpose of all the new piping. They said it was in case of a chemical or biological attack, but it is just to pipe in oxygen throughout the facility."

"Yes, split from the lake water with electricity. As long as we have electricity, we survive. Without plants or algae, the levels may never come back. Food is the next necessity. There will be a decade worth of MRE's for us. We let outside people in, that takes us down to months or even days."

"This is... you knew when you visited before?"

"I was sworn to secrecy and I was against adding this facility."

"Then why are you here?"

"Punishment, I think. Let's go see if we have everything we need to survive this apocalypse."

We went through the supplies, some of which were stored in rows of steel shipping containers. They would be difficult to access once the snow started falling. I wondered if a flame thrower would be a good addition to the final provision list. I got a text from the facility manager to come to the conference room for a briefing. I hoped the reason for preserving this facility would finally be revealed.

Sky Without Stars — Chapter 3

A select few that knew about the comet settled into the moderately comfortable seats in the conference room. Al Dresden was the Navy nuclear engineer that was put in charge of the Diamond Beach facility. He was a tough, no-nonsense bastard completely without a sense of humor. I'm sure you think that's the pot calling the kettle humorless, but I used to be a lot lighter in both demeanor and outlook. End of the world has a way of drilling that out of you, but Al never had one. Hopefully that would be a useful characteristic when the fan got loaded up with excrement.

"Ladies and gentlemen, we are forty-one days away from Zero Hour. As of right now we are on lock down. We'll get shipments in, but no one that stays after Zero Hour leaves without my explicit permission."

"Why?" Everyone wanted to ask, but only the dumbass from Maintenance had the lack of sense to ask.

"President will be addressing the nation tonight. After that every facility will go to code Orange. You go out, we cannot guarantee you'll get back in. Since you are critical to facility operation, you stay. More importantly, we are going to be stuck in here for years. I'd like to know which of you can't hack it before it's too late to find a replacement that can."

"We can't go home before the President's address?"

"No. Phone calls are allowed when you are off duty. Next piece of business is the super-secret reason this facility was added to the list." All eyes looked up at him, even though some already knew the reason. Al gave a long pause, like when they announce who won the competition show. Not going to miss that mind dissolving crap. He finally nodded, and a geek fired up the projector and started the presentation from a small tablet computer.

"Project Long Shot is exactly that. If humanity has any chance of surviving this cataclysm, it is up to us, here in this facility, now. The boats that rolled out three hundred miles of plastic hose in the lake last week did so for Project Long Shot, not for deep water intake. The two major problems we will have after impact are lack of oxygen and an overabundance of particulates in the atmosphere. After Zero Hour, nearly one hundred percent of this electrical power plant will be splitting oxygen from water and pumping it down into those hundreds of hoses."

I had read several of the papers that suggested this. The obvious problem was that the hydrogen produced by the oxygen splitting was going to find other oxygen and burn back down into water. Zero net gain.

"The hydrogen will be compressed down in reactor pods with particulates filtered out of the air. This will sequester most of the hydrogen and make the particulates far more soluble."

"Hydrophilic," an engineer corrected. "The particulates won't dissolve in the water, but they also will not just float to the top. In fact, most will eventually sink down into layers of sediment."

The presentation geek nodded with annoyance at the unnecessary clarification, then continued. "The lake will hold onto the particulates and eventually the skies will clear."

"That will take millions of years. We're talking ten to the fourteenth cubic meters of air. One compressor here will not be able to process one trillionth of that in a hundred years. At best it will slightly clear local air."

"That's why they are calling it Project Long Shot," I said to the math wiz. See, I do have a sense of humor. What do you mean that isn't funny?

"The oxygen bubbling up through the water will be doing something else. It will keep the ice open above it. Approximately forty square miles. Moisture will be carried aloft with the oxygen and that will precipitate out, carrying more of the particulates to the ground. The more particulates, the dirtier the snow. The less particulates in the air, the more sunlight will get through and hit that dirty snow. Computer models show a chain reaction that will take no more than eight years to get us to equatorial normalization.

"The equator is thousands of miles south. How long before we start to thaw?"

"The model is less clear on that. We could be in a full-blown ice age so it would be a long time."

"The last Ice Age lasted twenty thousand years," one of the amateur historians said.

Al grunted in annoyance. "Now you know the 'why' of this facility. We are the only chance the other facilities have of making it through."

I had already made the leap, but the others came to it fairly quickly. We aren't intended to survive with the others. We might be under a hundred feet of ice by the time the atmosphere starts to clear for the southern facilities.

"What about the fish?" I asked. Everyone looked at me like I was an idiot. "If we are keeping the ice open and the water oxygenated, the fish near here will survive a lot longer. I assume we have a way of harvesting them to supplement our food." They continued to look at me and each other. No one had thought about this contingency. "I'll take Everett down to get some fishing equipment."

"No one leaves," Al said. "We can requisition what you need."

He was testing his authority. That was smart as long as the request was reasonable. Starting lock down weeks early was not reasonable. He should have run it by the section heads first.

"We'll be back in a few hours," I said getting up and heading for the closed door. Getting kicked off the list was unlikely at this point, but possible. Certain death at Zero Hour seemed a lot better than slowly freezing to death up here. I expected his booming voice before I turned the knob, but it didn't come. I didn't look back as I walked quickly to Everett's office.

"We need to take one of the van's down to the city. He was quiet and compliant, his mind going through the various stages of grief. I let him drive since he knew the roads better than I did. "They're locking us down tonight. The President is making a speech about the comet. You think you're going to hang with us?"

"Not much choice in the matter."

"Going to need a little more positive response."

"Yes, I want to survive. Yes, I want my son to survive." I didn't bother telling him about the delay of the inevitable. Hope was our best ally at the moment.

"Good. The briefing I was just in says the lake will be open and oxygenated for the forseeable future. We need a way to pull fish out in extreme sub-zero weather. I can't imagine anyone wants to stand on the seawall casting a line."

"That's easy. The plant water intakes pull in thousands of fish an hour. There are separator tanks that gently guide the fish out of the flow and back into the lake. A lot of the fish like the trip, especially in winter because it is warmer water in the separator."

"So we just need nets to pull them out of the tank?"

"Nets on long poles. The tanks are enormous. Strong lights will probably pull them to the surface. They need more than open water. Algae and insects are in their food chain and those will be gone pretty quickly, right?"

"I guess. Anyone you want to say goodbye to in person? Today, before lockdown ends your normal life?"

"My daughter won't understand. Too young, like you said. My ex-wife wouldn't allow it since it isn't my day with her anyway. You ever married?"

"No," I lied. This was about his grief. I'd processed mine months ago.

"Best and worst thing a man can do with his life. Most important part is who you choose, or who chooses you. Didn't know that tidbit when it would have had some value."

"Everyone changes over time, so it is impossible to make those kinds of decisions with incomplete information. At least a dozen of my men chose to stay behind with their ungrateful, miserable wives. They literally spent half their day complaining about them, but when it came down to it, they chose death."

"I would probably choose the same if we were still married," he said after a long silence. "There aren't many women on the list, are there?"

"No hospitals, no midwives, no chance for children to grow up in a normal way if they did survive. Only the Presidential site has all that," I said with some disgust. The man intends his genetics to be the majority of what survives into the future. Some have called it the 'harem', the hundreds of young girls added to the list specifically to repopulate the species.

"Where is that?"

"Top secret. That is the place most vulnerable to massive overrun. If word gets out, there are very extreme measures in place."

"Like what?"

"You don't want to know."

"Nukes?"

"That is one of the options."

"Not for us, right?"

"No. Public probably won't even know we are on the list until it is too late. Where is the best place to get nets and long poles?" I asked to change the subject as we approached the town.

Sky Without Stars — Chapter 4

"**M**y fellow Americans..." Blah blah blah. I watched the address with far more cynicism than it deserved. I think I was just bitter about being stationed here, and then informed that we were cannon fodder for the rest to survive. He laid out the dangers of the comet impact, not much different than Morgan Freeman did in that Deep Impact movie. He notably left out the lack of oxygen portion of the brief. No need to tell them now that they were all going to die. People would start hoarding oxygen bottles thinking that would make a difference.

I was more interested in the reaction afterward. The world was ending. Women and minorities hardest hit. Most of the pundits were pretending they knew, but they were all indignant they had been kept in the dark. There were no journalists on the Ark list. When that news came out, they would turn on their fearless leader in a heartbeat.

I went back to my cruise cabin... that's what I started calling it after they showed us to the modular housing that was shipped in from a FEMA storage site. It was small, functional, and I was not looking forward to spending the next decade in it. Climbing the tower was going to be the best exercise available with everything snowed in so I started establishing a routine of stairs, push-ups, and stretching. I did some target shooting from the various overwatch positions, happy that my skills were still better than 99% of the rest of the world. It didn't predict the shooter's ability to hit a moving target, like a guy running toward the fence. It was even less likely a guy could hit a fellow American, their only crime wanting to survive the coming apocalypse. The smart ones will carry a child, making it impossible to pull the trigger. Of course, once they reach the fence, a head shot will be far easier regardless any passengers they are carrying. What will these guys do? Shoot the guy and leave the crying child there to starve to death? Better to put a round through both of them at a distance and let others know there would be no mercy transiting the open field.

This was not a post for caring, feeling humans. Psychopath Marines like me are what they need, but most chose home and family. After Zero Hour the need for psychopaths will be significantly reduced, so it is a delicate balance. There were many ideas about how to have the necessary security until it was no longer needed. The best would have been patrol robots, but unfortunately they didn't exist yet. My idea was to bus out the unnecessary troops the day before Zero Hour. They'd die in transit from a distant post like this. Feeding them to secure this place from no external threat only quickly shortened the time we could fulfill this mission.

The simple fact that the survival of a moral civilization is always dependent on men willing to use extreme violent force to visit immoral horrors on others has never bothered most citizens that benefit from the actions. They just don't think about it in those terms.

Sky Without Stars — Chapter 5

The first wave arrived in a long parade of vehicles. None made it through the outer concrete barriers. Some were damaged in the attempt, but I instructed my guys not to damage any vehicles as long as they stayed outside. This was so they could drive home after they gave up. I made sure they gave up. We were still five days from Zero Hour.

The first group emerged from the trees in late afternoon. Seventeen men, women, and children. Possibly family members of people inside the facility. There was a little wind, but it was at my back, not theirs. I transitioned my scope from one face to the next. Even with the magnification, they were small indistinct flesh-colored circles. I stopped on the obvious leader. She was carrying a toddler, walking briskly a few yards ahead of the rest. I waited. I knew the maximum range of this rifle and she had a few hundred steps left to live, and hopefully change her mind. Donny fired a warning shot from the other tower despite my explicit instructions. It was lost on them, a subtle boom seconds after the bullet went over their heads with a sonic crack.

As she stepped over the low barrier that marked my thousand-meter mark, I squeezed the trigger. My heart resumed after a brief pause and beat twice before her head exploded. She fell backwards at the feet of a man that had just cleared the barrier himself. He stood frozen for a few seconds, wondering if his name was on the next bullet. Only if he kept walking. Half of them had already started running back. The others were waiting in frozen panic. He knelt down and took the crying toddler from the headless corpse. They were conversing, weighing their options.

He turned and looked at me. I was hoping he made the right decision, but a second death would show them it wasn't a lucky shot. Donny sent another warning shot, or maybe just missed. It hit the low barrier behind them with a profound whack and they headed back toward the tree line. Two smaller groups emerged in a fast run from further west. The sun was on the edge of my vision. A little further and it would have degraded my ability to target them. None of them were distance runners, but they were dodging left or right evasively. They must have been watching the other group from the tree line.

I didn't wait until they reached the low barrier. I hit center mass on the biggest target of each group. One was winged, choosing to dodge right at almost the right moment. The others kept running forward having made their decisions before starting the run. Donny winged another and I dropped two more. We traded shots as the numbers dwindled.

The last one alive fell to her knees, then fell forward in exhaustion. I moved my scope to scan the trees. A few shadows watched. When I looked back, the woman was on her back. Her breathing seemed to have normalized. She was going through stages of grief, or making peace with her maker, or waiting for us to go to sleep.

Donny fired another warning shot, the dirt spraying over her face. I was going to have to smack him around about the waste of ammunition. She didn't react to the shot. I looked back to the tree line. Five shadows were talking and watching the woman. The gate guard reported that most of the vehicles had left. They would take news of the failure back to town. Either they would give up, or more likely they would try again at night.

The woman was up now, walking back to the trees, but angling toward the lake. The shadows were moving in the same direction through the trees. I think they were calling to her. When she reached the seawall, she just sat down, legs dangling. After ten minutes I stood up and let Donny know he was on his own for a few minutes until I sent Otto up to take my spot. I used the bathroom and glanced at my emails as I walked to the motor pool. I fired up one of the four wheelers and headed out to the seawall.

"Ma'am, you need to leave," I said after shutting off the noisy motor. She didn't respond so I circled around to her side. Her mascara streaked face looked pathetic. "Ma'am?" I asked in a more concerned voice. That got a response. She turned and looked at me blankly then back at the water. "I can give you a ride back to your car."

"What's the point? Dead here now or dead in a few weeks from starvation." Word still hadn't gone out about the oxygen depletion.

"What did you expect coming here?"

"A chance to survive with my son." For a moment I wondered if she was Everett's ex-wife.

"At best we only have a few months or years more than everyone else."

"You say that like it means nothing."

"Months in a dark frigid wasteland slowly buried alive by snow and ice."

"Alive with people you love."

"I can't stand most of them."

"You will. Once you only have each other. I could be whatever you need me to be. I'll do anything." She turned with what she hoped was an alluring smile. I tried to not reveal just how horrific she looked. Blood splatter from her friends, grass and mud in her hair, mascara pointing up at her bloodshot eyes.

"Go home to your family."

"Son is in there with you and my husband is dead back there. I got nothing left." She looked back at the water. I drew my pistol and ended her suffering. I heard someone cry out from the distant woods. Apparently, she did have someone who cared enough. I drove toward the tree line and that was enough to send them running from the devil they certainly thought I was. I drove to the bodies littering the field. I checked them for weapons. Only one small knife. Maybe they thought they had a better chance getting in unarmed.

I contemplated moving the bodies closer to the trees so the warning hit them before it became a decision for the sniper. I heard a screech and looked up to see a buzzard circling high. They would be well fed tonight. I left the bodies where they lie and drove out to the guard shack.

They were in an impregnable booth without any worry. I purposely did not post any locals out here. These guys were impassive jerks. One vehicle was wedged between concrete and entry steel posts which will make deliveries impossible. I used the four-wheeler to move it out of the way enough to get trucks through. I'd call the towing service to have them remove it permanently. There wasn't much point having the Sheriff arrest and prosecute the driver. They all had a death sentence coming, including the Sheriff.

I complimented the guards on their restraint and headed back in. I talked to Donny about his wasteful shots and saw something in his eyes. He hadn't wanted to kill. It was affecting him badly. I wasn't sure if it because they were unarmed, or fellow citizens, or fellow humans. I thought of the woman on the seawall and saw his face there. Executing someone in your command was a dicey proposition. Exile was better, but I would need him for a few more days in case they came in force.

I chose to spend an hour with the entire shooting team going over what we did. What worked, what didn't, and most importantly the psychology of someone running across that field. I spelled out my orders again and hoped they would stick better this time.

The next wave came at night and they were well armed. The guard booth took some direct fire and they responded with two explosive micro-drones. One hit the truck they had drove in and were hiding behind. The second hit the small group that had survived the first one and had gathered in a ditch a few miles away.

The towing company refused to take the smoldering remains of the truck so we used the fencing truck to drag it a mile back down the road. We added their weapons to our inventory after cleaning them.

The final attack came the night before Zero Hour. There were several big trucks led by a dump truck with a massive snow plow. We had been warned by a family member who wasn't bitter enough to have been left behind.

A pair of National Guard Apache helicopters met them on the road ten miles away. Warning shots were enough to turn them around. I wondered if having that snowplow here at the facility would be valuable and cursed the eggheads that had not thought of it. What else had they not thought of?

Sky Without Stars — Chapter 6 Zero Hour

We rode it out in the various gathering places that had been pressurized with supplemental oxygen. In almost every room was a digital readout of both Oxygen and Carbon Dioxide levels. Everyone looked over at the readouts periodically to reassure themselves that all was okay.

Several of the TV news stations had gotten the inside scoop and they had supplemental oxygen to continue broadcast after Zero Hour. Some watched the small screens in morbid fascination as the world around them ended.

I rode it out in the tower. It was too difficult to seal and supply oxygen all the way up there so I took a bottle of compressed air and a thermos of coffee. I felt the tower sway as the first wave of compressed air flashed by. Some thought the tower might be blown over. I had read almost every prediction related to the impact. I watched the oxygen meter slowly wind down the scale, bottoming out at one percent. I think that was a rounding error. I had put on my breathing mask as soon as it started to change. I occasionally took it off and inhaled deeply, wondering what the end of the world would smell like.

The horizon began to glow red. Although the sun was still high behind me, the blue sky began to darken. I remembered something about the color of the sky being refracted light based on the molecular makeup of the atmosphere. That's why the same sunlight becomes black as you fly upward out of the thin shell.

Over the next few hours, the meter went up and down like a yo-yo as the invisible waves of atmospheric disturbance swept around the world. It settled at one percent. I got lightheaded when I took more than one unassisted breath. I looked out at the lake and the water had begun shimmering. The compressors' hum had been a constant background noise since we ramped up to a full-scale test and decided to just leave it running. Most of the air bubbles were tiny by the time they made their way to the surface. It was just enough to make the normal deep blue of the water an almost prismatic white. Like a frozen surface that was alive. Would it be enough to save the world?

The sunset was spectacular as the atmosphere loaded up with all kinds of formerly earthbound matter. Some millions of humans had been flash vaporized into that ether. I knew it would be the last sunset of my lifetime, no matter how long it lasted.

I heard the feint grind of the elevator. I wished I had flipped the circuit breaker. Last thing I wanted was some chatterbox waxing ineloquent about the current state of the world.

"Commander asked me to check on you." I didn't say anything, hoping they would join me in silence. I just held up my hand with a thumb's up and continued enjoying the sunset. "Wow." Yeah, wow. Shut up. Please. "You think we'll ever see the sun again?" One more word and he'll cease seeing this one. He blessedly took the hint and just watched. He headed back down once the room became dark.

A few stars came out, but the haze was beginning to envelope the world. By morning the sky was a permanent dark gray, like the middle of an intense thunderstorm. I mean that literally because the lightning was nearly constant. This was not something I had read about in any of the scientific predictions. It wasn't followed by acoustic rumbles. It was what we used to call 'heat lightning' when we were kids. The name came from seeing the flashes on a clear summer night, not knowing they emerged from an intense thunderstorm raging just over the horizon.

This lightning came from a much shorter electrical path between polar extremes and the abundance of swirling ions in the air. We began wearing goggles on outside patrols. Once away from the facility a mile or two the particle density was negligible. The people that designed the atmosphere processor at the heart of our mission had built it to attract the particles from high up in the air, and it worked to our detriment. Air filters inside the facility had to be washed almost daily. With more time to plan, it probably would have been smarter to build the air intakes miles away. If we'd had time, there would be smokestack tower more than a mile high to pull in the worst of it. They did the best they could with what time they had.

Turns out the last-minute attempt to get into our facility was thwarted by the Comet itself. They thought that getting past the security in the immediate aftermath of the impact would be possible. Maybe they could slip in unnoticed? We found the vehicles a week later. Some just sat in their seats. Others used their last gasps of oxygen to crawl out of the car. No carrion eaters had nibbled on them because they were all dead as well.

We found them when I decided to make a run down to the city for supplies. Although we were no longer feeding the national electric grid, our power was still on to people hundreds of miles away. Freezers full of good food would be gratefully eaten over the coming months. It wasn't like the walking dead, where you had to worry about a zombie jumping out at you. The people were all dead and were going to stay that way.

Our oxygen masks did a good job filtering out the rotting smells of human death. That did you no good if what you saw made you puke. I never took any of those that lived here on these runs, because they would have undoubtedly seen someone they knew.

We identified the best locations for foraging and used paint cans from the hardware store to mark buildings that had been depleted. Shutting off electricity in depleted houses made the most sense, but some guys did not want to stay long enough to find the panel, especially if it was in a basement.

I always went on these external runs, mostly to get away from the facility boredom, but also to make sure the guys didn't try to create a black-market enterprise within the facility. All of them pocketed candy bars and other things that might be used to entice one of the few females into their bed for a night or two. I figured most guys would pair off into prison couples eventually. It was common enough in the Navy I'd heard until they started letting women on board.

Anyway, I kept the foraging runs orderly and safe. One of the biggest surprises that first month was finding a bear doing some foraging of her own. We guessed she had fallen into some sort of hibernation long enough for the oxygen level to normalize to the current two percent. She was definitely moving in slow motion and her winter hibernation was going to be permanent. I'd never had bear meat, but the idea of getting her after she had fattened herself on the discards of a dead civilization made me curious. I brought the hunting rifle along on subsequent trips, but never saw her again.

When the snow started falling, the foraging accelerated. We could store the frozen food outside now, but keeping the long road passable became almost as hard as the foraging itself. The commander finally put a stop to it when the heavily loaded van got stuck and it took hours to get it out.

There were calendars on the wall and all the computers displayed the date, but there was no real sense of day and night anymore. Daytime was only a slightly less black sky. The bubbling lake could be heard but not seen. The fish we pulled out of it was quickly diminishing. I had been hopeful for years of fresh food, but it looked like that was not to be.

My job slowly transitioned to keeping simple slights from becoming violent fights. I was the sheriff in a mostly peaceful town.

Sky Without Stars — Chapter 7

Morale hit a low in March. The biology expected Springtime. No amount of artificial light was going to fix that. If we'd had more time, a greenhouse might have been a good addition. About the only green in the entire facility was the military vehicles in the motor pool.

A fight broke out over a woman and Everett tried to step in and break it up. His failure demonstrated the reality of our situation. Asshole #1 was waving a butter knife as a weapon at Asshole #2. Everett approached, talking calmly and trying to deescalate things. Asshole #2 pushed Everett toward Asshole #1 as an offensive maneuver. Asshole #1 dodged with a shove and Everett fell backward over a chair to a hard tile floor. A small bump on the back of his head turned into a fatal hematoma and now his 11-year-old son was an orphan among strangers.

I took Asshole #2 out to the sea wall for a talking to and fed him to the fish. Asshole #1 was kind of important to the mission so I only roughed him up enough let him know the consequences of trying to butter someone. The woman he had been 'protecting' from Asshole #2 ended up committing suicide a few weeks later. I swear I had nothing to do with it. Being desirable to most of the men around might be a school girl fantasy, but it was a nightmare in this apocalyptic scenario.

The staff psychologist, who definitely had the hardest job in the place, admitted the woman had come to him about the near constant harassment. The worst offender, I'll just call him Asshole #3, got a quick public neutering at the end of my razor-sharp knife.

The commander was not the least bit happy with my punishment method, but the harassment of the other women stopped immediately. Asshole #3 tried to take me out a few nights later and I used the knife again. Some people think 'Marine' is just another type of soldier. We aren't. We are a swift certain end to the things that ail society.

The Commander communicated with the other sites and passed on what he thought we should know. I had my own back channel to the security people I had put in place before being sent here. There was little gossip of interest beyond the official. The lack of outsiders trying to get in made security an internal matter. Others reported conflicts, but none admitted to using the extreme measures I did. They all had better weather since they were in the south. Clouds blocked the sun, but they had gray days with the sun overhead. They were cold, but the rare snow melted quickly. My guess is the Gulf of Mexico buffered the temperatures.

They came up with a few ways to measure the cloud density using radar equipped spy satellites. There had been many fluctuations, but none could be attributed to our work here. Honestly, if that is true in one year, I'm giving up on this pipe dream and heading south. It would take a snowmobile to get down to lower latitudes, but a hand crank fuel pump would give me all the fuel I would need. If the machine died, I probably would too.

Anyway, cloud density was reported daily, and my back channel let me know I could trust the numbers. My fear was some idiot would start manipulating the numbers to give everyone hope. The biggest surprise was the explosion of fresh fish. Apparently, the eggs laid by the few remaining fish had no natural predators and thrived in the oxygenated water. Morale improved significantly as the outside air hit a two percent solid value. That told me what was left to be burned had burned. There were many natural sources for oxygen, but we were definitely making a difference on that front.

We churned out nearly as much oxygen every day as the plant mass of North America. Another two percent and we would be able to go outside without supplemental air. That was projected for next September. We'd still be falling into the depths of an Ice Age, but at least we could survive without technology beyond fire to keep us warm.

Sky Without Stars — Chapter 8

My first few trips into town I had everyone armed and on high alert for survivors. Survivalists with electricity from our power plant could have had an oxygen generator to stay alive with. After a month of no signs, I knew that anyone that wanted to survive, headed south fast to stay ahead of winter. What I didn't anticipate were Canadian survivalists heading south.

If there were any in the months after Zero Hour, they didn't come near our facility. It was mid-summer that they started emerging from their bunkers and making their way through the deep snow toward us. The first group made it all the way to the inner fence before Donny even noticed them. I heard the shots and sprinted up the tower stairs thinking it was hundreds. I had to physically pull the rifle out of Donny's hands to make him stop wasting ammo.

The seven figures were face down in the snow. Through binoculars I could see the big ice encrusted backpacks and the tails of their snow shoes sticking out of the snow. None were moving, but none had telltale red spots from Donny's fusillade.

"I'm sorry I let them get so close," Donny apologized. It was not totally his fault since we all focused our attention south and west.

"Hold your fucking fire. I'm going down to talk to them if any survived." I didn't hand his rifle back, just leaned against the wall and gave him the binoculars, a less than subtle hint to remain unarmed. I jogged down the stairs and went to put on the extreme weather clothing.

When I made it to the fence, I could tell they had moved, but remained prone as the motor of my snow machine approached. I shut it down and silence returned. I lifted my M-16 but held it across my chest pointed down and away from them.

"Sorry about the shooting. My guys are a little trigger happy with nothing to do all day." There was no response. "I promise you are safe as long as you don't point anything in my direction besides an angry scowl." The smallest of the figures was first to lift their head. Small woman or a large child, I couldn't tell. "Anyone hurt?"

"No. That guy is a terrible shot." Young boy's voice.

"I'm Colonel Marsh, in charge of security. We have fresh fish from the lake and moderately bad coffee. You can't stay long but you're welcome to come in and warm up for a few days."

The boy stood up and the others began moving. I kept watch for weapons, two of which were hunting rifles which stayed over their shoulders. The boy approached the fence.

"I'm Roy." He pointed back and named his parents and older siblings.

"Hey Roy. Did you cut the outer fence?"

"What outer fence?" he asked in return.

"Guess the snow has buried it. If you walk west along this fence, you'll find gates half buried in the snow. Pass them by and you'll eventually come to gates that we keep clear. I'll meet you there."

"You aren't cannibals, right?" Roy asked.

"Not yet," I answered with a grin that he couldn't see under my thick ski mask.

I left him guessing if I was serious and fired up the snow machine and headed for the motor pool gate. It was a good hour before the family made it around. I assumed a lot of arguing on whether to keep walking or come in was done on their walk. Both hunting rifles were held across their bodies, upside down with open bolts. I had left my rifle inside and two of my guards kept a friendly watch from a distance.

Inside the cool motor pool garage we began slowly peeling off our outer wear, the family helping each other. Many curious faces began appearing at the motor pool door, thrilled to have something break up the monotony of Project Long Shot.

"What is this place?" Roy asked.

"Nuclear power station. Where have you been for the last year?"

Roy looked to his father to answer. The man shook his head. I thought secrets were a bad sign and my hand twitched toward my holster before I stilled it. The man then started telling the story himself with a heavy Norwegian accent.

"My father was soldier in war. He tell stories of life without any of basic things. All my life he is obsessed with having supplies to last months, years, decades without social order. We all thought he was crazy since we never used any of things he hide in cellar. When he died, cleaning out these supplies, most rotted, rusted, and useless, I pitied man for wasted life. Then Europe economy get bad for men like me with aging skills in Petrol industry no longer desired by green economy. We have hard winter seven years ago and cannot get to market for almost two weeks. I look to these mouths to feed and wonder what I can do. My life was wasted in opposite way of father. I have no skills to prepare for being cut off. As family, we decide to learn so it not happen again.

"I get job in new oil fields northwest of here and we begin building an off-grid homestead. Not much worry about nuclear war, but we make underground bunker that can supply own oxygen in case war happens. The night before comet, Jeremy hear on long distance radio that all oxygen be gone after impact. I not hear this on news so I don't believe. Luckily wife believe and she made us go to bunker for impact.

"Very few of the voices on long radio returned after impact. Some lasted only days, a few survived weeks. They store oxygen, not have generator. Eventually no one is on long radio. Winter was very difficult, but we survived. We have food for maybe two more years, but none of us wanted to stay in hole in ground."

"Surviving is not living," Roy said.

The father nodded and then continued. "We began making short trips to find clothes for long distance walk. We have GPS coordinates for many places to find refuge, this one was closest."

"As I said, you cannot stay long." He nodded with grim acceptance. "Why don't you follow me and we'll get you a warm meal and you can tell us the story of your walk here."

They left all their outer clothes and packs and followed me to the cafeteria. Doris had set a table for them and the rare smell of baking bread filled the air. They told their story of walking from rural home to rural home, most with enough firewood to last through the winter. They would stay for a few days, eat what little food most homes had on hand, bury the dead if necessary, and then move on.

A ring of people sat nearby, enjoying new tales from the world outside. I sat at the table with them, asking questions when the conversation lagged, doing my best to get each of them to have their say. I studied all of them as their father told their tale. Not a one of them seemed the least bit hostile except the wife. She was road weary and not in the best of health. She was not as enthusiastic about the survivalist ethos that had helped them survive. She was working toward her demand to be allowed to stay.

The boys all seemed intelligent, hardworking, and most importantly, passive. They could all be useful workers once trained up. The daughter would be nothing but trouble. She was thin as a rail with no curves to speak of, but most of the gathered men looked at her like a big stuffed turkey on Thanksgiving. She seemed oblivious to it, but her momma wasn't and that might have been a good portion of her growing hostility.

Al emerged from the crowd after the food had been consumed. It probably would have driven him to distraction as he subtracted project sustaining provisions from the end date of his mission. He didn't look my way, which told me he was seriously PO'd that I had invited them in. He introduced himself and asked their intended destination in a way that was meant to communicate they should be heading there before nightfall.

After he left, I told them they had three days to thaw before they headed back out into the cold. It seemed to be what they were used to on their trek thus far. It gave me time to figure out if we were going to keep any of them to fill the vacancies that had been created, not the least of which were the aforementioned Assholes #2 and #3.

We found empty beds for all of them, the daughter and Roy sharing my cruise cabin. I had spent enough sleepless nights up in the tower watching cameras, it wasn't much of a hardship to go without a warm bed.

On the second night I took the father up to the tower with me. You might be wondering why I don't use his name. Truth is it was incomprehensible and I feel like a fool trying to guess at how it is spelled. I asked him if he'd want any of his boys to stay behind. He wanted good things for them, but even he saw the futility of our mission here. I got the impression he wouldn't mind leaving the wife behind, but we had no need for that kind of distraction in our delicate social balance.

I accepted his refusal and then I pulled out a map and told him what he'd find in the towns that we had plundered after Zero Hour. I'd been trying to figure out a way to help them get south easier without having to part with our precious vehicles. He'd been thinking the same thing, so we just went through the options knowing it was the snowshoe express for them. I gave him some information I probably shouldn't have about what he'd find down near the Gulf of Mexico if he made it that far, including GPS coordinates to the better facilities. Not the Presidential bunker, for sure. They'd be shot before they ever got close to that. I'd alert my security group of the possibility these poor bastards might make it there in a year or two. Hopefully they'd get some sort of vehicle once they got out of the deep snow.

I did drive them in the fence van as far as our front gate which we kept clear of deep and drifting snow more out of boredom than anything else. Probably saved them most of a day's walk, but they were likely six days from reaching un-plundered housing. Because of this I slipped them seven days of MRE's and a box of Clark bars I had stashed for a rainy day.

As they walked into the dark gray morning I sat in the warm idling van and wondered if I would be following them. Oxygen was up to four percent now. Nobody knew exactly why, but they assigned us the credit. Equatorial particulate density was down to forty-five percent on average, so it was possible some ocean algae had survived and was starting to spin back into production. It would be a long time before anyone took a boat down and did their science shit to find out. There was a nuclear aircraft carrier that was readied for post-apocalypse use in a San Diego dry-dock, but most believed the skills necessary to run it would be long dead by the time the air cleared.

Roy and his family finally disappeared from sight so I drove the van back through the gate, locked it up, and headed back to the garage. I think they might have tried to steal the van from me. Once they saw the depth of the unplowed road, they realized how futile such a gesture would be. We would be watchful for the next few days that they didn't try to come back and steal the snow machines. I think they hoped to find brand new ones at the

Moraine Ski-Doo dealership I told them about. It was on a hill that might be kept out of the deep snow by the strong winds. Finding good gas would be difficult in deep snow, but not impossible.

I fell in my bed in exhaustion, having not slept much during their visit. There was a distinct foreign smell of Roy and his sister. They had showered, so there was perfume of the soap, but something else. I tried not to think about it as I sank into a long, deep sleep.

They did return two weeks later. I was watching the video feed when I saw the snow machine with three figures on it pulling up to the front gate. The clothing was unique enough that I called for Anatoli to hold fire. He hadn't seen them yet, probably sleeping up in the heated sniper hide on the other tower.

When I saw the father pull out wire cutters and start snipping the fence, I questioned my earlier assessment of the man. I assumed he brough his two oldest sons back and they were going to make a run for our supplies and snow vehicles. It made no sense. Then one of the boys shouldered an AK and I knew these weren't members of Roy's family. After a big hole was cut through the fence, they drove the snowmobile through and resumed their charge toward our buildings.

"Anatoli, weapons free. Kill the vehicle first if you can."

"You don't want to talk to them?"

Anatoli had visions of the family come back to stay and the skinny sister joining him in bed. He'd share if he had to. "Don't think it's them. Most likely they got ambushed," I said. Anatoli swore in Greek words I didn't care to understand. They might as well have killed his bride on the way to the honeymoon. He wasn't as emotional or wasteful as Donnie. He'd take his time and put one round each through the bridge of their nose. No chance for interrogation.

We'd never know what happened unless I went out and took one of them alive. In my experience, a man willing to heft an AK-47 had decided to wage war and not stop until they had what they wanted. I watched the video as the vehicle rode up and down across the frozen waves of the open field. When they passed the thousand-yard marker the driver's head kicked back into the one behind it. The snow machine veered in one direction and the driver fell off in the other. The middle man was holding his neck with one hand and reaching for the handle bar with the other. The one in the back eventually fell backward off the seat.

"I think I got all three with one shot," Anatoli said after all the bodies stopped moving, the middle man having driven a slowing circle as he bled out from the neck wound.

God bless the Barrett rifle company and their fifty-caliber finger of death. "Good shooting. I'm going to investigate. Keep an eye on the tree line for me."

I went to the snowmobile first. It was still idling. The middle man was pitched forward on the handlebars. My flashlight illuminated a coating of red blood from his fatal neck wound. I killed the machine and pushed him off onto the snow. My fear that I made the wrong call dissipated when I saw the scraggly bearded man's face. Definitely not a member of Roy's family. However, it definitely was one if their unique parkas. I drove over and found the driver with no identifiable face.

Could have been the father pressed into service against his will. I stopped next to the AK guy and shut off my motor, but left the headlight on. I kicked the body over to look at his face and recognized him immediately. Not Roy's family. This man had been in the Noah project. Was he a courier of some sort? No chance they would have sent a messenger this way without some kind of radio communication. I unzipped the coat and began searching every pocket. I found nothing of value aside from two energy bars and a pack of cigarettes. No lighter. One of the other guys must have it.

I stood to go check the pockets of the other two when I heard a groan. I looked down and a small woft of steam came out of his mouth. When I rolled him over again, the bullet hole that had gone through the back of his coat had also penetrated his shirt, but there was almost no blood. I pulled the collar down and found he was wearing a military grade flack vest. Worthless against a fifty-caliber round unless that bullet had been slowed as it passed through two other human bodies. The blood on the outside of the coat must have been from the middle man. I zipped him back into his coat and dragged him around to the snow machine. I lifted him onto the back of the seat, facing the opposite way with his chest and head on the utility rack in back. I used a bungee to strap him down. I drove over and quickly went through the other pockets, finding the lighter on the driver along with another pack of cigarettes. Middle man had two milky ways and one of my precious Clark bars. Anatoli met me in the motor pool garage with the entire sniper team.

"Who's watching the tree line?" I asked gruffly.

"Keith is in tower one watching the cameras," Anatoli said. "You bring this guy in because he survived?"

"Afraid you only got two kills with that shot, but won't surprise me if this one is paralyzed from the impact on his vest."

"He had a plate?"

"I didn't strip search him out there, but he's still breathing. Is the doc on his way down?"

"Yeah, woke him up. He said to give him ten to get some coffee."

"Let's carry him up then," I said, grabbing an arm at the shoulder. Three others grabbed a limb and we lifted him up to our shoulders and started walking up the loading ramp. Anatoli held the doors and we wove our way through the building and up the fire stairs to the third floor.

"He just kicked. I think he's waking up," Gerald said behind me. We doubled timed it down the hall and got him onto the gurney in the doc's examination room. He was starting to move, but there wasn't any consciousness as far as I could tell. We started pulling off the cold weather clothing and layers of inner clothing.

Anatoli wrapped his knuckles on the chest. "Ceramic plate in front. Must have been what saved his spine." We rolled him and found a matching plate in back as we pulled the bullet proof vest off him. "Lucky SOB," Anatoli said in disgust.

"His name is Sgt Logan Meyer of the United States Army," I said, his face now fully uncovered.

"You found his ID?"

"No, I worked with him at the start of this Charlie Foxtrot. He was in charge of munition procurement for all the sites. Not sure what happened to him after I got shipped up here."

"Messenger?"

"What is the point if we still have satellite comms?"

"What do we have here?" the doc asked as he walked to the sink to wash his hands.

"Took a hard strike to the back from one of my bullets. High end vest saved his life," Anatoli said, once again regaling us with his feat of marksmanship. Once he was done, I sent him back up to the tower to finish his overwatch shift. The others went with him hoping for more action that they could get a piece of.

The doc started his examination with a pen flashlight in both eyes, then worked his way down, cutting off clothing as he went.

"Recent knife wound, maybe a week old," he said when he cut away the jeans. I hoped it was Roy's father getting in his final blows to protect his family. "This looks like ordinary sewing thread. He's lucky he hasn't died already from infection." Not exactly a miracle since there are plenty of antibiotics out there to be found. "Help me roll him over."

We both saw the enormous black and blue contusion on his upper back. "Jesus."

"Could be spinal disruption. If he's bleeding into his CSF, there's nothing I can do for him."

"No way you can wake him up for a few questions?" I asked. Doc just shook his head. Probably didn't want to waste any precious meds on him. "Fine. Let's put the restraints on and have someone watch him."

"You know where I'll be if you need me." The loud 'you woke me up for this?' subtext came from his grumbling tone. He was a world class doctor without most of the modern diagnostic equipment of a modern hospital. I would have rather had a field medic, but I wasn't in charge of personnel.

I opened a drawer and pulled out the Velcro restraints and began tying his wrists to the bed rails. After doing the same to his ankles I covered him up with a few blankets. I went to my room and returned a few minutes later with a war novel I'd been reading.

Sky Without Stars — Chapter 9

"Colonel Marsh?"

I jumped awake and the novel on my lap fell to the floor. I looked at the restrained Sergeant on the hospital bed as I leaned forward to retrieve the book.

"Sargeant Meyer. Why were you cutting through my nice fencing?"

"I had to get here to tell you important information."

"Sat comms would have been faster and safer for all of us."

"No... they don't know I'm... Where's Joe and Hector?" he asked looking around.

"Frozen fifty caliber donuts. You knew we had sniper overwatch."

"Didn't think you'd still be on high alert."

"You would have known if you asked the people you stole the cold weather gear from."

"Stole? We found the stuff in the Ski Doo dealership."

"Just the outerwear? No bodies? No blood?"

"No. You knew the people?"

"They left here two weeks ago, headed to that Ski-Doo store on my recommendation."

"We ain't seen no one for months. Joe and Hector are dead?"

"Very. What are you doing here, Sargeant?" He didn't answer, just seemed to be processing the deaths of his friends. He didn't seem at all concerned that he was tied to a hospital bed. "You are going to join your friends if you don't start talking."

"Colonel, I..."

I waited and watched. It seemed like he was trying to spin up a new story to match the facts. "Truth or consequences, Meyer."

"This is the only facility left."

"What do you mean?"

"Everyone else is dead."

"I just talked with Dexter yesterday."

"It isn't him."

"Who is it?"

"Computer generated."

"Not possible. Even if it was, it doesn't explain why you are here."

"It's the only facility left," he repeated.

"The value of your presence, especially after you cut into my nice security fence?"

"You need to know what happened."

"Yet you still haven't told me."

"I need to use the bathroom," he said, finally pulling at the restraints.

"Who stabbed you?"

"What?"

"Stab wound on your leg. Who stabbed you and who sewed it up? Doubt it was Joe or Hector."

"You don't believe me, do you?"

"You haven't said anything. Spill or I'm going to start hurting you for hurting that family."

"Your friends are fine for now, but you need to let me go."

"You just admitted to lying to me. Why would I believe anything you said now?"

He started to talk but my fist crossed his lips before his words. His unconsciousness returned immediately. I went to the communication room and sent a message for Dexter to call me.

"HEY DEX, WHO WON TAGFEST?"

"Blue team totally cheated," Dexter said angrily.

"How do you cheat at laser tag?"

"I don't know, but three of my shots were dead on and didn't register. Wanted to pull out my forty-five and show them I don't ever miss at that range."

"You'll get 'em next week. Hey, do you remember Logan Meyer?"

"Sargeant pushing bullets out to the sites, right?"

"Yeah. Do you know what happened to him after I got deported to the great white north?"

"I think he was going to Scorpio for the long haul. We had our munitions early so I didn't interact with him much after he got us that container of M4's. Those came in real handy that final week. Why?"

"Who do we know at Scorpio that I can get discreet answers from?"

"Nobody I know. Why discreet? He owes you money or something?"

"He showed up here last—"

"He left Scorpio!? To go up there!? What the fuck!?"

"My thoughts exactly. From what he said I'm not sure I can trust official channels, so I wanted to make a back channel query first. None of my guys went to Scorpio because—"

"Secret Service handled all the security there. What did he say?"

"Nothing that makes sense. Claims all the other sites are gone, that Diamond Beach is the last site standing."

"That's crazy. How could you believe that shit?"

"Don't believe him, but I also don't exactly trust leadership—"

"Since they sent you to Siberia. I assure you, Walrus, that we are nominal here at Coronado. I can guarantee Libra and Hercules are five by five as well."

"I appreciate the reassurance. I'd just like to know why he is here."

"Maybe he tried to hit on one of the Harem girls and they booted him."

"Not likely since Logan is DADT (Don't Ask Don't Tell). I think he's out of it right now because we killed one or two of his boyfriends. Probably should have withheld that info until I got his story."

"Killed them?"

"The three of them came in on a single snow machine, cut a hole in my nice fence, and then made a sprint for the buildings across the no go zone. My sniper put a fifty through the drivers nose and it continued through the middle guy's neck and terminated on Logan's Ceramic back plate."

"Almost three kills with one shot?" Dexter asked reverently.

"Almost. Kind of wish he got him too because I hate mysteries."

"Wouldn't be official business. Must have been excommunicated and thought you would be the only one to take him in. You always were an old softie."

"Foxtrot Yankee Alpha," I said slowly, keeping my laugh inaudible. "I guess I just have to see if he comes to his senses and gives me the truth."

"He had to have connections to get assigned to Scorpio. Finding those connections might answer some of your mystery."

"If only I had access to the NSA database."

"I'll ask around quietly if anyone knows anything."

"Thanks, Dex."

I hung up the phone and wondered if any of that conversation could have been generated by a computer. Not a chance. That meant that Logan was wrong about the other facilities. Why would he claim that? If he was kicked out of the Presidential compound, wandering the wasteland could have driven him batty. I went back up to the doctor's exam room and found Logan talking at the deaf Marine I put in the room to guard him. He wasn't completely deaf, just enough that it was easy to ignore the ranting prisoner.

"Colonel, you need to listen to me."

I waved the Marine to head back to his off-duty pursuits and looked at the bruise I'd created on Logan's face.

"How long were you at Scorpio?"

He seemed surprised I knew his assigned destination. "Never went there."

"Why?"

"They reassigned me to Cadillac the day before I was to travel."

"How long were you at Cadillac?"

"Five months. I left because everyone was dead."

"How did they die?"

"Some sort of flu. Swept through fast. One day everyone was sick, the next they were dead."

"And you were the only one who survived."

"No, there were eight of us. We figured it was one of the vaccines they gave us.

"You didn't bother telling any of the other sites what happened?"

"Of course we did. They told us to stay there and bury the dead. Had to be bio war so no FN way we were going to stay. When we got to the Hoover Dam, no one was home."

"You went in?"

"No way in, but no one answered the door. We figured they were all dead too."

"So you headed here instead of the coast sites?"

"No way they'd be able to get a bioweapon here. You were already frozen out."

"So you drove most of the way?"

"Hector is a genius with mechanical shit. We drove in the Hummers until the snow got too deep. We got a big snow cat from a ski resort and used that most of the way across the Rocky's and plains. Eventually, we lost a track and no ski resorts in Kansas so we had to start walking. Venus grew up in Alaska so he helped us make snow shoes. It was slow going and two guys just gave up and headed south. Eventually, we got far enough north that we found a snowmobile dealer and Hector got four machines with makeshift fuel trailers. They didn't last long sucking in the crappy air. Filters clogged and no way to wash them out. Venus froze because he wouldn't share a tent. Joe, Hector, and I just kept going, knowing this was the only place we'd be able to survive."

I sat there watching his face as he spun out this tale. His face was constantly shifting as if to find the next piece of bullshit that fit the circumstances.

"You have to believe me, Colonel. We just wanted to survive. Are Joe and Hector really gone?"

"Yes."

"I'm really thirsty."

I just got up and left the room and went to find the doctor. I sent him back to tend to his patient with strict orders that he remain restrained. Logan wasn't a physically imposing guy, but soldiers are by definition trained killers. I went to the communication room and found the contact number for the Cadillac site. I knew two of the security officers that had been assigned there. No one answered the phone. Not unusual, so I sent emails for them to call me.

Emails got to individuals at a site immediately since most carried phones. They couldn't make calls with them. They were mostly used for games and streaming movies through WiFi. Our system never worked that well, probably all the radiation shielding, so I didn't bother carrying a phone. I waited ten minutes, more than enough time for either of my security officers to get the email and get to the commo van.

Site Cadillac was on the lake created by the Hoover Dam. It was considered the best source of clean drinking water so they were tasked with keeping the intakes clear for the underground pipelines that fed many of the sites, including the ones on the SoCal coast.

"YOU GET MORE OF THE story?" Dexter asked when the call connected.

"How's your water supply?" I asked.

"Clean and reliable. Should I be worried?"

"No drop in flow?"

"Not that I've heard. What's up?"

"You talk with anyone at Cadillac?"

"Never."

"How about the Dam?"

"My LT has a brother up there. They play chess over the phone every week."

"Logan says he was sent to Cadillac and everyone died four months in of some kind of flu. The eight survivors went to the Dam and no one was home. They decided Diamond Beach was the least likely to succumb to the bio-weapon that killed their friends, so they headed this way."

"No way Cadillac dropped out and no one noticed."

"What if they did notice and just kept a lid on it?" I asked.

"Meaning we're going to lose our water eventually."

"I still think he's full of shit, but Umar and Fender were at Cadillac and I'm getting nothing back from them."

"Send someone from the Dam to check it out?"

"Only if they have Level 5 bio suits. Better to just ask if anyone at the Dam is in regular contact."

"I'll get the LT on that right away. Don't want to be drinking seawater if I can help it.

"The Ohio class tied up in San Diego harbor can churn out water and electricity if you need it. That was always in the backup plan."

"Good to know. Anything else?"

"No. Email me either way."

"You're carrying your phone now?"

"Until this mystery gets solved."

"DOCTOR, CAN I TALK to you outside?" I asked after listening to his interaction with Logan.

He came out into the hall and I closed the door. "What's the prognosis?"

"Seems like he'll pull through. More worried about the knife wound."

"Mentally?"

"Seems a little detached and evasively passive. Definitely not to be trusted."

Doc was a shrink as well, so I'm glad he agreed with my assessment. I nodded and opened the door again.

"Colonel, I can tell you've confirmed my story about Cadillac."

"Just talked with Fender," I lied. "Says you were never there."

"Bullshit. Fender was one of the first to drop. Must be some kind of computer generating his voice."

"To what end? I think your head is full of paranoid delusions. We don't have a padded cell for you, so you better start figuring out what truth you have left to tell."

"I am telling the truth," Logan said with exasperation, pulling at his wrist restraints for the first time since I knocked him out.

"So you have my friends tied up somewhere? The two missing from your story are with them and will do bad things to them if you don't report back to them in time?"

"I never met your friends. You just seemed obsessed with them, I only said what I thought you wanted to hear."

"Is that what this Cadillac story is? Just telling me what I want to hear?"

"No, it's the truth."

"Then why didn't you start with that?"

"I did. I could tell you didn't believe me and you kept threatening me. It was the only leverage I seemed to have. I knew once you heard my story you would believe it."

"What happened to the other two that left Cadillac with you?"

"Other two?"

"You left with eight. Two went south when the snowcat died. Vinny froze because he—"

"Venus."

"Right, Venus. Three of you cut through my gate. That leaves two."

"Ike and Rico went toward Vegas from the Dam. That was probably the smart move."

"Definitely smarter than cutting through my fence. How did you find the Ski Doo dealer?"

"Snowmobile tracks lead right to it," he said.

"Tracks?

"Whole bunch of them. At least three or four. Figured it was you guys, but the tracks only went south."

"That seemingly inconsequential tidbit may have just saved your life."

"Your friends?"

"Most likely. If they found better snow gear, they might have left behind what they had."

"What they left was way warmer than anything on the racks of clothes."

"Not if they put on three or four layers. So, Sargeant, now that you've made it here, what do you expect me to do?"

"You have to let me stay. I can do almost any job here."

"You can run a nuclear power plant?"

"I can learn. I'm really smart."

"Maybe you can, but we have all the people we need."

"I can—"

"You left what you called a bio-war target and came all the way here to possibly that last place with human life, not once thinking you might be carrying that infectious bioweapon with you? Does that sound 'smart', Sargeant?"

"No..."

"You may have single handedly ended the human race by coming here."

The flu did go around a week later. Very mild, and it happened often enough that it was impossible to know if Logan brought it in with him. I thought about going down to Moraine and confirming his story about snow machine tracks headed south. I preferred to just believe that Roy and his family were happily motoring down through Iowa now.

Logan turned out to be nearly useless in any job since he had been a leader, not a worker, for the last decade. I made him come out and fix the fencing on the gate he had cut. Then I took him to the place where his comrades fell and left him there with them.

Sky Without Stars — Chapter 10

L ocal oxygen levels shot up to ten percent in less than a week. No one had an explanation. Well, everyone had a theory, but none of them made sense. It slowly tapered back down to seven percent over the next few months, but any improvement was a light at the end of a very long, dark tunnel.

We were now three years into project Long Shot. Many of our underwater hoses had formed leaks. Although this was their purpose, having one large leak instead of a hundred small ones, meant the air spent less time in the water and therefore less particulates were removed. The way we knew this is by seeing large bubbles erupting from the surface miles out in the lake. This may seem like bad news, but the fact that we could see these bubbles miles away meant the particulate density was dropping and more sunlight was getting through. The satellites reported no such change in density.

We didn't bother questioning the scientists who looked at the satellite data. We just chose to believe we were making a difference, even if it was just local.

The fish continued to show up in our filters, but less every month and the taste of them began to be offputting. I decided to make a run down to Moraine for the express purpose of procuring seasonings. Garlic, paprika, oregano, hot sauce, anything that could turn the sameness of the daily foodstuffs into something palatable, if not outright desirable.

My other purpose was to investigate Logan's story about snowmobile tracks heading south. There was at least a foot of snow added by that time, so it was impossible I would find any such evidence. I knew the only real evidence I would find was the human remains of Roy and his family. I didn't find evidence in either direction.

You probably think I was a monster to have ended Logan's life so callously. I'd probably agree if he hadn't lied to me. I have a serious problem with liars. Probably grew out of dealing with politicians and bureaucrats during my military career. When lives are on the line, the most innocuous of lies can get thousands killed. A good liar can plant a seed of doubt into a chain of command. If you can't trust the leaders, you can't trust the orders. If you don't trust the orders, you have to start thinking about what you are doing. In a battle, wasting brain cycles on whether you should advance or retreat because someone up the chain wants the target attacked for personal reasons will paralyze you.

Logan was typical of supply officers. They had to do things that were outside regulations to get stuff done in time for it to matter. To me the most important thing was results. If he said the bullets would be delivered and they didn't show up, no amount of excuses mattered to those that died for lack of ammunition.

There were other reports of survivors showing up at facilities down South. Even the best survivalists did not have endless supplies of food. Most expected to be able to supplement on animals and plants after the bulk of humanity disappeared. I paid attention to such reports hoping Roy would find safe shelter at the end of their long hardship. Others listened for the same news in hopes such a trip could be successful.

There is a term in general usage that denotes a psychological condition most deadly to enclosed societies. Cabin fever. Scientists have probably studied it most intensely since we began sending men into space for long durations, like the fake trips to the moon.

No, I know we went there, I just like tweaking those who doubt pretty much everything any government has ever said or done.

Our facility is enormous compared to the International Space Station, but it is completely shut off from the world with no way to go elsewhere. I admit my spice run was pushed to actuation by my own need to get out of the bubble. Our best ally against cabin fever was entertainment. We had an extensive library of movies and TV shows created before Zero Hour. Some people took to creating stage plays either of original material, or re-enacting some of those movies.

The second-best ally was companionship. If there was a better balance of men and women, we'd be overrun with babies we couldn't possibly care for. Many discreetly paired off for short periods of time to satisfy the urges that build up. I had made it clear that all such things had to be explicitly consensual and they all knew the consequences.

There was no room for large scale sports, but every week they emptied the cafeteria of tables and chairs and played kickball, dodgeball, even soccer was poorly attempted. Live sporting events would have done a lot to quell the male appetite for competition. It is said that war replaced hunting in the agriculturification of humanity, and sports replaced war. This isn't as true as it sounds, but it does reveal a biological need for men to cooperate toward the goal of killing.

I saw firsthand a lot of killing in my life, usually the aftermath. It was a necessary evil only because some amount of poor bastards thought that they could improve their lives with violence. A moral military is formed to show them that it does not.

My seemingly swift decision path to violent ending lives is not for the one I kill. It is for all those that stay alive and harbor thoughts of improving their lives with violence. People can remain civilized only if the consequences of not doing so are a clear and present danger in their lives. I am that danger. That is why they sent me here. Feel free to hate me, but this place would have devolved into a tyranny of the strong over the weak within a few months. Since the most important people in the facility were the pencil neck scientists, project Long Shot would have failed completely by now.

ANOTHER SPIKE TO TEN percent happened, and then it fell back to four percent. One step forward, a little more than one step back.

The good news is the air is clear enough that we can see the edge of ice our air pumps are keeping open. The bad news is that this edge is much closer than the original fifty miles we started with. If the ice closes completely, our job here is done and we have failed.

The eggheads tell me that the air bubbling under the ice is still effective and is likely exiting through small vent holes. What they avoid telling me is that the particulate clearance rate has been steadily trending down and we are looking at a ten-thousand-year ice age.

"Colonel, we need you in the conference room." The words came from a young man named Sullivan who had started the project as a radiation technician but had busied himself learning nearly every job in the entire plant. Some joked that he could run the place by himself if everyone else left or died. I was in the middle of a novel I had read when I was in high school yet had almost no memory of the plot.

I put the book down and stood and stretched. I looked at all the monitors one last time before joining Sullivan in the elevator. When we hit the ground floor and the doors opened, I asked "What's this all about?"

I could tell he was excited about something, but not whether it was positive or negative. "It wasn't my idea, so I'll let them tell you."

'Idea' seemed like a positive word. The conference room has the usual brain trust that went over the latest data from the plant and what was provided from the science collection sites down south. I took the only open chair and Sullivan went to the other end of the table and sat on a window sill. They all just looked at me expectantly.

"What?" I finally asked.

"Colonel, we have come up with an idea that might increase our chances of succeeding."

I just sat there waiting.

"It is not ecologically sound and could present a worse long-term problem than the one we are in."

"Just spit it out," I said.

"They want to take the spent fuel pellets and use them to melt the ice," Sullivan said when no one spoke up.

"What does this have to do with me?"

"Colonel, you control what happens outside this facility."

"My snipers won't shoot you if that's what you're worried about."

"Someone needs to take the pellets out onto the ice," Sullivan said.

Ohhh... They need me to take lethal radiation material out onto ice that has unknown thickness because of the warm oxygen bubbling up from below.

"Denied," I said standing up and heading for the door.

"Wait!"

I turned back at the door, hand on the knob.

"Colonel, we need to do something."

"You want someone to die of radiation poisoning in the act of making the Great Lakes a radioactive wasteland for fifty thousand years."

"If we do nothing..."

"I remember idiots saying that about the Earth's temperature going up a degree every hundred years. Keep those brain cells churning until you find something reasonable to try."

I left the room and heard feet running after me. It was Sullivan. He rode up the tower elevator with me in silence. He followed me out of the elevator and walked to the windows. I sat in my chair and lifted the book, but didn't open it. I knew Sullivan had something to say. It took almost ten minutes to come out.

"I know it is a bad idea, but the ice is closing up. I think we have less than a year."

"That's why they named it Project Long Shot."

"They're going to do it with or without your help."

"Over my dead body."

"If necessary. You don't have to be the one that goes out. I'll do it."

"Always amazes me how stupid smart people can be. You are not thinking logically or rationally. You are thinking desperate thoughts toward taking desperate actions against an overwhelming enemy that has no chance of losing. You are literally a Kamikaze pilot flying toward an enemy ship after you have lost the war. You might as well drain the coolant tanks and let this place go China syndrome. The explosion would take out far more ice than the pellets you plan on dumping out there."

Sullivan was quiet so I opened my book and started reading.

"Explosion," he said. I ignored him. "What kind of explosives do we have here, Colonel?"

"Nothing big enough to make a difference out there. At best we could crack off a section and it would float into the middle of the open water. You know exactly how much energy it takes to melt even a small amount of water."

"Are there any nuclear weapons nearby?"

"I don't have that information, but best guess the closest is at the SAC base in Grand Forks, North Dakota. Hundreds of miles of deeper snow than we have here. You'd need a nuke just to get at the bunker where they are stored. If they were smart, they would have removed all the triggers from the site before Zero Hour so no one could ever use them."

"What about—"

"Sullivan, I am not going to brainstorm with you. If they want to kill me to do this stupid thing, they are welcome to try. I hope you are smarter than that."

Sullivan left and I resumed my book, though my mind sidetracked through all my knowledge of military munitions and how they could be used to destroy ice.

WE FIGURED OUT WHY we were getting spikes in oxygen. The amount we pumped under the ice was collecting, and then it would release in one big burp. This also tended to clear out the particles from nearby, which explained the other things that were happening.

They never did try to kill me to implement their hairbrained scheme. As with most men that want to live more than they want be thought correct, they were cowards. Sullivan came up with a better idea. The outflow of water from a nuclear power plant was never to be more than one degree higher than the intake. Some EPA asshat probably came up with that arbitrary number. This meant that the plant could only produce so much energy per hour based on how fast it could flow water through the cooling towers.

We had been operating well below that since our power requirements no longer included millions of electricity customers.

Sullivan, having worked in every area of the plant, started to see outside the narrow focus of the others. He suggested they intentionally heat the outflow of water as well as the oxygen being pumped into tubes. This would burn through our nuclear fuel faster, but we had a few hundred years supply on hand. They even got to use their spent fuel rods, though in a far safer way. The containment pools that were naturally cooled by the ambient air became part of the water heating system. As temperatures rose on the outflow to ten degrees above intake, far more fish were drawn to the outflow as well as taking the ride through the system. They didn't taste any better, but more fresh protein meant a later end date of our food supply.

Within two months the far edge of the ice slipped out of sight over the horizon. The burping stopped completely and the air became denser with particles. Sullivan was voted employee of the year and rumor has it Kendra, the kind of woman no men fought over, gave him his own special thank you for saving the project.

I haven't said much about the five women at the facility and that is mostly on purpose. You'll remember the three Asshole incident and the subsequent suicide of the poor girl at the center of it. The five women that remained were not substantially less attractive, but they wisely tied themselves to men that had either physical or political power over the other suiters.

Two were already married at Zero Hour, having met and married at the plant before anyone heard about the comet. To say Kendra's personality was abrasive would be like saying George Clooney was likable. A lot of Hollywood rich tried to buy their way into our sites. A few young starlets did make it into the Harem down at Scorpio. I'd like to think Clooney took it like a man standing on the veranda of his Lake Como Palace.

Anyway, Kendra was more manly than most of the nerds that worked at the plant. Whether she was a lesbian was beyond my interest, but she never consorted with anyone physically on a regular basis. The other two were married, but the husbands were not brought into the Ark. I knew a lot of men that chose to stay with their families and in some cases I understood it. They were military and had some experience with survivor's guilt. For Janice and Oleanna to choose surviving over their husbands was not something I expected. Janice had two children, but they were grown up. Oleanna was doing fertility treatments for two years without success. She didn't make her final decision to stay until a week before Zero Hour.

I didn't want either of them to be included, but it wasn't my decision. Unattached women were the worst kind of unknown to have in a closed social group. By definition all of the men became unattached at Zero Hour whether they had someone in their life or not. I watched the other military branches experiment with women in combat platoons to very mixed results. I never saw it improve the readiness or effectiveness of any fighting unit. Regardless physical or mental capabilities, men just treat women differently. They are more protective in a big brother kind of way and they are more reckless in a show-off kind of way. No matter how ugly, they were perfect tens in a foxhole. Put simply, it is always a distraction.

Janice and Oleanna were a constant distraction. Everyone grieved their losses for a month or two, but then courtships began. The two women knew the power they had and they knew how to use it for the most part. Once I established the consequences in the three Asshole incident, these two women became the Queens of the facility.

Some might call them sluts, or whores, or empowered feminists. I just wanted them to stop causing problems. They both eventually picked a man and stayed exclusive for the most part. The other men still attempt surreptitious courtship, but I haven't heard any problems coming from it in a long time. Feel free to wonder if I did any courting myself. I'll never admit it in this or any other forum. Not because I'm ashamed or anything. I'm just not the type to kiss and tell. You should know that since I never talk about my personal life.

Why? Because it doesn't matter. This story isn't about the thoughts and feelings of a grown ass man in a crappy situation. This is the end of the FN world here. Probably less than ten thousand humans left. That sets us back two to three million years. Sure, we have better technology. But this comet is no different than the pride of lions stalking us on the East African savannah. Our chances aren't good. I'm telling this story so you know how we made it... or why we didn't. Final act has yet to be written.

WHEN I GOT OUT OF BED feeling like absolute feldercarb, I didn't bother showering. I dressed and headed for the Doc's office.

"Morning, Colonel. You look like you have the latest bug." I nodded and pulled off my shirt so he could listen to my lungs.

The 'latest bug' had killed two and incapacitated five. I'd helped one of my sniper's walk here two days ago, so I knew the necessity of quick examination, and the likelihood I was going to be the next to catch it. Oscar had waited until it made him too weak to walk.

Doc gave me a shot of steroids and had me breathe deep on pure oxygen for half an hour. I didn't feel any better, but I knew the theory behind his treatment. Antibiotics were getting scarce, and only the most critical people got them. I went to write my inactive status on the duty whiteboard and headed back to bed. Three days of misery before I started to pull out of it, but I knew I would because I woke up regularly. Not waking up was a bad sign for this 'latest bug'. I had to wonder if it was some mutated form of the bug that killed Cadillac. Logan was an idiot to come here. Could have been a Canadian bug that Roy's family brought, but I doubted it.

It was a week before I could climb the stairs of the tower without stopping every other landing. As far as I could tell, I hadn't passed it on to anyone. Socialization was tamped down and eventually it burned itself out. I wondered if that would be how I went out. Five or ten years from now, when my body's immune system was weaker, would I catch some bug and just fade out?

Not a great way for a soldier like me to finish out his service. Most successful military men have a bit of suicidal tendencies inside them when they join. I never wanted to die young, but I didn't care too much about living either. I had friends that did crazy shit on skateboards, bikes, motorcycles, whatever they could use to give them that adrenaline rush. Some became more cautious after first broken bones, but others charged harder toward the spectacular end.

I wanted my death to have some meaning, a purpose that made the world better. Plenty of opportunities presented themselves on various battlefields, but I fought to stay alive just a little bit better than my enemies. Now my only foes were aging and microbiology. Because of the fish, starvation wasn't likely to be our end. Perhaps if the skies cleared enough, outsiders could become a problem again.

If the plant had a catastrophic failure, we'd freeze to death within a few weeks. I had contingency plans for heading south with anyone interested in risking the frigid journey. Fighting to stay alive for a few more years in a dying world seemed anti-climactic. Freezing here or out there didn't present enough of a difference to matter.

After that would be either endless nothingness or an afterlife in which I'd pay for my sins. The idea of life continuing on in a different form was intriguing, but I sincerely hoped there weren't

72 virgins there waiting for me. That might have appealed to the teenager I left behind in boot camp. Having a woman that knows what she wants and wants what she knows is the only definition of heaven there could ever be.

I was raised with religious respect, but never really believed all the fantastical stories. It would feel like a self-serving lie to try to believe it all now on the off chance it was all true. The chaplain they sent to help everyone here was a very good man as far as I could tell. He did his best to help with the general crisis of losing all their loved ones in the blink of an eye.

The Doc was generally more successful helping people actually heal and get beyond the survivor's guilt and hammering sameness of post-apocalyptic life. The chaplain found himself becoming the most useless of survivors, and ended up answering the afterlife question for himself deep into the second winter. He didn't see that his skills would become more important as time passed. Existential questions tend to crop up as one's existence becomes tentative, or even precarious. The most important question was why would a deity allow this to happen to us? One only needs to look at all the political scandals and recall the story of Noah to put it together.

Like I said, a 'meaningful death' is what I had been aiming at, but such a thing seemed impossible now. It was doubtful any of the billions that blinked out at Zero Hour had a death anywhere approaching meaningful. How valuable would a sacrifice be now? One in ten thousand humans left on the planet. Even if we clear the skies, how many child bearing women are still alive? Will they be forced to churn out children from random fathers in order to maintain population levels and genetic diversity? Picture thousands of years of oppressive patriarchy to return humanity even to a basic level of sustainability. Is that worth the effort?

If there is a deity, it seems like the intent was clear. We were weighed, measured, and found wanting.

Sky Without Stars — Chapter 11

I spent much of my stateside service in southern California. I couldn't tell what the magnitude was as some people claimed, but I knew what an earthquake felt like. I was in the tower reading when I felt it. The tower probably exacerbated the effect of shifting, rolling, and heaving. It was stronger than the biggest one I had been near the epicenter of, a five point six. Alarms went off and the lights dimmed. The reactor had automatically gone into safe mode, so we were on the dwindling steam that remained in the system.

I was 99% certain this area was near no known faults. That was confirmed in later meetings. Therefore, this was likely not an earthquake but another impact of a large meteor. Supposedly someone was watching for more near earth objects, but it wasn't like we could do anything about them. I would have been surprised if some had reported one out to us.

The other option was that an extremely massive earthquake occurred very far away. The closest I was aware of was the New Madrid fault down near St. Louis, more than five hundred miles south. For us to have a six, they'd have to be in the nines. I had a brief fascination with earthquakes in my youth and my current guess was in no way scientific. There were no sites any nearer to the New Madrid than we were, it likely would not affect anyone directly.

I suspected meteor and wondered if I would ever be able to get the truth of what happened. In the communication room, there were messages from nearly all the sites about ground shaking to a greater or lesser degree. This was still fairly common in California, so they put the info on the net without any panic.

Site Scorpio was going Apeshit. They were demanding answers even though they had the links to most of the nation's scientific equipment, including all the seismometers that were still on line. I saw no useful information as the chatter devolved into speculation. I went to the reactor control room to see if the plant was being restarted.

Through the windows I saw the concern on all the faces. Then the announcement went out.

"Brown alert. Shut off all non-critical equipment." This repeated ten times in a prerecorded loop. This meant that they were switching over to the diesel generators. There was no chance they were going to fire up. They had twenty feet of ice and snow on top of them. The majority of the facility was going into darkness. Only the emergency lights that still had good batteries would give any illumination. We drilled all this when we did a refueling last year, so everyone knew to find a place to sit or lie down, or find a good flashlight if they wanted to stay mobile.

The inside generators in the motor pool were well maintained and were already running. They provided power to all the critical plant systems. We had one month of diesel left for those if they ran at full power. I opened the door and went to the group of managers trying to figure out what was happening.

"One minute to blackout," was the next announcement. Anyone that was stupid enough to have gotten in an elevator was in for a long, claustrophobic wait.

I listened in on the discussion which was full of technical jargon about the control rods and actuators. Shit wasn't working like it should because it was old and not well maintained. This power plant was supposed to be run by four times as many people, each group commuting in for their shift and then going home to their families.

These poor bastards have been on-call for four years straight. Luckily the plant is very low maintenance, so it is doable – unless there is a major problem. I pictured some short straw guy was going to have to go in and turn something manually as their life force slowly faded to nothing. That's the Hollywood version. Reality was three guys going into a completely safe area and replacing a big ass electrical switch with a new one. It would take a few hours as long as the new one worked.

They were wasting time arguing about how to test it. I finally spoke up. "Test it in place. Let's go." I went with to help carry tools and hold one of the big battery powered work lights, since the area was poorly lit under blackout. While we were down in the bowels of the plant, we felt small rumblings.

"Not a meteor," I said confidently. They looked at me. "I didn't know if it was an earthquake or meteor. Those were standard aftershocks." As they worked, I gave them a short dissertation on earthquakes.

"Could be the super volcano at Yellowstone," one of them said. That punctured my balloon. It certainly could be and would explain why they felt it in California.

"Might as well be the Death Star."

"Why?"

"Super volcanic eruption will dump a brand-new load of atmospheric particulate, especially on us since we are downwind of Yellowstone."

"At least with the Death Star some humans will survive this," I said.

"Can they be considered humans if they are from another galaxy?"

"Looked human to me," I said.

The power was back up to normal after three hours of down time. Most of us congregated in the cafeteria and watched Star Wars.

It wasn't a volcano, and the geniuses at Scorpio told us ten days later that it was a nine-point-five near Memphis.

Sky Without Stars — Chapter 12

S atellites reported a temperature rise of seven degrees at the equator. We weren't in contact with anyone down there to confirm it, but our outside temperature had bumped up a degree. That didn't mean much when fluctuations were from minus ten to minus eighty. Oxygen had stayed above ten percent almost two weeks now. This wasn't local burping. This was planet wide. That could only mean the ocean algae were back in business.

Everyone was beginning to hope.

You probably think I'm going to throw in the next twist, a reversal of fortunes that make everything look grim. Nope. All signs showed slow, steady improvement. Even the particle density began to drop. Someone finally admitted that the satellite they were using to measure it was looking at the wrong light frequency.

Humanity had a good chance of making it through this.

You sense the 'but' coming, right?

If we are the reason for the improvement, then we have to maintain our facility until we run out of nuclear fuel. There is nobody venturing out into the dark frozen world to bring us more. That also means nobody is bringing us more food. Finally, no one is coming to replace us. They can't get here any easier than we can get there.

You see it has been snowing for five years now. Our facility is in a shallow bowl of a forming glacier. At the peak of the last Ice Age the glacier here was over a mile thick. Even if Southern Florida eventually becomes sunny again, that does not mean the ice here will not continue advancing. Sorry, double negative. It is very likely the ice will keep accumulating for thousands of years. The heat we produce will keep us out of it, at least until an avalanche buries us completely.

Humanity may survive. We won't.

Earlier I was talking about a meaningful sacrifice. This is what it looks like.

Sky Without Stars — Chapter 13

The last of our communication satellites shut down. The ham radio still works but it is rare that we get any useful information from it. This has had a very negative affect on almost everyone. Getting atmospheric updates, regardless how static and unchanging they were, gave us a feeling of mission. Our local measurements would have to suffice, but I wondered how long those instruments would last.

People went through the motions of their day, but I felt the powder keg of tension growing. My personal threat of violence carried far less threat to those who began to accept the doom. I cursed the pastor for leaving us. The most uncomfortable of my duties was going to each of the women and imploring them to not be alone when outside their personal spaces and to lock their personal spaces when they were there.

Instead of watching camera feeds that only saw the ice that was closing in on us, I patrolled the corridors with my holstered Beretta. A smile probably would have done more good, but I had no smiles left. Being cut off from outside communication had hurt me more than most. Few people here had anyone to talk to at other facilities. I did, and checked in with a lot of them weekly, if not daily. I was expecting bad things from others because I wanted to do bad things myself.

When it happened, I was surprisingly temperate. Two of the nerds got in a fist fight over the last candy bar one had been saving and the other was accused of stealing and eating. They weren't physically capable of hurting each other, so that wasn't at issue.

Normally I would have offered to cut open the alleged thief's stomach for evidence of the theft, but the statute of digestive limitations had passed.

What I did was have the two of them sit down in my office and we just talked about our feelings. End of the world, the end of our branch on the genetic tree of life, the value of a stale five-year-old candy bar when nothing in life had much meaning. Maybe honing a nine-inch combat knife while we talked put a certain spin toward a peaceful resolution. I can't really be sure.

Certain people were becoming unnecessary for the mission. In truth, few of us were indispensable. No such culling was in the plan, but as the ice wall began closing in, it might be merciful to thin the herd a bit.

IN AUGUST OF YEAR SIX the air temperature over our reactor building broke through the water freezing barrier. Whether it is thirty-two degrees, or zero for you metric freaks, we hit a balmy thirty four degrees for more than an hour before the medium gray sky went dark again. The sun was breaking through. I could only imagine the benefit of this might have at the southern sites. There had been no communication from anywhere on the globe.

During the past winter, the surrounding snowpack had finally exceeded the height of sniper tower one. The heat of the reactor building kept us in an inverted cone, so avalanche fears were still on the low side. The only side of the bowl that was open was the one the looked out on the bubbling open water of Lake Superior. When the ice gets another ten feet deeper, we'll have no warning of an attack. Only a nutcase would try to come up here, let alone attack, but my job was anticipating threats and eliminating them. They could get to the edge undetected, slide down into our bowl with a one out of ten chance of being seen. Breaching outer doors would be child's play with explosives.

The only answer was sunlight. We needed constant sunlight for months to start melting the ice. Decades away. Decades after we run out of food.

Sky Without Stars — Chapter 14

If the human race does survive, this microcosm of society would be a fascinating source of long-term isolation study. I'm not a sociologist, so I'm not going to record anything approaching a scientific description. There have been many ups and downs, mostly tied to the hope for survival as certain atmospheric indicators fluctuated.

Most of us have accepted our end, still about five years away when the food runs out. That number kept getting pushed out while the fresh fish lasted, but it's been two years since we've had any of those. The number also gets extended as people opt out of this misery. Part of our initial medical supplies was an abundance of painkilling doses. Fentanyl gained its notoriety as an additive to the common street drugs to enhance the potency, but in retrospect was more of an intentional poisoning of the supply. A very small amount was capable of rendering a person pain free and spiraling quickly down into a cessation of life functions. That was the main legal use of fentanyl, an end-of-life pain suppression.

In our medical inventory was a supply of painless bullets to the head. It was considered better than slowly starving to death. Nobody openly talked about cannibalism, but it was one of the folders in the file cabinet on how to do it safely. As if gaining a few months of life to extend this futile project was worth crossing that line.

I wondered how I would take my final exit. Going to sleep and never waking seemed like a nice way, but only for a coward. I dream of ending my failure with the honorable method employed by warriors throughout time. A bullet to the head while standing at attention. I thought about doing this on my watch tower, hoping in that last moment I might once again see stars in the sky.

I didn't wonder if I would be one of the last to go. The depression that hit most never touched me. The futility of the mission that seemed so obvious to all never bent my will toward despair. I saw my place here as an act of greed. He put me here so I would not supplant his leadership when things became darkest. He knew that was my way, and I was better suited to the outpost of last resort. If I survived this I would seek him out and thank him for giving me this mission. Then I would take great pleasure in beating him to death for taking away my nice San Diego office ocean view.

Sullivan was the most optimistic, constantly coming up with hairbrained ideas on how we could melt the snow, now compacted into a rising glacier, and make our way south to enjoy the fruits of our labor. Particle density was still dropping. Far more than could be explained by our work here. The planet was healing itself. The equator probably had enough consistent sunshine that photosynthesis would resume in earnest. Humanity would likely survive.

Temperatures here did not improve. The snow continued to accumulate, our glacial bowl growing ever deeper. Our numbers dwindled as each person found their existence untenable. When we passed the ten-year mark, we were down to ten percent of the original workforce. It was enough to keep things running, but much of the larger maintenance tasks went undone. Doing a refuel was beyond us now. The plant output slowed as the material burned through half-life after half-life. The open water in the lake was slowly closing as well.

Our food supply would now outlast our electricity generation capability. We would freeze to death in darkness before starving. Wonderful news.

Sky Without Stars — Epilogue

"What are you reading?"

"Looks like a diary of someone who worked here. A Marine that ran security."

"What happened to them?"

"How should I know?"

"Read the last page, you idiot."

"I don't want to spoil it. You find anything of value?"

The food is like a hundred years old, but it's been in the deep freeze all that time.

"Eighty-seven years. The last entry is dated fifteen years after the impact."

"They survived fifteen years under the ice?"

"Maybe longer than that. The guy writing it wasn't the lone survivor. There were five still alive."

"Then you did read the last entry."

"Guilty as charged. You find anything else besides frozen food?"

"No bodies if that's what you're asking. There are some unused fuel pellets, but they'll need reprocessing because they're so old."

"How many?"

"About four hundred. Barely worth the trip up here."

"Better we get to them before the scavengers do. Either they'd know what they had and used them for bad or they'd crack them open and spread the radiation."

"You knew what we'd find here."

"My grandfather served with the man that wrote this diary. Communicated on the radio with him until everything went dark. I half expected his descendants to be here running the plant."

"No one had food to last that long. The ice must have covered them inside a decade."

"It was a bowl until the plant powered down. Lake ice was open too. These people gave their lives to keep the human race alive."

"That's why you came up here?"

"Mostly. The stories Gramps told me growing up always made me curious."

"I'd prefer the big payday you promised."

"You know what was worth more than gold before the impact?"

"Plutonium."

"Even more valuable were the stories of heroism."

THE FLOW

SCI-FI * SPACE TRAVEL * TECH WAR

PG-13 7000 WORDS 28 PAGES

The Flow — Chapter 1

"Approaching exit point, sir," the computer voice said calmly. "Hostile activity?" Captain Miles asked, looking at the graphical display in front of her.

"None detected."

"Then I guess we have to stop." She pressed two buttons on the touch screen, one authorized the AI pilot to execute the skid at the proper time, and the other was an automated prompt that told the passengers to return to their seats and buckle in. It only took one unrestrained skid to make them heed this warning.

"Skidding in ten centons... mark."

Dierdre Miles tugged at her own shoulder restraints to verify they were nice and tight. The attitude display began to turn off the main course line and the ship began shuddering. The microsecond reactions to control made this maneuver far beyond a manual human operation. Many of the pioneers died trying to navigate the Flow.

What is the Flow? Imagine a river that flows in a circle. Not possible on a planet with only one direction for gravity to flow - down. Now imagine this river flows around the galactic core. The closer to the super massive black hole at the center, the faster the river moves. But it isn't really movement as you understand it. Physical movement is constrained by physical laws, well described by Newton, Einstein, and McGonagle. Acceleration and deceleration on the scale that occurs in the Flow would spaghettify the human body. It is better to think of it as jumping in the river in one place, and a few seconds later surfacing hundreds of miles downstream. The longer you stay in the Flow, the farther you travel. The closer you navigate to the galactic core, the faster you travel, but that usually means traveling a farther distance. Simple. At least now it is simple. Thousands of drones disappeared into the Flow while the engineers tried to figure out navigation. All of that is automated now.

"Captain, have we arrived?"

She didn't look up from her display which was slowly scanning in every direction for other ships.

"Two more hours to the colony," she said when the sweep produced nothing of immediate concern. She pressed several touch buttons to change screens and begin standard acceleration toward the ring that encircled the planet.

"Wonderful news, Captain. Our prayers have been answered."

"Yes, they have, Padre. Thank Him for His assistance." She said this without a trace of the dripping sarcasm she felt. They were poor folk, but they were among the most reliable in paying for transport of their goods and beliefs to the distant corners of the Milky Way Galaxy.

The Flow — Chapter 2

"Welcome to Sagan Station, Captain Miles."

"Don't 'Captain Miles' me, Gundar. Do you have the lithium you owe me?"

"Yes..."

"Yes, but?"

"Not all of it."

"I'm not unloading your new cargo until you at least pay for the last two shipments. I'm not a charity."

"I just need a few more weeks."

"I'm not sitting around for a few weeks. I guess we'll see if the next colony wants the supplies."

"Dierdre—"

"Don't you dare call me that!"

"Sorry..."

"How much do you have?"

"Twenty-five kilos."

"You only have twenty-five after three months and you think you'll have two hundred more in three weeks? I think I'm going to remove this station from my regular route."

"You can't... we won't survive."

"You should have been self-sustaining decades ago. I didn't mind that failing when it was profitable for me. What's really going on, Gundar?"

He looked around and lowered his voice. "The Zels. They are back every month now."

"I thought the Marines cleared them out."

"They have a Flow Ship now."

"Where the hell would they get that?"

"I don't know, but they... they said they'll kill my family if I don't give them what they want."

"So... they stole my lithium? You expect me to believe any of this?"

"If you stay, you'll see for yourself."

"You think I'll protect you. Why would I risk my ass for you?"

"It's the only way you'll get your lithium."

"You're a weasel and a coward, Gundar. Give me the twenty-five kilos and I'll give you one of the pallets. I'll even let you choose which one. Have the two hundred for me the next time I stop or it will be the last time I do it. I find out you're lying to me and my face will be the last one you ever see."

Dierdre pushed him aside and continued down the walkway into the space station. She went to the least expensive of the three bars and ordered a beer. The robot bartender poured a tall mug from the tap and handed it to her. She walked into the crowd at the back.

"Hey Dee, long time, no see."

"Got anything to haul?" Dierdre asked.

"I might..."

"Legal," she clarified.

"Always..." he lied. "You ever go to the Omega colony?"

"Hell, no. Even the Marines are afraid to go there."

"I can guarantee safe passage."

"You going with me?"

"No—"

"Then your 'guarantee' means nothing."

"Pays five thousand kilos."

"Gundar says he's only got twenty-five to give me..."

"Yeah, the Zels take everything they extract."

"Then where are you getting the five thousand to pay me? Don't you dare tell me they have it on Omega."

"That's the deal."

"They would just take my cargo and ship and... you already send someone else?"

"No..." She was pretty sure he was lying.

"Is that where the Zels got their Flow ship? You send some newb to Omega without warning them?" He didn't say anything. "What's the cargo?"

"Just one passenger."

"What idiot wants to go there?"

"I didn't say she wanted to go there."

"Now you're human trafficking?" She finished her beer and stood up. "I should have known better than to come back here."

"It isn't like that, Dee. She's a fugitive, a terrorist. Warlord on Omega wants to personally carve his pound of flesh from her. You'll be a hero bringing the bitch to justice. A whole new lucrative market would be open to you."

"Illegal market. I don't do that and you know it."

"Just because you like to go home every once in a while. Why did you come to me if you aren't at the end of your rope? Listen, just do this one job and then enjoy some extended time on Earth with the profits. I'm sure you'll find some good, legal clients there."

"Why don't they just come and get her themselves?"

"Marines shoot first when they see one of their ships. They stay on their side of the galaxy now."

"Even more of a reason for them to take my ship."

"They won't if you let them believe that you are working for them. You can go where they never could. Five thousand kilos, Dee."

"I'll think about it."

"Wheezer died. His kid sold me his business and settled down on the colony with a fat widow."

"You telling me you're the only game in town?" Dierdre asked.

"For now. Zels scared away most of the competition."

"They don't scare you. It wouldn't surprise me if you hold their leash."

The Flow — Chapter 3

"Bishop, your newest sheep settling in?"

"Yes, Captain Miles. Thank you for delivering them safely."

"Anyone headed home?"

"No, Captain Miles. They have all committed their lives to this place. There might be a few more passengers to bring here on your next trip."

"I'll be sure to inquire with the temple before I leave Earth. Any communications you want me to deliver?"

"A few are writing letters home if you'll be in port for another day."

"Two days."

"Grand. I'll have them sent up on the elevator tomorrow. Good morrow to you, Captain Miles."

The screen went blank and she knew the prisoner was her only option. She hated seeing the smile on Figo's face when she walked up to his table two days later. He knew she was out of options. He lifted his enormous bulk out of the chair and mini hopped in the low gravity over to a door in the back of the bar. He opened it and led her into his office.

"Bring her up," he said into his wrist communicator.

"Easier to just have them take her to the ship."

"I want to say goodbye to her," he said, sitting in a similar wide chair behind his desk.

They waited for more than ten centons. The door opened and one of Figo's thugs pushed a box on a cart into the room.

"What the hell?" Dierdre asked.

"Hello, my dear," Figo said when the thug opened a small hatch. It was just a defiant stare that looked out at him. "You are going on a little trip. You better start making peace with your maker."

"Diaso condraso," the woman in the box growled.

"Perhaps hell is where we'll meet again, but Kindrasa assures me your trip there will be far more... interesting." His eyes flared with delight at the imagined torture she was going to endure.

The woman spit ineffectually, obviously weak and dehydrated. The thug closed the hatch.

"How long has she been in that box?" Dierdre asked.

"Since she put one of my men in the infirmary two days ago. You'd be wise to leave her in there until you get to Omega."

"You're even more of a monster than I thought. You better hope these friends of yours treat me fair or I will end you."

"Threats are meaningless here. You would have no better chance of hurting me than this poor girl. However, Kindrasa will reward you for bringing him this... prize. Safe travels, Captain Miles."

Dierdre got behind the cart and pushed it out of the door held open by the thug. She weaved among the bar tables, drawing curious stares. She rode the freight elevator up to her ship sitting on the box. The gravity here was almost negligible, so she used handholds to maneuver the box into her open cargo hold. She left it hovering and spinning slightly above the deck as she prepared two cargo nets. Pushing the box between them she tied them closed.

She reached through the netting and released the latches on the box and pulled the lid off. The woman sprang from the box weakly, finding herself similarly trapped in a different kind of cage. Dierdre dodged the arm the reached out to grab her, or strike her, she couldn't tell. She went down to the deck and retrieved a large container of water. She lifted it up and hooked it to the cargo net.

She unraveled the nozzle and extended it toward the feral eyes that glared at her.

"The net is temporary until we are in the Flow. Then we can talk about your situation."

"The situation is you are dragging me to the gallows. If you think I will do that passively, you have no understanding of human nature." She said this between long pulls on the water source, her voice improving with each intake.

"Only way I am going near Omega colony is if you hijack my ship and take us there." Dierdre could tell there was no trust of her words. It didn't matter at the moment. She pushed off and pulled her way up to the command deck.

The computer began the startup sequences. Dierdre checked her fuel levels. She'd had thieves drain her tanks while docked and she wasn't interested in being stranded in deep space. The levels were lower than she preferred, but she had no money to buy more. Five hours later she was in the flow and on course to the next colony on her route. She had eaten her nutribar as she worked, so hunger returned to the bottom of her current need priorities. After one last systems check, she locked out all the controls. If the woman got free, she wouldn't be able to take control of anything. If she killed Dierdre, she would spend the rest of her life in the Flow.

The Flow — Chapter 4

"Eat slowly," Dierdre said, pushing the wrapped nutribar toward the net. The woman grabbed at it, her physical strength and dexterity now back to a higher level. She tore open the package and devoured it in seconds. "You'll regret that. Try to vomit toward the ventilation intake." She pointed toward the back of the cargo bay.

"Where are we going?" she asked after drinking more of the water to wash down her food.

"I have three more colony stops before I am back to Earth. Fear that more than the Omega?"

"Their fascism is so much more polite."

Dierdre nodded. "I'm Captain Miles. All the ship controls are locked out. If I let you out of this net, are you going to make my life miserable?"

"Probably."

"You want to stay in the net?"

"Better than the box."

"What is... Figo implanted a veritas chip?"

"Yes. Couldn't tell a lie if I wanted to."

"He's a monster, but I guess honesty is better than wondering at this point." Dierdre went to the edge of the netting and started releasing the ends. After five were unhooked, the woman could easily get out of the net. Dierdre kept an eye on her, hoping she would stay passive. She slowly pulled herself out of the net and tumbled forward in a slow spin.

Dierdre couldn't understand what was happening until she heard the retching. She pulled herself over and gently guided her toward the vent intake. The airflow pulled in the semi liquid mess and began processing it into the recycler system. After the spasms stopped, they worked their way up to the passenger deck.

"You can have this cabin unless I get a paying passenger. You ever use zero gravity cleansing?"

"No."

"I can tell you how, but it's easier if I just help you the first time."

"You a lesbian? That why you're being nice to me?"

"No, and no. I'm being nice because I am a nice person."

"Religionist?" she asked, unzipping the one-piece containment suit. The smell that came out made both of them turn away in disgust.

"Wait," Dierdre said, not wanting human waste to fill the air of the cabin. She guided the woman into the small cleansing unit and closed the door. She touched a few buttons and a moist airflow began flowing past their bodies. "Now you can take it off." Dierdre helped her push the suit down and saw all the fresh bruises along with numerous scars. She pulled a shower sprayer from the wall and activated the water spray, checking the temperature with her hand. She aimed the stream at the woman's back and began rubbing the skin with a soap scrub pad.

The smell was overwhelming until they had washed down her legs and rinsed out the containment suit. The water was filtered and recirculated continuously. Dierdre purged the system and a new batch of fresh water began flowing. She helped her wash her hair and then began the drying cycle.

"Thank you," she said as Dierdre helped her pull on a new containment suit. "Been a long time since anyone has been nice to me."

"That why you became a terrorist?"

"That's what he told you?"

"Fighting the Omega's is hardly a terrorist act. You a runaway?"

"I was never Omega. They grabbed my parent's ship and killed everyone I ever cared about. I killed my way out of slavery. That's why they want me back."

"Makes a lot more sense than what they told me. Doesn't make it true. What was your parent's ship called?" Dierdre saw the flicker of evasion in her eyes as she searched for a name. "I have compassion, but I'm not an idiot. Omega's don't kill a ship full of free labor. Make the next story you tell the truth or you are going back into the cargo net."

"You wouldn't believe the truth."

"Maybe not, but I'll never believe lies. Guess there is no Veritas chip after all. Get some rest and when I give you food again, eat slowly." Dierdre left the room and pulled herself into the cargo area. She gathered the cargo nets and stowed them, then checked on the temperature sensitive containers.

The Flow — Chapter 5

"**A**pproaching exit point in ten centons."

Dierdre approved the commands and then unbelted and went to make sure her guest was properly restrained as the automated voice made the standard warning speech. She found the woman in a passenger seat looking out the window.

"Where are we stopping?"

"Not Omega. This is a small military outpost. They monitor the traffic inside the Flow."

"I thought ships are undetectable from outside the Flow."

"They are. The military has microdrones that ride the Flow and catalog what goes by, then reports what they see to the outside."

"That sounds like classified information. Are you sure you should be telling me this?"

"It was a secret until a whistleblower blabbed about it all over Earthnet. Guess you haven't kept up with newsfeeds. I'm just dropping off some supplies. If they see you, they'll want to know who you are. If you are in the system, they will process you. Are you in the system?"

"No," she said with some difficulty.

"Thought so. Stay scarce."

"So you can get your bounty?"

"I do not profit from human flesh. Never have, never will."

"You'll turn me into the authorities for free?"

"I maintain my operating license by doing the right thing, always. It is so much easier than playing shady games and hoping not to get caught."

"CAPTAIN MILES. WELCOME back to FOB Tesla. What do you have for us?"

"Five pallets, two are frozen."

"Hopefully some edible MRE's are in there."

"I have ten pounds of lobster tails if you have a few kilos of lithium to spare."

"Tempting, but the supply officer is not open to lithium trades. Earth credits?"

"Can't convert those to lithium. I run out of lithium and I'm out of business."

"And we run out of food and ammo."

"Marines will bring you what you need... eventually."

"Let me see what I can do. Who's your passenger?" Diedra looked at him with concern. "Saw her looking out the window as you were docking."

Dierdre sighed. "I actually don't know. Paid to take her to Omega against her will, but there's no chance I'll go there and survive."

"Companion?" he asked hopefully.

"She's more likely to cut it off than fondle. Can we keep it just between us?"

"For two of those lobster tails?"

She nodded the concession. "Do you know anything about the Zels having a Flow ship?"

"We haven't seen any unregistered traffic, so it must be stolen or... leased. We've heard whispers of the Galactic Emperium looking to expand. Zels might be playing nice with them."

"Wonderful. My last stop before Earth is in their territory."

"I could see if there is a Marine ship nearby to protect you."

"For the rest of the lobster tails?"

"No... would you have any Champagne?"

"Two cases. They're worth ten times the lobster."

"Just a bottle and I'll make the call."

"Got a new girlfriend?"

"Not yet..."

"Fine. Two bottles IF you get me cover. Give me a couple splooges I can mark their ship with if they try to board me in the Flow."

"Pleasure doing business with you, Captain."

"YOU TELL HIM ABOUT me?"

"Nothing to tell. Don't even know your name. He wouldn't even know you were on board if you weren't staring at him through the window."

"Pi."

"Nutribars are all we have."

"No, my name is Pi. My parents were mathematicians."

"Nice to meet you, Pi. Call me Dee. Got a safe place to go?"

"They killed my entire family. Not all at once. They used us for mining on some rock way out in Virgo—"

"Lithium?"

"Some of that, but mostly this toxic black stuff they called—"

"Kalinide"

"Yeah."

"Frak. A flow ship and Kalinide weapons. Someone needs to drop a nuke on the Omega colony once and for all."

"Ever wonder why they haven't?"

"Way above my pay grade. So you got no place to go?"

"I'll go anywhere and do anything to kill those—"

"How's that working out for you so far? That's right, lives in a box. One person can't do anything in this verse, certainly not against an organized criminal enterprise."

"You have a ship. Together we could—"

"Stop right there. I'll go out of my way to take you somewhere safe, but I am not a weapon of your suicide vengeance mission. And before you think you can take my ship from me..."

Pi turned away and disappeared into her quarters. Dierdre went up to the bridge and began her preparations for leaving.

"Captain Miles, I have sent the encryption codes you will need to contact the Marine ship Panther Paw. I recommend skipping the stop if you cannot make contact."

"Thanks. See you in about eight spins."

Dierdre authorized the ship to depart and head for the Flow entry point. The ship reported a forty-five minute maneuver to save fuel since they were now below half. She studied the systems along the way looking for a safe place to leave Pi. She started an onion query to obscurely search for Pi. It took several centons but the results finally spit out she was nineteen years old now, reported missing twelve years ago with her colony ship. It had been headed for Babylon Prime.

Dierdre had never been to Babylon since it took a lot of fuel to get back to the entry point. It was four habitable planets orbiting a red giant. Most thought it a bad location but once there, it was relatively safe from unwanted intruders. She asked the computer to estimate fuel consumption for a visit there after the next stop. While the numbers crunched, the rest of the information came in. Pi was in the Earth crime system as a terrorist. The details of her crime were classified and not available, but that made it impossible for her to go to Earth with Pi onboard.

"PI?" SHE DIDN'T ANSWER. She was curled up in the bed facing the wall. Dierdre sat on the bed and rested a hand on her shoulder. Pi didn't move. "You're in the system. I can't take you to Earth unless you want to enter the rehab system." Pi shrugged as if her fate no longer mattered to her. "Your ship was headed to Babylon Prime. Did you know anyone there? Maybe distant family that could take you in?"

"My brother had a job offer. He was going to support us while my parents set up a school. No, I don't know anyone there. I'll just get off at your next stop."

"The next stop has about forty years of terraforming to do before anyone can settle down there. Unless you want to be a companion, there would be no place for you there."

"That a civilized name for whore?"

"I guess."

"That's what I was at the mine, though I never got paid for it. That's why I didn't die with my family in the mines."

"I'm sorry that happened to you. I'm sure your parents knew the risk of transiting uncontrolled space."

"We had two Marine escorts!"

"Two men?"

"Two Destroyers!"

"That wasn't in the records I found."

"They left us for a fake distress call."

"That wouldn't be protocol, but neither is an escort. Was there someone important on the ship?" Pi turned and gave her a malevolent stare. "I know your family was important to you. I meant was there an ambassador or maybe retired military?"

"There were thousands of people. They took our ship without firing a shot. My father kept telling us the Marines would find us and rescue us but they never did. I doubt they even looked for us. I went looking for them. I wanted them to pay for what they did to us."

"Is that why you are in the system as a terrorist?"

"Why would Omega's want me for hunting down Marines?"

"They wouldn't," Dierdre said.

"Exactly."

"Tell me what happened."

"Doesn't matter since you won't help me."

"I won't help you kill people, especially Marines. I will help you find a place to live a normal life."

"Normal life. No such thing left in the verse. You are either a wolf or a sheep. Eat or be eaten."

"Ten centons to entry point," came through her wrist communicator. Dierdre squeezed her shoulder and helped her to strap in, then got up and headed for the bridge. As she buckled herself in, Pi pulled herself into the rarely used co-pilot seat. As they were pulled into the Flow, Pi let out a childish whoop. Dierdre looked over with a smile.

"You should trade me for your lithium," Pi said.

"That will never happen."

"They'll take over the Flow eventually. Better to be on their payroll than in their way."

"What makes you say that?"

"Earth doesn't care about us out here. They only care who provides the lithium."

"And their local supplies are almost depleted. You know part of my decision to not go to Omega was because there is no way they would part with three kilos for a person, let alone three thousand."

"They have millions."

"Nobody has millions."

"I've seen it. Almost blew it up."

"You can't blow up a metal. All you'd do is spread it out."

"Kalinide will trigger fusion."

"Not unless it is refined."

"Exactly."

"You almost ignited a million-kilo fusion bomb?"

"Millions. Plural. Would have been a hell of a fireworks show. Kindrasa's brother stopped me at the last minute and took the brunt of the explosion. Would have scorched the entire planet and everything in orbit, which included five Marine destroyers and two battlestars.

"In orbit?" Dierdre asked in confusion.

"Fueling up."

The Flow — Chapter 6

"Approaching exit point, sir," the computer voice said calmly. "Hostile activity?" Captain Miles asked, looking at the graphical display in front of her.

"Extensive ionization detected."

"Solar activity?"

"Not likely. Stay in Flow?"

"No, but be ready to dive back in immediately."

"Critical fuel levels, Captain."

"I know. If the Ganges terraform has been attacked, we need to alert... Earth."

"Preliminary message composed, Captain."

"Send along with our official logs. Pi, can you come to the bridge, please?"

"What's up?"

"This next stop might be dangerous. Get strapped in the copilot seat. Good, now, you see that panel on your left?"

"Yes?"

"If I say 'dump it', hit that big red button three times. It should turn orange and then green. If it doesn't just keep hitting the button until it does."

"What does it do?"

"It will eject the cargo for the terraformers."

"They have to come out and get it?"

"If I'm dumping it, that means they are probably dead and I want to put as much debris between us and whoever killed them."

The computer counted down and they felt the turn and deceleration as the streaking stars became still. "One Helenic transport ship docked. Distress beacon activated indicating repelling hostile boarders. We have been detected, canon rotating toward, fifteen seconds to reentering the Flow. Commit?"

"Commit. Launch a splooge pack. Get ready to dump cargo, Pi." Dierdre watched the attitude of her ship as it turned back toward the Flow entry point. "Now! Dump it!"

"Splooge impact positive. Five seconds to reentry. Canon firing. Deflectors discharging, sixty percent, twenty five percent, entering Flow."

Pi didn't bother with the 'woop' this time as the ship slid into the Flow.

"Hit the close door button, Pi. The white one at the... that's is. Damage report?"

"The ejected cargo took most of the laser fire. There is a small breech in the lateral thruster array. Two astro-mechs are working on it. Cargo bay four is failing to pressurize— "

"Leave it down, don't waste atmo."

"Aye, Captain. No further damage detected."

"Run a complete diagnostic."

"Send post-contact message?"

"Yes, include the splooge pattern in the report."

"What is a splooge pattern?" Pi asked.

"Splooge is a package of radiactive waste that is nearly impossible to clean off your hull. We just marked that marauder for the Marines to hunt down and destroy. I know, I know, they might be working with the Marines, or at least the scumsuckers that are allied with the Marines. The universe is upside down."

"We should head straight for Earth," Pi said after a long silence.

"Not safe for you there. I'll take you to Babylon Prime—"

"Captain, we do not have enough fuel—" the computer voice started to say.

""I know. I'll take you to Babylon Prime unless you have somewhere else you would prefer."

"Ragmot Seven."

"Never heard of it."

"It is the mine that killed my family."

"I told you, I'm not—"

"There's no one left there. I blew it up."

"Then why go?"

"To get some Kalinide. There might be some lithium if you don't mind reconcentrating it."

"I will not bring explosives on board."

"You just delivered ammo to that outpost."

"You know what I mean."

"Then just leave me there. I'll figure out how to get it to Omega Colony."

"How much lithium?" Dierdre asked after running through all other possibilities.

"At least a hundred kilos."

"You are just saying that because you know that's how much I need."

"And you won't find it anywhere between here and Earth."

"Captain, the splooge pattern has entered the Flow. Shutting down external transmissions. Recommend altering course."

"Agreed. Where is this Ragmot system?"

"Head into the core. It's on the Epsilon ring."

"I've never gone that deep before."

"That's what he said," Pi said, with a sideways smile. "Third quadrant."

"Course plotted, Captain."

Dierdre reviewed the course on her screen and then hit the approve button.

The Flow — Chapter 7

"What am I looking at?" Dierdre asked.

"Appears to be the remains of an orbital mining platform, Captain, there are one-hundred-seven droids attempting to rebuild it."

"No humans?"

"None detected. Thorium radiation levels are multiple times lethal."

"Scan for Lithium."

"Scanning."

"Pi, how exactly did you do this?"

"Lolo was a soft headed guard that liked to rape me above all the others. One night I killed him and used his keys to get into the Kalinide refinement facility and set a delayed explosive. Then I stowed away on one of the transport ships and was in the Flow before it exploded. I had no idea it would do this much damage."

"How many of your family were still alive when this exploded?"

"I hadn't seen them in years. I have no idea. I assumed they were all dead... or wanted to be as much as I did."

"Approximately four hundred kilos," the computer said. "Recovery time would be ninety hours, twelve inside the lethal radiation area."

"Any of those repair droids down there capable of lithium collection?"

"Most of them. Do you want me to hijack them?" Dierdre laughed at the twinge of excitement in the computer's voice.

"First look for any sentry droids and power them down."

"No sentry, but there are seven proximity mines. It will be easy to avoid them."

"When we're done with the droids, send them in to detonate the mines."

"Droids have been reprogrammed. New recovery time will be sixty-one hours with all the lethal work done by the droids."

"Commence," Dierdre said as she pressed the necessary buttons on her control screen. "You got me my lithium, how do I get you what you need? Looks like all the Kalinide in the sector ignited to cause this damage."

"One of those proximity mines would do the job."

"You want a pressure suit to go out and grab one?" Dierdre asked sarcastically.

"Can't your computer shut it down?"

"That would be a sentry drone. A mine is not programmable, by design. It is there to destroy trespassers, not guard an area. Let's say you get some explosives. What's your plan?"

"Blow up Omega and any Marines in orbit."

"You intend to swim there in a pressure suit?"

"You exit the Flow, push me out the airlock, and head back in."

"They stop you again and now I have a price on my head for assisting a terrorist in a terrorist attack."

"Leave me here and I'll figure something out."

"Leave you in a lethal radiation zone with no life support."

"I can die here, with my family."

"Survivor's guilt. Believe it or not, I understand that."

"If I may make a suggestion, Captain?" the computer asked.

"Go ahead."

"I can easily convert one of these droids into an autonomous explosive device. Since it has Omega encoding, it will pass through security grids unimpeded and do what the young Miss wishes."

"Remind me never to get on your bad side," Dierdre said. "How do we get the droid there?"

"Still working on that part of the plan, Captain."

"Pi, have you ever had lobster?"

"COLLECTION COMPLETE, Captain. Yield was four hundred eleven kilos. Seventeen kilos of Kalinide were filtered out. I have processed it and added it to the droid with the highest remaining propellant. I can leave it here to possibly attach to any ship that comes to investigate or I can put it in cargo bay four for deployment."

"Put it in the cargo bay and set course for the Proto system."

Once underway she pulled out the encryption code for the Marine ship Panther Paw. She typed a message and waited for the response. It came an hour later and she altered course toward them.

After following the instructed vectors, Dierdre took up position behind the enormous, heavily armed ship. She followed it into the Proto system and docked with the space station orbiting the mining planet of the self-proclaimed Galactic Emperium. She saw no reason to be concerned for her security during the cargo offload and even picked up new cargo headed to Earth. She wasn't sure if it was the looming Marine ship that made everything smooth, but she was happy to have an uneventful stopover.

Her ship headed out of the system to the entry point. She thanked the Marine commander and headed into the Flow on her way back to Earth on a very slow trajectory. After checking everything one last time she went to her cabin to finally get some restful sleep.

The Flow — Chapter 8

"Good to see you again, Commander Higgins. You get demoted to cargo inspection?"

"Welcome back to Earth, Captain Miles, I hear you have cargo from the Emperium."

"You might want to scan it for Trojan Horses."

"You haven't heard? They just announced a new treaty. You are the first ship to carry goods in a new era of peaceful space exploration."

"New era? Not a chance with the Omega Colony... What?"

"Vaporized. They bit off more than they could chew. Ambushed and destroyed three Marine ships so the Marines went and nuked them."

"You don't say. Guess I can spend less time looking over my shoulder in the Flow. Any luck tracking down that Zel ship I tagged at the Ganges?"

"That was reportedly in dock at the Omega complex when it went up."

"And I thought they were blood enemies. Pretty soon the Flow will be safe for everyone. Any cargo headed out?"

"As much as you can haul."

"Then I'll get them working on cargo bay four. Sprung a leak at the Ganges dustup. You stay safe, Commander Higgins."

"You too, Captain Miles."

Dierdre supervised the unloading of the cargo and had them top off her lithium tanks. She flew her ship up to high Earth orbit and docked it in long-term storage. She enjoyed the view all the way down the space elevator to the ground station in Nairobi.

Secretly attaching Pi's improvised bomb to the Panther Paw had been the computer's idea. It would only detonate if the ship went to the Omega colony. Apparently, that was their next stop. She felt bad about the crew, but if the Marines were now colluding with the Omega Colony, the galaxy had taken a serious detour into corruption.

The hyperloop transport was far less scenic, diving through the bedrock under the Mediterranean. In the Paris station, she transferred to the magnetic levitation bullet train up to Amsterdam.

Emerging from the station she finally smelled the familiar springtime essence of home. Walking through town brought back all the memories of childhood.

"Dierdre, you're finally home."

"Yes, finally."

The Flow — Epilogue

"Welcome to Jamanar colony. Passport please." She held her hand up to the scanner and held her breath. He scrutinized the digital readout, looking up several times to compare the photo with the face. "Ahh, a citizen of the Emperium, very nice. How long do you intend to stay, Miss Squared?"

"I have a job interview at the shipyard, so maybe a long time."

"Wonderful, wonderful. Emperium citizens are the most reliable immigrants we have." He pushed his button to imprint the electronic passport stamp. Best of luck with your interview, Piastora."

"All my friends call me Pi."

DADDY LOVES YOU

SCI-FI * BIO-TECH * AI * PARENTHOOD

PG-17 4800 WORDS 14 PAGES

Daddy Loves You

D addy loves you. I say it too much. Started as a genuine emotion whenever her tiny face looked up at me with innocent hope. I didn't expect that when her mother first told me of her existence. Back then I had a career that I loved and a wife and child could only take away from that.

The good thing about the human gestation period is that one has months to get one's mind right. Took me most of the seven months I had available. She knew what I initially wanted when I asked her the indelicate question.

"HE IS NOT A CHOICE!" was her emphatic, indignant answer before I got to the end of it.

Back then she was certain it was a boy. Maybe she thought I would be more enthusiastic about marrying her if she could provide me with someone to carry on my family name. She didn't know I had changed my surname to escape the negative association. It wasn't a story I told. It came out long after Sara was born when Ally insisted we do genetic testing.

I won't burden you with it, since it is completely irrelevant. Well, not completely, but I don't want to talk about it. Don't even want to think about it.

Daddy loves you. It is now a mantra. It reminded me who I was to her and why I was doing this. Cleaning up excrement never entered my decision process during the pregnancy months. Even the diaper pail we got at the baby shower didn't trigger the revulsion I should have been anticipating. It was the second night home from the hospital that Ally refused to get up. I had changed a few diapers, but they were simply wet. The toxic mess that confronted me that night was the closest I ever came to leaving. I did my best, choked back my vomit, then brought my beautiful daughter in to suckle on her mother's perfect breasts.

Those breasts were the trap.

Exotic filming location. Hopeful young starlet in her first studio movie. Even the boner shrinking butch lesbian 'intimacy coordinator' couldn't diminish the joy of seeing her shyly disrobe, over and over, until the director had exactly what she needed. No acting was necessary to convey my love for them. Ally of course interpreted that as love for her. After seven hours of painstakingly choreographed simulated sex in front of four cameras and three dozen crew, she came back to my hotel suite and we had the best sex of both our young lives. That was a bit of acting on both our parts. She was conquest number seventy-two for me and I was number three or four, depending on whether you counted the blowjob she gave the producer to get the role.

We went our separate ways after two weeks of filming. She took my brushoff like a champ, not the common tear or expletive filled parting. When she showed up on set two months later, I was well into a steamy affair with the real star of the movie.

Memories of her perfect breasts did their magic and I found myself naked and sweaty in the studio trailer less than an hour later. That's when the plot thickened with her pregnancy reveal. My range of emotions were tempered by post coital bliss, thoughts of a lifetime with this body were not unpleasant.

Ally was on her best behavior while the trap ensnared me and we married in her fifth month. To say it went downhill from there is an understatement.

Daddy loves you. My mantra. I say it hoping to see recognition in her eyes. I often do, but it is imaginary. I finish cleaning her private areas and putting the fresh diaper in place.

I get new clothes out and dress her. I talk to her about why I chose that particular outfit and tell her how beautiful she looks.

"You never tell me I'm beautiful anymore," Ally said one night when I arrived home after midnight. She assumed I was sleeping with my latest costar, but I wasn't, not for a lack of trying. It surprised me that Ally never got jealous. Not about other women. She got angry at almost everything I did, but not that, knowing I'd never stop, Ally would feel compelled to leave me. I honestly think she didn't care.

"I'm sorry Ally. You are the most beautiful woman I have ever known." It was the truth, mostly, when I met her. Less so know, but lines and sags happen so gradually.

"Do you have anything left for me?" she asked. It was a challenge to prove my words. Was she beautiful enough to arouse me after an eighteen-hour shooting day? Or, in her mind, after an evening with a younger, wilder woman.

"I'm not fucking Tina," I said defensively even though she hadn't accused me of that.

"That isn't what I asked." She sat up and pulled her nightgown down. They weren't perfect anymore, but they still did the trick. I showered her with compliments as my exhausted body did its best to convince her I was faithful and attracted to her. She faked her enjoyment and I was asleep twenty minutes later.

Daddy loves you. I lifted my daughter Sara into her wheelchair and secured the straps that held her in place. Though usually passive, the possibility of leaving her bedroom often sparked anticipatory excitement. This was probably because outside the safety of her bedroom, she couldn't be left alone. I tried not to think about it, but lying in bed, staring at the ceiling for most of the day had to be as close to hell on Earth as could be.

It was better when her mother spent time with her. Ally would talk constantly, ever hopeful for the cure to Sara's affliction.

Affliction. Probably not the best word to describe Sara's condition but it is the one that holds primacy in my head.

It started almost two months after a round of vaccinations. Not close enough to be accusatory, but it made the most sense to us. Especially after reading all the vax horror stories they don't tell you before you submit your child. A few twitches here and there became brief seizures. Only once when I was bouncing her on my lap trying to make her laugh did I see an episode begin. Her smile became a grimace and her eyes rolled up out of sight. Then the body shakes began and all I could do was hold her until it stopped. The screams of pain, fear, and whatever else she felt was the hardest part. I felt bad for Sara, but eventually I had to just walk away for my own sanity.

Daddy loves you. I wheeled her into the kitchen and fed her. She didn't like or dislike anything, just hummed as I spooned the different foods in. Her biology knew she needed nutrients, but she was indifferent as far as I could tell.

Ally tried all different kinds of diets, hoping one would provide the miracle cure. High carb, low carb, high fat, low fat, megadose vitamins, in endless combinations. Nothing made a difference because the damage had been done. The seizures stopped, but so did the brain development. She has brain activity, but it is comparable to a toddler. A toddler without any ability to communicate.

Sara has the ability to move every part of her body, but she does not do it by verbal command. I get her arm moving in a direction and she will continue until it hits the limit of motion, I push it in the other direction and this is her daily exercise. We do this in her bed, then again in her chair. Once a week we do it in a pool with a vest that holds her head above water. We did it more often when Ally was here to help.

I guess I need to stop avoiding the subject. Ally left us a few years ago. She grew to hate the futility of Sara's existence. She started... no that isn't fair. We both began edging toward conversations about how she we would be better off with her maker. If I had been a moral, religious man, Sara would not have been conceived.

It was during one of those post seizure screaming hours that we looked at each other. We never said it out loud, but murder was most definitely on the table. We knew the breaking point was approaching. I was holding the screaming Sara and Ally just nodded. It was permission to end the suffering for all three of us. It was up to me, the indifferent, disconnected, narcissistic man.

I began going through the options. A quick snap of the neck or four minutes of suffocation. Either would end with me in jail. Prison. Not sure if it is a phobia, or just common sense, but I didn't want to spend the rest of my days in a concrete cell.

Ally saw my incompetence and held out her arms. She would take care of it. Then she would take care of herself. A woman's right to choose. The horrible thought that came to me was Freedom. Freedom from all obligations. Parenthood, marriage, all of it. I almost handed Sara over to the executioner. Then I laughed. It was so inappropriate at the moment, but I couldn't help it. Ally was in her nightgown, somewhat lowcut and revealing. I looked beyond her extended arms and couldn't help thinking – 'there is a shortage of perfect breasts in the world, it would be a shame to damage yours'.

That line from the Princess Bride saved Sara's life. I put her down and literally tore that nightgown off and raped Ally until her screams were louder than her daughter's. That became our scream session coping mechanism. Animal sex while Sara cried.

Daddy loves you. I really do. Most of my life now is taking care of Sara. I've given up acting for writing. I've given up womanizing for... writing. Probably a good thing with the whole MeToo movement sweeping through the business. Some might find the life I live now as a boring monotony. It most definitely is that. Luckily, I can escape into my made-up worlds. I don't need the money, but I still submit my screenplays to see if anyone will buy them out of pity. I've read a lot of bad scripts, so I know just how terrible mine are.

I wash Sara's face to remove the food that didn't quite make it inside. She often whines because it is so often followed by a return to the bed. This tells me she is capable of memory as most dogs who learn the patterns of family existence. Her whining is one of the more useful indicators of preference. They usually stop when I wheel her chair toward the front door.

Taking her for a walk is usually a hassle because humans like to recognize and approach famous people. I'm not as famous as I was, but new generations are continuing to discover my entire catalog of movies and TV shows.

If it is just a greeting and a selfie, I wouldn't mind. With Sara it becomes question after question devolving into pity for a has-been and his afflicted daughter.

Sara's whine stopped because I wheeled her chair toward the garage. We are going for a ride Sara. This was more often than not a trip to the doctor, but she had not internalized that negative association. Her head pivoted side to side as we passed through the kitchen and laundry.

I rolled her down the ramp and beside her van. Lifting her into her special back seat was easy, she only weighed sixty pounds, but her whine returned as I strapped her limbs with Velcro, her head bringing the highest pitch whine. This was only for physical safety should we get hit by a truck. I stowed her chair in the back and climbed into the driver's seat. Her whine disappeared when the radio came on. Maybe it was still there, but I couldn't hear it over the classic rock beamed down from a satellite.

I watched Sara's eyes as we drove. They were constantly darting as new things flashed by her window. Television can also give her new things to look at, but she seems not to care about those.

Daddy loves you. You alone have taught me who I really am. Perhaps you made me a better man since who I am is quite different than I was at your conception.

Sara's whine returned as I pulled in front of the big white building. When I opened her door, I could tell she was fighting her restraints. She hated doctors more than her lonely bedroom. She had a good reason.

Her mother did little more than alter her diet. Doctors poked her with needles and locked her in big scary machines. Today would be no different. Well, hopefully, a little different.

"You must be Sara. I'm Kahna."

Sara was a good lens on seeing the human soul. Strangers talked to her as normal, then slowly understood there was no value in the communication. Those that wanted something from me would often pretend to care about her. Most people were frustrated by the lack of response. It was rude, disrespectful. They had to fight their natural reflexes, which was all but impossible in the bastion of narcissism that is Los Angeles.

How are you holding up, Sam?" Kahna asked with seemingly genuine concern.

"Honestly, I'm a little scared, Dr. Winston."

"Don't you dare Dr. Winston me." She hugged me. I think it was less to assuage my fears than it was to tell Sara she cared about me too. I would have thought the gesture useless, but the whining stopped.

Kahna knelt down in front of the chair and took Sara's right hand in both of hers. "My favorite story when I was a girl was

Tabillia Myorca, which was not dissimilar to Alice in Wonderland. Today, Sara, you will begin a great adventure of your own.

Back when Kahna first approached me, I rejected her proposal after only a few sentences. A few lonely, boring nights later led me down an Internet rabbit hole about her, her laboratory, and her patient testimonials. Sara would be far beyond anything Kahna had tried before. It was still many months before I agreed to hear the rest of her proposal.

Ally talked to every doctor and charlatan promising a cure or just a new and different therapy. They heard my name and assumed a bottomless pit of gold to draw from. They are most of the reason Sara hates doctors. My increasing resistance to them is why Ally left us. One of the reasons.

Dr. Kahna Winston was the leading researcher in controlling missing limb prosthetics with the interpreted impulses of the brain. Soldiers demonstrated their restored humanity, wielding arms that could do almost anything their missing appendage could have done.

This was after painstaking training of course, but the technology improved consistently as they learned more about how brains worked.

Sara was missing no limbs. It was the other end of the connection that was missing. Kahna believed she could fix that. I doubted it because I knew Sara. Kahna said she didn't want any money, but knew any small success would open my checkbook.

Daddy loves you. I said this as I changed her clothes into an accessible gown. Sara was about to be covered with all manner of sensors and wires.

It was going to be days, maybe weeks of learning every impulse and nuance that was Sara. I would be there to wash, feed, hold, and care for her throughout it all.

Kahna never asked me the important question. What did I expect if she succeeded? Sara loves Daddy. Just once I wanted to hear it, feel it, believe that the silent girl truly felt that. To know that these years had not been a waste.

Realistically I know her brain is both damaged and undeveloped. To think she could comprehend any words, let alone put them together in a meaningful sentence, while understanding the complex meaning of love, is complete fantasy. At best they figure out how to repair her brain and then she begins the learning process of an infant, toddler, and everything that follows.

Daddy loves you. All evidence to the contrary. Three technicians were applying adhesive patches, some with wires, most without. Her hair was already cut short enough that her scalp was accessible. I did this a week ago during our normal grooming routine. We always kept it short for simplicity, but I cut it way down at Kahna's direction.

I held Sara's hand as they did their work, talking amongst themselves and to the team monitoring the sensors outside the room.

When they finished, they added a mesh cap over Sara's head. This was a different type of sensor. Kahna had explained patiently several times before I agreed to this intrusion.

"You may begin, Sam," Kahna's voice came through the overhead speaker. I began our normal physical therapy routine, moving her arms, legs, head, back, fingers, toes. The dentist chair movability made this process easier, and I thought about asking for the supplier's name.

After the third time through, Kahna came into the room. We did a fourth round with her mirroring all my movements on the opposite side of Sara's body. Then Kahna gave me things to mirror her.

The entire time, she talked to Sara in a way that was similar to how I had been talking to her. After about thirty minutes, the technicians came back in and began removing most of the sensors. Some would stay on during sleep, which is why they were wireless.

We went to the small hospital room and I put Sara on the bed and began washing her, mostly to remove the adhesive that was left behind. The rest was for the toxic diaper that she had filled during the tests.

"You are an amazing father, Sam," Kahna said when she sat down for dinner in the small cafeteria.

"Hardly."

"Most would have put her in an institution."

"Thought about it. Still do every time she has diarrhea."

"You could afford a nursing service."

"I did that for a while after Ally left. They just couldn't do it with the tenderness she deserves."

"And you are an amazing father for providing that for her."

"Did you find anything useful?" I asked to change the subject.

"The AI is digesting the data. There is just too much information for the human mind to comprehend. It can take days just to process a few minutes of data, but we upgraded our computer system in anticipation of this project."

"Now that you have seen her, spent time with her, you know there is no hope."

"Her verbalizations tell me there is more going on in there than she is capable of communicating explicitly. I believe that is something we can build on. Sam, I'm not... I wish you could trust me."

"She wouldn't be here if I didn't."

"To a certain level. You still think I want your money or your name to raise our fundraising status."

"I'm just tired. Not just physically. It is just so much harder..."

"Alone. I know exactly how hard that can be. I won't bore you with my story, just know that I will do everything I can do to help. Would you like to go home? I promise we will take excellent care of her."

I hadn't had a night to myself in... "No."

"As I said, amazing father."

I shook my head. "Let me know if you find anything in the data. I'm going to cry myself to sleep." I took the tray and emptied it into the trash. There was a recliner in her room. I changed into sweat clothes and made myself as comfortable as possible.

Daddy loves you. The smell was strong enough to wake me up. Sara usually had to lay in her filth until I woke in the morning. She was a mess so I stripped her down and began the extensive cleaning process. A nurse came in with a change of sheets, letting me know for sure someone was actively watching the camera.

I couldn't get back to sleep so I pulled out a novel and began reading. Kahna came in around five.

"I meant to ask about the camera," I said.

"It is a critical part of the data collection. If you want a private room..." Kahna said.

"Tabloids would pay top dollar for a video of me washing a naked young girl."

"There are far sicker people in the world than those that live in hospitals. I was just going to talk to my team and saw you were awake. Would you like to come with?"

"They start this early?"

"Most have been up all night. This is the next leap forward in our work. I barely slept myself."

Kahna led me down the hall to a conference room. I made myself a sweet cup of coffee and sipped while their jargon filled conversation built in excitement, but went way over my head.

"Sam, this is amazing news," Khana said.

"I might agree if I understood any of the words you used."

"Sara appears to respond to different words consistently."

"Respond?"

"Brain patterns emerge from the chaos as we repeat things. That's why we did so many versions of the same movement. It's the best way to find the necessary patterns for controlling a missing limb. We are trying to reverse the process, assuming language was not a factor since this happened before she began speaking. The AI found consistencies in the repeated words."

"So today you want to focus on spoken words?" I guessed.

"Yes, but even better we can interpret her responses into words of her own."

"No." I looked around at all the bright eyes at the table. I was still asleep in that uncomfortable chair right now dreaming about the thing I wanted most.

The video began to play. It was my back as I began our physical therapy. I moved her arm out, then back, then up, then back. On the screen was a mass of electronic lines, the waterfall of data that was being collected.

"Daddy Loves You," I said on the video. My mantra when I didn't know what else to say.

"Daddy." It was a computer-generated voice, electronic and male. The technician paused the video, made some adjustments and rewound. The electronic voice was now feminine and child-like. It was the same word repeated over and over, but only as an echo when I said it. As with most mantras, I said it a lot.

My mindless describing what I was doing, such as "I am moving your beautiful arm up... and down" did not evoke an electronic response. When I got to her toes, she repeated "piggy" after every time I did. I could almost imagine her giggle as we went from the market to we-we-we all the way home.

"Remarkable," Kahna said.

Nothing changed through each repetition of the therapy. Daddy and piggy over and over. Then Kahna joined me for the mirroring exercise.

"Momma," began repeating. The electronic voice stayed flat, but we could all see the brain patterns increasing in intensity. Then a word was added. "Not. Not Momma. Not. Not Momma." Over and over. Then her head turned toward me. It was barely noticeable but I was focused on her face in the video. The video me was lifting her leg and moving her foot, so I didn't see it then. "Daddy. Where Momma."

"Stop," I said, standing up. "What kind of game are you playing here?" I was angry. I was tired. I had started trusting them. I had started to believe. I had been played the fool too many times. I stormed out of the conference room.

I was pulling the remaining sensors off Sara when Kahna entered the room.

"Sam."

"I thought you were different," I spat, pulling the gown off and reaching for the bag of her clothes. Kahna moved to the other side of the bed and helped me dress Sara.

"Sam, I'm just as surprised as you are. I swear, we did nothing to manipulate that data."

"Bullshit. The word 'Momma' never came up in our session. How could she repeat that... how could you interpret that if it was never recorded?"

"Sam. Please. Stop."

I refused. I got the wheelchair from the corner of the room and moved it to the bed. Kahna just watched as I lifted Sara in and strapped her down. I knelt down, tears streaming down my face, looking into her vacant eyes. I heard the electronic 'Daddy' repeating over and over. I wanted to believe it so much.

"Sam. Our AI has scanned the brain patterns of thousands of patients through millions of hours of thoughts and speech. It pretty much has the entire English language on file. There are variations from person to person of course, but some things are the same in every brain.

"I don't believe you," I said, and I turned and pushed Sara out into the hall and headed out to our van.

"ALLY?"

"Hi, Sam," she answered with great sadness.

"Where are you?"

"New York. Off Broadway play. Wretched dialog but... how is she?"

"No change. I miss you."

"That me is long gone. I barely recognize myself anymore. Do you need something... besides the obvious?"

"I'm sorry I didn't have the..."

"No, you were right. It was just too hard..."

"I know. When you get a break... come home."

"I can't," Ally said.

"I need you to... help me..."

"What?"

"All those doctors. All those charlatans. You never stopped believing..."

"Until I couldn't take it anymore."

"Couldn't take me complaining about the wasted money."

"It wasn't... only that. This run ends in October. I only get one day off and I can't do a coastal round trip in that short of time. Whatever this is... if it isn't physical... can't we just do it over the phone. If you want out, I won't—"

"No, Ally. I don't want out. I want you to come home. I did something... turned out to be another charlatan... I think. I'm far from objective."

"What?"

"I can send you the video. I don't..."

"What video? Sam, what is going on?"

"I just sent you the link. I don't trust them... I just want so much to believe."

I heard her playing it in the background. The endless daddy and piggy. She stopped it exactly when I did.

"What the hell, Sam? This some kind of guilt trip?"

"No, Ally. They are the top in the field of reading brain patterns and converting them into prosthetic movement."

"How much did they take you for?"

"They haven't asked for any money."

"Yet. What do you want from me, Sam?"

"Watch the rest of the video."

I listened as she played the video. 'Where Momma' turned into 'Miss Momma', then 'Need Momma' I could tell Ally was crying. Maybe I was trying to guilt her into coming home. The call ended without another word. I felt terrible adding this burden to her life.

An hour later her flight information came through. Sara and I picked her up at the airport the next morning and the two have been inseparable ever since. Charlatan or not, Ally bought into what Kahna was selling.

Ally now has a special hat that converts her brain signals into words. There is as often as not a random word that makes no sense in the context, sort of like Tourette's without the profanity. They are working on actuators for Sara's arms and legs, but she is a long way away from walking. Mostly her hand is used to push away food she doesn't like. I'm told this is common for toddlers.

As a family, we are a work in progress. I guess what inspired me to write down this story is what I will end it with. Last night, after another horrific diaper change, she finally said the words I've been waiting to hear. "I love pineapple, Daddy."

It was close enough for me.

THE CONFESSION

SCI-FI * BIO-TECH * TECH-WAR

7000 WORDS 35 PAGES

The Confession — Chapter 1

"Bless me Father, for I have sinned. It has been seven years since my last confession."

"Why such a long time, young man?"

"I have been overseas in a place that does not look kindly on those that practice the faith."

"Are you a soldier?"

"Yes, but I have never worn a uniform. My life is enshrouded in secrets I must keep, but I have done things that weigh heavily on my soul."

"Our savior will forgive any transgression as long as you are truly repentant and you have no intention of continuing them."

"I am done with that life whether I want to be or not. I do truly regret almost everything I have done in the name of my country."

"Everything?"

"Yes, Father. Had I known where it would lead, I would have chosen to continue my studies and become a doctor like my mother."

"There is still time, my son."

"No, there isn't."

"Are you ill?"

"No, Father... I guess I should just get to my greatest sin and save us both the dance of discovery. Within the next five years more than ninety percent of human life will disappear from this Earth. I am the sole cause of this. Last month I had the opportunity to stop it, I chose not to."

"I have no way to comprehend your words. What you describe is not possible."

"I assure you it is. I know this is beyond Our Fathers and Hail Mary's, but is there any chance my soul can be saved from eternal damnation?"

"You are talking about the deliberate murder of billions of people. This is not something I am prepared discuss with you. Any chance this is some kind of prank, or a delusion perhaps?"

"I wish it was, Father. It was foolish of me to come here. Thank you."

"Wait."

"Father?"

"I'm thinking. Tell me your story and I will do my best for you."

The Confession — Chapter 2

"Nathan, do you have a minute?"

"Of course, Professor."

"There are some friends who would like to discuss an opportunity with you."

"Friends?"

"We went to Yale together many decades ago."

"Sounds like influential friends."

"In certain circles. Meet me in my office in five minutes."

"Will do, Professor."

"What was that about?" Marcy asked as he approached her at the back of the lecture hall.

"He wants me to take over teaching the class since I know the material so much better than he does."

"Right."

"He wants to discuss an opportunity. Hope it is another scholarship. Medical school is going to bankrupt my parents. I have to go to his office now. Hopefully it is quick, but in case it isn't, I'll just meet you at the apartment."

"I have to go to the library, so I'll meet you at O'Reilly's."

"I thought you gave up drinking for Lent?"

"Jesus will forgive me for one beer."

"Not exactly how it works, but I'm glad you are trying."

"I'd convert to whatever religion you wanted as long as I get to have you in my life forever and ever. Bye, my love."

"Forever and ever," Nathan repeated quietly to himself as he watched her hips sway down the hall.

"NATHAN ZACHARY, I'D like you to meet Robert and Edward," the professor said in introduction when I entered the office.

"Nice to meet you." Nathan shook their hands in turn and they waved him to an office chair.

"I'll leave you to your business," the professor said as he stepped out of the office and closed the door behind him.

"Professor Salzman tells us you are headed for medical school."

"Yes sir."

"Is there anything that one could offer to dissuade you from that path?"

"I... don't understand," Nathan said.

"Your records indicate you would be a fine doctor, perhaps even a surgeon of high renown. There are dozens of others in your class that will do nearly as well, perhaps better. It is a common use of a keen mind like yours. If you are accepted, one of the other dozens will not have the same opportunity to pursue medicine."

"What my friend here is saying is that there may be a better use of your intellect and the other unique life skills you have accumulated."

"Such as?" Nathan asked.

"What are your thoughts on global terrorism?"

"Central Intelligence?" Nathan guessed.

"Yes, we represent certain entities that have a need for bold young men with... you look disappointed."

"I thought this was about a scholarship. The last thing I want to do is chase ghosts around the planet for this fascist president," Nathan said.

"We were not aware of your distaste for the current political party in power."

"Not much difference between the two parties. Politics is for polarizing and destroying social groups, nothing more. It is a game best not played," Nathan said.

"It may surprise you that we agree completely, but perhaps for slightly different reasons."

"Nathan, this best of all imperfect government systems is not made for the modern world in which it finds itself. Our organizations seek to minimize the damage of our temporary... 'fascist' politicians that lie their way into office and proceed to monetize their positions."

"Wait, are you saying you work outside the government's control?"

"We work in the shadows of government power."

"What plans do you have for me? A Mitch Rapp assassin or a Clancy save the world by Sunday shadow force?" Nathan asked.

"The fiction you have read is not all fiction, young man. They are stories that are told in a way to desensitize people in case details leak and, perhaps in your case, they work toward recruitment."

"You guys are serious?" Nathan asked.

"Far more than we wish we had to be. Today we just want to get a sense of whether you are interested in exploring the possibilities."

"That's simple. Hell no."

"Can we ask why?"

"I want to save lives, not take them. I will be getting married to Marcy, the mother of my future children, and live a normal life in wealthy suburban comfort. I will not be running around the world putting out fires started by the apes that lie and buy their way into political power.

"Medical school and residency will be the next six to ten years of your life before you can start building that suburban picket fence. You will have extensive debts and an uncertain economy that may convert to socialized medicine before you can make your fortune."

"Better than risking my life..." Nathan said.

"The reality is nowhere near the knife-edge danger portrayed in fiction. That is done simply to sell the books."

"Give us ten years and you will have ten million dollars, adjusted for inflation, to enjoy your wealthy dream life with the lovely Marcy."

"Ten million?" Nathan asked.

"Tax-free."

"Not much of an enticement if I'm not alive to spend it."

"Unlikely outcome, but Marcy will receive the money if unfortunate events occur. You will likely be in more physical danger during your residency at an inner-city hospital."

"Thank you, gentlemen, but I cannot possibly consider your offer regardless of how much money you offer," Nathan said after a long pause.

"If you change your mind, you can contact us through Professor Salzman."

"I won't."

The Confession — Chapter 3

"Ten million?" Marcy asked.

"That's what they said. Adjusted for inflation," Nathan said.

"That would make staying home with the children easier."

"I'll be a doctor in private practice by then. You'll never have to work again."

"Still, it would take a lot of the worries about our future off the table."

"You want me to give up medicine?"

"Let me see... married to Doctor Kildare or James Bond. Difficult decision."

"That's not funny. We have a plan."

"I'd get to chase more career opportunities for ten years instead of supporting you through medical school and residency. I don't think you understand the pressure I'd be under."

"You never said anything. I thought this is what you wanted too?"

"I want what is at the end of the rainbow, not the thunderstorms between us and that pot of gold. They said it isn't dangerous?"

"I wouldn't be a doctor."

"You could still study, probably get your degree online from wherever you are in the world. You said you wanted to travel."

"I doubt they'll be sending me to Paris or Fiji."

"We could go there together after things are..."

"I can't believe you are even considering this."

"I consider all the possibilities. You should do that as well. You can tell them you are considering it without giving up your medical school slot. You have all summer to make a decision."

The Confession — Chapter 4

"Welcome, gentlemen. You all know why you are here. What you don't know is why you are here." The nine recruits all looked at each other in confused amusement. "We are on the precipice of Armageddon. That may sound like grandiose hyperbole, but I assure you it is not. Can any of you tell me what that word means?"

"Hyperbole?" Nathan asked with a smirk. The instructor just looked at him with expectation. "Armageddon is a Biblical term that describes the end of the world brought about by demonic forces. Have we been recruited for a Holy war?"

"Is there anyone in this room that does not believe there are 'demonic forces' at work in the world?"

"I guess I'm the only one," Nathan said as he looked at the others. "There are no supernatural forces at work on the side of good or evil. It is simply men choosing to follow moral or immoral drives."

"I agree completely. We all have our demons and you may find out just how under control you have them. You all have higher than average IQ's, exemplary physical condition from successful participation in team sports. You also all have extensive education in science, and specifically, biology. My task force will be the thin green line between flourishing humanity and a bioterrorist Armageddon. Your country needs you to focus all your efforts on this task. More importantly, your family, friends, loved ones, they are all in the crosshairs of the next catastrophic attack."

"I'M NOT SURE WHAT I expected, but it wasn't that. Where are you from?" Nathan asked, sitting down with a tray of food in the almost empty cafeteria."

"Colorado. You?"

"Ohio originally, but I was at NYU when they approached me for this. You're premed as well?"

"Molecular biology. Hoping for a cush pharma job when I get through school. I'm Phil."

"Nathan. Is this what you expected?"

"The woman told me most of it to talk me into putting college on hold."

"Why do they need college kids? Why aren't there veterans with decades of experience?"

"There are. They need manpower to get ahead of the problem before it's too late."

"The guys that recruited me told me nothing."

"They probably would have if you played harder to get."

"I didn't want it at all, but my fiancé got all hot at the idea of marrying James Bond."

"I doubt we'll be carrying guns."

"Not gonna tell her that. You think we have any chance at stopping this?" Nathan asked.

"If we don't, nothing else really matters."

"I guess. At least the food is top notch. I always figured the AI robots would do us in."

"COVID took less than five months to get around the planet. If that had a real payload, we're talking half the planet gone inside a year."

"Payload?"

"It was engineered to stay respiratory and just take out the already infirm."

"You've been reading too many conspiracy blogs."

"We sequenced in our lab. It had all kinds of manufacturing artifacts never found in nature. They've done a good job suppressing that information, which is probably a good thing. People would go apeshit if they knew it was a weapon."

"What good is a weapon that kills your own people too?"

"Depends what war you are talking about. Because of the one child policy and mass foreign adoption of female babies, there is a huge demographic time bomb in their near future. Not enough young hands to feed old mouths. You create a weapon that wipes out those old mouths and your problem is solved.

"No way."

"Easier than the way commies used to do it. Russia is in a similar demographic boat for different reasons. Only the Mormons and the Muslims have growing young populations."

"Shit."

The Confession — Chapter 5

"What's it like?" Marcy asked.

"Long days in a classroom with lots of physical conditioning. A lot like what I've been doing most of my life," Nathan said, lying down on the bed and putting the phone on speaker.

"Weapons training?"

"Definitely," he lied.

"Tell me all about it."

"That really turns you on?"

"I'm a Texas girl. Unless you want to tell me about your rodeo riding, guns are the best way to my... heart."

"I'm not going to stay."

"Why not?"

"I can't tell you, but I just know this isn't what I want to spend the next ten years of my life doing."

"When are you coming home?"

"I'll give it a few more weeks. The... weapons training is a lot of fun."

The Confession — Chapter 6

"Nathan, you're sure you want to drop out?"

"As I told Eddie and Bob, I want to be saving lives, not chasing bad guys around the world."

"You'd be saving far more lives on this mission."

"Maybe, but we'll never know for sure. You have good people here and I have the utmost confidence they will keep us safe," Nathan said.

"What if I told you your participation is crucial to the mission?"

"I'd call you a liar since nobody knows the future."

"I wish that were true. I guess your mind is made up."

"It is. What... nevermind. I'm going to pack and say my goodbyes."

"I'll have a car waiting for you outside the apartment building to take you to the airport. Best of luck in medical school, Nathan."

"MARCY?" NATHAN ASKED the empty apartment. He set his suitcase down next to the open bedroom closet. All her clothes were gone. He dialed his phone and left her several messages. He called mutual friends. No one answered. He finally called his best friend. "Keith, have you heard from Marcy?"

"Never had a conversation with her without you there. You know she doesn't like me that much."

"Apparently, she doesn't like me much either. She moved out while I was in Virginia."

"I thought you were both moving soon for med school. Maybe she just went ahead of you."

"Without telling me? We talked every night I was gone."

"If you're asking me to explain a woman to you, I have genuine doubts about your sanity. You want me to come over?"

"No, I'm just gonna crash. Talk to you tomorrow."

"HI, NATHAN," MARCY said.

"Where the hell did you disappear to? I've been out of my mind with worry. What's going on, Marcy?"

"My name isn't Marcy. It's Linda. I was hired to screen you for the program."

"Hired? Screen?"

"I guess I was wrong about you. I really thought you were the perfect candidate."

"That's why you pushed me toward it. You slept with me, said yes to my marriage proposal. All part of the job you were hired for?"

"I would have married you and happily had your children, but only if you were the man I needed you to be."

"James Bond."

"Yes, Nathan. They told you what is at stake and you walked away to cure the sniffles in upper-middle-class brats. I can't tell you how disappointed I am."

"Where are you?" he asked.

"Not in New York."

"If I go back, will you..."

"You think they'll take you back after quitting?"

"That isn't what I asked."

"I don't know. I'd have to see that you're really serious and committed to it."

"Would you do it because they pay you to?"

"Are you asking if it was all an act?"

"I guess."

"You are fucking clueless. Goodbye, Nathan."

He sat staring out the window. Even with the scholarships and his parents paying tuition, he wouldn't make it through medical school without her support.

The Confession — Chapter 7

"Your target is a non-descript building in the center of Moldovastan."

"That isn't a real country," Nathan said.

"Just shut up and listen. The four of you will do a night parachute jump to this field two kilometers away. You will covertly penetrate the target and destroy any biohazards you find."

"That's the Sandy River reservoir. I recognize the shape," Nathan said.

"What about hostiles?"

"Terminate anyone that is armed. If there are lab personnel with valuable information, wrap them for extradition."

"Will we have a Moldovastanian translator with us?" Nathan asked. "I'll shut up."

The four men jumped off the rear ramp of the cargo plane into the black abyss of the moonless night. They popped their chutes just five hundred feet above the ground and almost immediately landed with a roll.

"Lima Golf Pandora. Target secure. Payload neutralized. Six moving to extraction point Delta," Nathan said in a no-nonsense voice through the satellite radio link. The exercise concluded in a conference room after two hours of simulated prisoner interrogation.

"What did we do wrong?" Nathan asked.

"Only minor deviations from expected performance. I believe your team is ready to be real world deployed."

"You have targets? Real-world targets?"

"There are a lot of unknowns in the world right now. We need you to help us change that."

"Let me get this perfectly clear. We need to kill people to get into these labs, but it is unknown whether or not these labs necessitate the killing to get into them?"

"Would you rather they kill billions of people because we were too timid to act."

"Timidity or Helter Skelter. No middle ground."

"You volunteered for this duty."

"That's debatable. I want to see the underlying intel before I go in."

"That could compromise our sources if you are captured."

"How convenient," Nathan said with disgust.

The Confession — Chapter 8

"Marcy?" Nathan asked, seeing the familiar silhouette by the window in his small dark apartment.

"Linda," she corrected. "You did well on the exercise today. I knew you were the right man for the job."

"Why are you here?"

"To see if there is anything salvageable in our relationship."

"You mean the relationship that you invented with your lies, *Linda*."

"Yes, that one. If you weren't a man I could respect and love, you would not have been right for the program."

"Bullshit. I was an assignment for you, nothing more. I want you to leave."

"If you wish. I was 'assigned' to befriend you, that is all. The sex and future life planning was all your doing. I really could see having your children. If we make it through this, I hope you will understand why I did what I did and give me another chance. Bye, my love."

"On to your next assignment?" Nathan asked jealously.

"Yes. Her name is Margaret. It was a mistake getting involved with you, I see that now. I won't be recruiting men for a while." She opened the door to the apartment and paused, hoping he would change his mind. He didn't, so she left and pulled the door closed behind her.

The Confession — Chapter 9

There was no need for a midnight parachute drop on this first mission. The owner of the bio research company was giving them a tour herself. She believed the four team members to be with the CDC doing a random inspection.

Nathan stayed at the rear of the tour, barely listening to Shane asking his probing questions as they stood in front of a series of windows. Nathan did his best to observe the space-suited people inside the various level clean rooms. He was also spatially mapping each lab within the building on a notepad, looking for dead zones and hidden rooms. There were several small ones that likely housed the air handling equipment that contained and destroyed any microorganisms that might escape during their testing.

The supposed assistant to the CEO kept trying to see his writing, so he made sure he hid his drawing and notes, finally giving the guy a dirty look at a more brazen of his snooping attempts. This did not stop them, only made him more careful. There were likely cameras aside from the obvious ones, so it was unlikely he would maintain complete secrecy. The attempted snooping was telling him more than the lab tour. They were concerned with this inspection.

The tour ended in the same conference room it began in.

"I need to go into laboratory seven," Nathan said.

"I'm sorry, that will not be possible," the CEO said.

"I'm certified to level five containment and I know lab seven is where we'll find what you are hiding. If you would prefer, we lock down and empty the entire facility now. We have the authority."

Nathan was bluffing since they weren't even with the CDC. The other three team members looked at him with barely concealed confusion.

"Hiding? I don't understand. What is it you think we are hiding?" The CEO asked after making a knowing glance to her assistant. He headed for the door.

"You stay here, Grant. You know exactly what is being hidden and I am the last person that would reveal just how much I know about your work here. If you would like to detail it to us now, we can forego the direct inspection. No? Since Grant also seems to know at least some of it, perhaps he would be willing to tell us to avoid prison."

"I truly have no idea what you are talking about," the CEO said, looking down at her phone.

"Messaging someone to destroy the evidence will only serve to lengthen your sentence."

The CEO held up her phone. "I was just looking at a text from my daughter. She always lets me know when she arrives home from school. May I respond?"

"I guess we are going to lockdown," Randy said, finally joining in on the ruse. He lifted his phone and pretended to search for a number.

"Wait," Grant said. "Emptying the labs, especially two and three, will set us back months if not years. We are on the verge of important discoveries."

"Why are you telling me this and not her, Grant?"

"She doesn't know about lab seven. No one does."

"I beg to differ. Who exactly are you?"

"I work for a private syndicate of wealthy entrepreneurs. That is who funds this research. I promise you there is nothing dangerous going on here."

"Forgive us if we do not take your word on that."

"I have a feeling I should call my lawyer," the CEO said.

"Good impulse, but hold off until Grant finishes telling all of us what is really going on here."

"Transmogenetics," Grant said after a long hesitation.

"Never heard of it," Randy said, pulling out his phone to search the term. "Appears Google never has either."

"That's just what they call it. It's research into epigenetics to turn off and reverse aging."

"With live organisms?" Nathan asked.

"Retroviral strands that deliver payloads to specific types of cells."

The team spent another thirty minutes squeezing more information out of Grant before they decided it was time to shut it down.

"How did you know it was lab seven?" Randy asked Nathan as they suited up to go into the containment environment.

"Wild guess."

"Seriously?"

"Grant got less nervous after we moved on to lab nine. There was something about the way he behaved that told me he wasn't her subordinate. The lab was way too nice, too well funded for the research they said they were doing. It just didn't add up. But, in the end, just an instinctual wild guess."

"Glad you were there. I was totally snowed by her openness and transparency. I guess it is easy if she doesn't know the dirty secrets."

"She knew that it was happening, just not what."

"Are we torching the lab or just the work?" Randy asked.

"We need intel. This can't be the only place doing research like this. We need to know what to look for in other labs. I'll go through the papers, you do your computer voodoo."

The Confession — Chapter 10

"**E**xcellent first mission, Nathan," "You knew what they were doing there before we went?"

"Of course. We're a part of the consortium that funds it."

"So... it was just another test."

"In a manner of speaking. Team two found nothing. You have very good instincts. Marcy was right about you."

"Linda."

"Yes, her name is Linda, Nathan. By recruiting you, she may be just as responsible for saving the world as you are."

"Saving the world? I doubt it. Who's to say a large reduction in human population would not be a good thing?"

"You think genocide is a good thing?"

"Not genocide. Elimination of the weakest. Modern technology has allowed a lot of suboptimal genetics to propagate into the future. Humanity has survived many plagues. I'm not saying I want millions to die. I'm just not as convinced as you that we wouldn't be better off as a planet with a natural reduction of population."

"It wouldn't be natural. These are bioengineered organisms we are looking for."

"I guess."

"What is really bothering you?"

"Nothing."

"You feel like you were coerced into this life?"

"Not at all," Nathan said sarcastically.

"I think it is her. You feel betrayed?"

"How will I ever trust a woman again?" Nathan said after a long thoughtful pause.

"In this business, trust is something we can't afford, regardless gender. I suggest you focus on the mission and leave romantic relationships to the future."

"I'm going to be a monk for ten years? I don't think so."

The Confession — Chapter 11

"What's that?" Randy asked. Nathan was at his desk, bent over a very thick textbook.

"Advanced Anatomy & Physiology," Nathan said, putting his finger on the page so he didn't lose his place.

"Upgrading your murder skills?"

"I should be finishing med school right now."

"Yeah, I would have been pulling down big bucks at Pfizer by now. Of course they don't pay you to shoot people in the face."

"Glad I haven't had to do that."

"Rumor is we're going into a hot zone."

"How hot?"

"Flaming Cheetos."

"You're a wuss if you think those are hot."

Randy mouthed the name of the country. Nathan's eyes went wide. "Might want to take a run through the shooting house, James Bond."

Nathan looked at his watch, pulling his placeholding finger off the text book. "Three o'clock?" Randy nodded and moved on to spread the word to the rest of the team.

The Confession — Chapter 12

"Insertion will be a HAHO glide of about seventy kilometers from the northern border. You'll still have thirty miles to cross to the target." The Colonel turned to the faces watching his slide show of terrain and maps. "Questions?"

"We waiting for favorable winds?" Nathan asked.

"Fairly constant crosswind is expected. If it turns against you, you'll hoof the difference. Target is very remote and underground." The slide showed a small home in the middle of a very empty surrounding desert."

"Barracks underground too?"

"Report says only four guards, rotated in every two weeks. Sixteen technical staff and probably six for cooking and cleaning.

"Why do they rotate the guards?" Randy asked. "Seems like a security risk."

"They stay on the surface since there is only one way in and out. It is thought they don't even know what they are guarding."

"How deep is our source?" Nathan asked.

"Can't share that. We deploy to Germany tonight. Go time is in two days. Any questions?"

"I say we nuke the site from orbit. It's the only way to be sure," Nathan said with a gravelly voice. The rest of the team chuckled.

"B2's will be overhead if you find any xenomorphs, Mr. Zachary." The Colonel smirked and nodded as the team filed out the briefing room door. "Nathan, hold back a minute."

"Sorry for the—"

The Colonel shook his head and waited for the room to empty, then closed the door. "Your gallows humor doesn't touch what proper military men spew before a tough mission."

"I wish they were going instead of me, sir."

"You're the best shooter in training, but you never pull the trigger in the real world."

"Against everything I believe."

"You won't have that luxury on this one. Pretend you are shooting paintballs if you have to.

"Only four guards? Randy will insist on taking them all by himself."

"These are not civilized lab techs. They are fanatics building a ninety-fiver there."

"Why wasn't that in the briefing?"

"You can tell them in the air, not before. I wasn't kidding about the B2's. Once you confirm it, we are going to thermite sterilize it all the way down."

"I'll confirm it right now and stay stateside."

"That would be optimal. Need you to terminate everyone, avoiding an intentional release, and leave all the doors open on the way out."

Nathan nodded. "If there is a release, we call in a tomb raider?"

"No one expects you to commit suicide. Stay in your suits, get clear, and decontaminate. We'll put you in isolation until you are clear."

"A ninety fiver isn't going to lay down and die, even if we come out into desert sun." Nathan walked to the window. "I should go in alone once they take down the guards. Less chance it gets out."

"You expect to go from zero kills to twenty?"

"How deep is the source?"

"The project was identified by electronic penetration. Progress beyond construction is unknown. The location is from the sister of one of the cleaning staff."

"Not deep enough to nuke it without investigation. If I find level three or better containment, I'll assume they are up to no good?"

"Your instincts are better than anyone in the service. Bet you'll make a hell of a doctor."

"If I make it out alive. Thank you, Colonel."

The Confession — Chapter 13

"Thirteen clicks to the target," Randy said as the others buried the parachutes.

"Wouldn't sound as unlucky if you used miles," Nathan said.

"Eight miles."

"Must have been a hell of a tailwind."

"Allah is on our side," Randy said. Nathan snorted and slung his rifle after checking the action. He nodded and Randy led them down the plateau onto the flat desert. They ran at an easy pace, separated by twenty meters to minimize a mine getting two of them. The satellite command gave them updates on the target. The only movement was the occasional guard or two going outside for a smoke.

"RASCAL SIX, YOU ARE go for penetration."

Randy and Tyler were already in the shadow of the house, peaking through the windows for a better feel of the inside situation. Two men watching a black and white TV, guns out of reach. Tyler tried the doorknob and found it open. He pushed through quietly and Randy took down both with a silenced round to each head. The remaining four converged on the front door as they cleared the rest of the house. The other three guards were sleeping and remained that way.

"Seventy-two virgins for five, please," Randy whispered through his mike. "Let's suit up."

"I'm the only one going in. You guys pull back to the ridge."

"Not a chance, Cochise," Randy said.

"Intel says there is a ninety five percent mortality bug down there. Good chance I'm not coming out. Get back to the ridge now."

Randy turned back to the other four. "Leave your charges and get back to the ridge. East side is upwind." They hesitated, not wanting to be cowards. "We fail, you're in reserve," he said to make them feel better. They unpacked the various explosives and slowly backed out of the room, then the house. Randy was pulling on his suit as Nathan set the charges on the pressure door. He helped Nathan get suited up and they seal checked each other.

As they waited for the detonation Randy tested the suit-to-suit communication. "Why didn't you tell us it was a ninety fiver?"

"Afraid you guys would chicken out."

Randy snorted just before the explosive popped. They both had to pull on the door to get it open. There was another pressure door at the bottom of the stairs, but it wasn't locked.

"Mad Max?" Randy asked before opening the door.

"Fury Road," Nathan said to confirm that everyone was to be put down with no chance to respond.

"Stay behind me and eat my leftovers."

They moved into the hallway and raised the silenced pistols. Nathan held onto a loop in the back of Randy's pressure suit and watched behind them as they walked. He heard two pops, but didn't turn to look at what Randy had shot at. A shadow darkened a doorway at the end of the hall. Time slowed down.

Nathan searched his heart, still not sure he could pull the trigger. It was a woman. Might as well have been Marcy... Linda. He saw his future with her disappear as the woman tried to make sense of the balloon man in the hallway. Nathan squeezed the trigger as her brain finally understood what she was seeing. A red blotch painted the wall behind her as she collapsed like a rag doll. It was close enough to a paintball gun that he did not spiral down along with his immortal soul.

"Going left," Randy said.

Nathan watched the end of the hall for two more seconds to see if anyone would come to the fallen woman's aid. He felt Randy turn and seemingly shoot continuously. Nathan spun around to Randy's right side and aimed at the closed door on the right.

"Clear. Breech right," Randy said.

Nathan finally let go of the strap and pushed the door on his side open, it was a lab with two men, their backs to him. Two more spots appeared on the walls.

"Clear. Nathan said.

They went down the hall to the next pair of doorways and repeated. After clearing the empty room on the right, he swung his gun back down the hallway. The woman's body had been dragged out of the hallway.

"Rear threat," he said and sprinted back down the hall. It was another woman on her knees shaking the fallen woman with a hole in her forehead. He put two in each and made sure the room was empty with no exits before running back to Randy. He had two or three bullets left in the magazine so he changed to a new one as he ran. Randy was waiting at the single closed door at the end of the hallway.

Nathan looked into the labs on the left and saw the carnage Randy had wrought. At least eight bodies. He was glad he hadn't come in alone. He grabbed Randy's strap and tugged twice to indicate he was ready. Without the suits, they normally used a hand on the shoulder, but that wasn't always felt if the suit was over inflated. The suits also had a narrower field of vision which made these situations more precarious.

Randy pulled the door open and they both went through and stopped quickly. It was a gowning room for the level five lab. The two men in the isolation suits could be seen on the other side of the airlock. They could not see the Americans watching them.

"You see any other stairs?" Nathan asked.

"Has to be more to this place, right?" Randy said, reading all the controls and signs. He pointed at the red one high on the wall. "Think it is connected?"

"I hope so," Nathan said. Randy raised his arm and flipped the clear lid cover and depressed the button. Nothing happened. "Try holding five seconds." All of a sudden, alarms lit off and lights began flashing The two men inside the chamber looked out at them. Randy waved and gave him the finger. They tried to get to the door but it would never open again.

Jets of propane shot out of the floor and ignited, burning everything and everyone in the chamber. Similar fire jets were hopefully burning in the exhaust vents to add to the UV Burn zone outside the biofilters. Randy just shook his head and laughed.

"Rascal Green, Lima Five is toast," Nathan said after keying the master communication system.

"Copy Rascal. Clear structure and signal safe distance.

"We forgot the graham crackers," Nathan said, turning back to the labs. It was nerve racking searching the rooms with the loud siren and flashing lights. They began a thorough search through drawers, cabinets and refrigerators for any information on what was there. The way it was set up indicated anything dangerous was kept in the level five chamber and was now atomized.

They found no more connecting rooms or stairs going down to lower levels.

"There had to be sleeping quarters," Randy said.

"Why would they hide those?" Nathan asked, trying to imagine where they would be. "Second lab from another part of the house?"

"We better find out quick."

"Rascal, Code Orange," Nathan said several times until control acknowledged.

"Yankee inbound, ten mikes."

"Did you just call in an airstrike on us?" Randy said, closing his bag of collected materials and slinging it over his shoulder.

"Told you this might be a one-way trip."

Randy swore as they sprinted up the stairway and quickly searched the house for another set of stairs.

"I don't see anything. Let's get out of here."

Nathan took ten seconds to picture the entire facility in his head. "Go, I have one place left to check."

Randy grumbled and followed him back down the stairs. It was a dead space between the two labs on the right. He pulled a rolling bench to the side and saw the hidden doorway.

"Rascal Orange Hold Five Xray Foxtrot."

"Yankee inbound, twelve mikes."

"They must have run in here and closed the door while we were clearing the first two labs," Nathan said.

"Just a room or a stairway?" Randy asked.

"Bomb shelter," Nathan guessed, pulling out two hand grenades and put fingers in the rings.

Randy grasped the handle and then pulled the door. He shook his head. Locked from the inside.

"Rascal. Third level is inaccessible. Likely reinforced."

"Understood rascal. Clear the zone."

Just then the hidden door burst open and a man swung out an AK 47 and started firing on full auto. Nathan pulled the pins and tossed the grenades on the floor at the gunman's feet as he pushed

Randy into the hallway. The gun was emptied in seconds and had climbed the wall in the inexperienced man's hands. The calls to his maker were cut short by the grenades.

Nathan jumped up and pulled Randy to his feet. Both saw the flashing red indicator on their visor that indicated their suits had been compromised. Nathan swore and tossed two more grenades into the now-open steel door and they sprinted for the stairs up and out.

"Rascal, Clear house. Five mikes to cover."

"Yankee inbound six mikes."

Randy pulled off his suit as he ran. "What if we're contaminated?"

Nathan looked over and saw Randy was covering his microphone and talking in a loud whisper. If so, their mission had failed so badly that they may have caused the Apocalypse. Nathan slowed to a walk. Randy didn't want to stop, but he turned and waited, walking backwards.

"Tyler and Pinball, suit up, ASAP. Jax and Gino leave your suits and run like hell north."

Jax keyed his mike. "Reverse that. We're already bubbled in case you needed us. Move your ass before the sky bats guano fire all over you."

"You sure?" Randy asked.

"Yes. My fault we went back. We'll be fine. Two weeks in quarantine will be a nice vacay."

"You'll get to finish that giant textbook," Randy said, turning and sprinting toward the two pressure suited teammates.

"Yankee inbound. Zulu in ten."

They only had their suits half on but dropped to the ground behind the low ring of stones.

"That was anti-climatic," Randy said when the whistle ended with just a thud. He started to lift his head.

"Stay down," Jax said, verifying his laser designator was still visible above the dust. "That was just a bunker buster. It punched a hole through every level. Here comes the fire, baby."

The inferno was intense and fried everything in the area. Randy and Nathan finished putting on their suits and tested the seal. Gino sprayed the suits down with decontamination chemicals and then Randy sprayed the other two to be extra safe.

"Please tell me we aren't walking ninety kilometers in bubble suits," Randy said.

"Not quite that far."

The Confession — Chapter 14

"So you think you should have stayed and let yourself be incinerated," the priest asked Nathan in the confessional.

"I know I should have. If it was just me, I would have. Of course, if it was just me, I would have died before the lab was breached. They never should have sent humans in. They should have trusted their intel and nuked them from orbit."

"Do you believe the plague would have been released eventually?"

"We'll never know."

"If you were infected, how are you still alive?"

"It wasn't me. It was Randy. They only quarantined us for three days. He took it home and died a week later. His wife took it to Chicago for a sales meeting while he watched the kids. Hasn't made the news yet, but it will. I figure it will peak in three months and then taper off over the next three years. With my luck, I'll be among the five to ten percent that are immune and I'll have to suffer for decades with perhaps the greatest sin any man ever committed."

"According to the book, all sins are forgiven, and you were hardly alone in committing this sin."

"I guess not."

"What about Marcy?"

"The priest wants to know how the love story ends?"

"I am fond of happy endings. Oh, that didn't sound quite right..."

"Thank you for the laugh, Father. See you on the other side, hopefully."

THE SEARCH

SCI-FI * SPACE TRAVEL * AI

PG-13 28000 WORDS 106 PAGES

The Search — Chapter 1

"Captain's Log, Sol date four thousand, two hundred eighty-seven point nine eight three-three. Acceleration continues on course to galaxy M227. All systems nominal, all personnel are within normal health parameters. The ore refinement from the last asteroid collection is complete but below standard, barely within limits," Tarlo Habaf paused the recording and sighed. She had a clear recollection of starting this journey and the enthusiasm she had for it. After more than two hundred and forty Earth years without any results, there was no enthusiasm left. She clicked the record button and continued her standard report. It would be broadcast at the end of the artificial day cycle along with all other relevant information. At this distance, it would take over a hundred years for the information to reach Earth. Her ship could make that distance in a small percentage of that time, but returning now would be pointless. The messages coming the other direction had slowly and steadily become grimmer, and those were from a century ago.

At first, she had censored the contents before she relayed them to the crew. That ended long ago. She rarely even read them herself anymore. The communications officer, Fedort Mu, was instructed to alert her if anything interesting came through.

Over twelve hundred so-called 'Goldilocks zone' planets surveyed. Only 8 with living organisms. None with multicellular life. None with more than trace amounts of oxygen in the atmosphere. Only one with significant amounts of water, and that was frozen. The next twelve hundred would be the same she was certain. As long as there was a chance, she had to continue the mission. Going back, she would likely find an Earth just as lifeless.

Tarlo was following the path earlier charted by a hyperlight space probe that identified Goldilocks planets but was traveling too fast to assess life signs. There were eleven other ships doing the same work she was, all on different headings away from Earth. An occasional report from one of the other ships found its way into their antenna receiver. Tarlo was evidently on the most fruitful of the discovery missions.

"Captain Tarlo."

"Yes, Mr. Uhar?"

"With the latest debris ejection, we will arrive at M227 in precisely twenty-three Sol years.

"Your use of the word 'precisely' fails in many dimensions."

He ignored her comment and continued his report. "At present burn rate plan, there will be seventy-four percent fuel capacity remaining after deceleration. There are five intervening stars that we are using for gravity assist. The likelihood of debris interference is calculated at point zero-zero-two percent. If there is nothing else, I will return to my pod for immersion."

"Nothing else. Enjoy your dreams."

"I will, sir. Thank you, sir."

Tarlo had only immersed once and that was before she left Earth. The hyperrealistic dream state that was induced had been described to her as a utopia. She found it the opposite. The thought chilled her so she refocused her mind on the gravitational routing display.

Satisfied the course was optimal, she pushed away from the console and stood up. The one-tenth gravity produced by the mass magnifier in the center of the ship gave just enough pull to allow a light-footed walk, not unlike those first moon landing videos.

Everything had padded rails in case the artificial gravity went offline.

Two hundred and seventy years old was not an unusual life span in modified spaceflight with plenty of time in the hyper-sleep immersion chamber. However, Tarlo was barely forty. Due to the relativistic form of light-speed travel, the clock wound much faster on Earth. Someone at NASA had thought it a good idea to root life on the ship to that on their home planet. The twenty-three years to reach the edge of the next galaxy was simply ten days in Tarlo's life. She reasoned the NASA people had insisted on communications being conducted in Earth time to impress upon the crew how time-critical their jobs were. Find a habitable world before time runs out for billions of people on a dying planet.

When Tarlo first tested for long-endurance spaceflight, she had no wish to leave family and friends forever. Her boyfriend was testing, and if she wanted to stay with him, she would have to go along for the ride. They both passed and committed to the training. NASA liked couples for the missions since it promoted stability. Unfortunately, the boyfriend was enjoying that 'stability' with several other crew members, as well as prostitutes. He contracted JFV, an extremely nasty and incurable venereal disease. Tarlo had escaped the infection that several other crew members got from him due to a strange quirk of fate. Tarlo's mother had been a bio researcher, and without her daughter's knowledge, regularly brought home unproven vaccines and inoculated all of her children. One of those inoculations was similar enough to JFV that it spared her, and helped set Tarlo on this seemingly endless mission to the far reaches of the universe. The betrayal, the humiliation, and the resulting anger spurred Tarlo into committing to the first launch to get as far from him, and the dying planet as possible.

Back then it was expected that the first two hundred Goldilocks planets would produce at least five worlds to choose from. The round trip would be two years of her life, about twenty years in Earth time. As the mission elongated and news of her family and friends dying off began to pop up on the feed, she became detached from that former life. Now ten generations had been born and died off, and Earth would be unrecognizable to her. Her sincerest wish was to find a new technologically advanced world where she could settle down and make new friends. The countless hours she spent as a child watching all forty-seven Star Trek series made her believe that the small corner of the Milky Way galaxy was teeming with advanced civilizations. Nothing could have been further from the truth. This was her fifth galaxy, and not even multicellular life was to be found.

Tarlo lay on her bed and opened her entertainment module. The inside of her eyes transformed into a holotheater. She chose a new program that had come in on the feed, the latest in a series of immersive mystery simulations. She played an observing character, watching the mystery unfold. Despite solving the murder long before the inept detective, it was still funny enough that she let it play out to its natural conclusion. She napped until late afternoon and then went to the mess hall for dinner. The food ten years into a two-year mission was understandably lackluster. Recycling can only go so far. The algae harvest the previous year had introduced plenty of fresh new protein into the system, but a few of the crew had developed allergies and subsequently died.

"I have a good feeling about M227," Jink Tavist said as he pushed the food cubes around in flavor sauce on his plate.

"You had a good feeling about M114 and the Pluton Cluster," Tarlo said.

"But this one is both spiral and hyper-spherical. I think some planets are affected by long-term core radiation."

"Black holes don't radiate."

"True, but I've been working on my slingshot theory. The same way we use gravity to assist acceleration, I believe the core can pull radiation around and fling it in concentrated directions."

"Why have we never detected that on Earth?" Tarlo asked, partially intrigued.

"Because the Milky Way isn't hyper-spherical," Jink said as if that was explanation enough.

"If this is the case, do you agree with the approach path we have chosen?" Tarlo had a feeling this was where he was going.

"If we deviate to quadrant four section beta six, we'll find far more evolved species, maybe even sentience."

"What is different about that area?" She asked, picturing it in her head.

"The spiral arm has a fat knuckle. I believe this would not only block core radiation, it could filter out non-plasmodial tetrions."

"I love it when you talk dirty," Tarlo said with a smirk, reminding him that she was not a theoretical physicist.

He sighed, knowing she didn't want an explanation. "Trust me, it will be the best place to start the search."

"I'll look into the course and do the calculations. Have you looked for peeprads?"

"There are no detectable signs of intelligent radiation. We are still more light years away than our human civilization has been radiating, so they could be at least as advanced as us."

"Only if they had their own Roswell technology to accelerate them." The comment came from Lendal Spinaker, a mechanic and amateur conspiracy theorist.

"Don't start with that crap, Lendal," Tarlo said dismissively.

"You think these engines were built by humans?"

"I know it. I was in the factory while these were put together."

"We may have assembled them, but the technology was from somewhere else."

"I hope we meet them soon and I hope they have better food. I swear this cube tastes like I ate it last week," Jink said.

"Hot sauce. That's what we need. Burn the taste buds beyond recognition and anything is delicious," Tarlo said.

"Is it true about the asteroid yield?" Lendal asked.

"Afraid so. We'll need to mine again soon."

"We won't need that because we will find exactly what we need in M227," Jink said confidently. "Has Yanti said anything to you?"

"About what?" Tarlo asked, playing dumb.

"You know."

"If I did, I sure as hell wouldn't tell you. She hates your guts and wishes you were dead," Tarlo said with a sardonic smile. Lendal laughed hard enough that she started coughing uncontrollably. Tarlo went to her and pulled the inhaler out of her pocket and held it ready. "Breathe slowly."

"I swear, I didn't cheat on her. Ranqor is just trying to make trouble."

"Who are you trying to bullshit. I have access to the video logs. I know exactly what you did with him, and frankly, it was disgusting," Jink knew he had done nothing sexual, but also knew Tarlo could manufacture a video of him raping Einstein if she wanted to. "Give her a few weeks. Try immersing until it blows over. I'll wake you if I find anything."

"She thinks that is cheating too."

"You can turn off the sexuality component of the immersion pod," Jink looked at her like that was the dumbest idea ever.

"That would be like going to Disney Nepal and not riding the K2 slide," Lendal wheezed out between inhaler hits. Jink nodded in agreement.

"Find someone less insecure if she doesn't complete you."

"Is that an order, Captain?"

"The only order I'd give is celibacy. Don't mess with me."

"Yes, Captain. Let me know about the course correction and I'll get to work on a dive plan," Jink got up and left after pushing his tray into the cleaning slot.

"I don't know why you put up with Ranquor's troublemaking."

"Keeps it from getting too boring around here. Didn't you watch soap operas growing up?"

"Never. I was more of a space ball enthusiast. Played until I got injured my junior year of Academy."

"If you had stayed healthy, I wouldn't have you on this mission. That cough seems to be getting worse."

"I'm fine. The medbot says it is the chemicals from the refinery leaking into engineering. I think it knows something and doesn't want to tell me until Doc Pember wakes up."

"It probably would lie so that we cannot use it as an excuse to head home. Meet me in medlab tonight and I'll scan you myself."

"I bet you will."

"Still not that desperate, sweetie," Tarlo said after she choked down the last protein cube. She pushed her tray in the cleaning slot and made her way back up to the bridge. She zoomed in on sector beta six and saw a dozen blue dots indicating goldilocks zone planets. Had she a telescope powerful enough to see the planets, she would only see what they looked like centuries ago. One of the most interesting fields of study that would result from her work was interpreting all the variables of the goldilocks planets to determine why they succeeded. Orbit too slow, or the tilt too large, and half the planet becomes ice bound, eventually absorbing most the water. Spinning too slow or fast and solar energy is not absorbed properly. No moons, no tides. How much effect do tides have on the flourishing and diversity of life?

Tarlo backed the display out and looked at the goldilocks distribution. Other sectors had more possibilities, but she did trust Jink's intuition more than others', more than her own. She backed out again to see the current course. She moved the galactic entry point and the computer did comparative trajectories until it found the faster and also the most fuel-efficient path. Rarely were they the same, so three sets of numbers were displayed. If she decided to make the change it would have to be done in the next twelve hours, which meant pulling Uhar out of immersion to do the navigation. He would spend the first hour complaining about the change, and then the next four grumbling about it.

Tarlo looked at the plan for surveying the systems as well as the exit for the next galaxy. Doing sector beta six first was among the least efficient of paths. She saw the many possibilities, and ultimately rejected the change. They would enter the sector when previously scheduled, approximately twenty-five days after galactic insertion. She sent Jink the bad news, and ignored his further pleading. He could wait twenty-five days. The years it would cost Earth were irrelevant at this point. She looked at the status board and saw an unusual red light.

"Mr. Dopple, why do I have a red light on array forty-four and no log entry about it?"

It took a minute for the response to come back. She had already switched on the video feed and saw the young man wiping sleep from his eyes. "I'm sorry, Captain, I must have dozed off," She was impressed that he didn't bother lying. "It looks like a power fault. Main bus B had an over voltage and safeties were engaged."

"Internal or external?" Tarlo already knew the answer.

"Definitely not internal, the auxiliary bus would have blown first. I'm showing a directed nanometer wavelength pulse just before the spike. None of the other arrays see it so it must have been razor thin..." she watched him on the screen pouring through the immense amount of data. "In both axis, sir," He added as an afterthought.

"A rifle shot?"

"Um, yes, I guess so. Diagnostics are good so I am bringing the array online,"

She saw the red light go out. "Well done, Mr. Dopple. Were you up late playing Harbin's Folley again?'"

"Yes, sir. I'm really sorry."

"No worries. The ship would have blasted an alarm loud enough to wake the dead if it had been a critical fault. Resume your duties. Bridge out."

She made her entry in his personnel file. It regarded his honesty and quick diagnostic ability, not his nap. She had napped herself only a few hours before. It was an accepted practice in intergalactic space. She had promoted the boy after his department head had been launched into space following his algae allergy demise.

Tarlo sat in the chair with the best forward view. She watched as a neutron star flashed by. She could almost feel the gravitational assist it provided, though that was extremely unlikely. It was more likely the curvature of the course that she felt. She thought it strange when encountering these stars out in the middle of nowhere, so far from the social safety of a galaxy. She looked at the trajectory of the neutron star and it seemed to have no connection to any galaxy. Had it formed in the early days before there were any galaxies? The computer was collecting and storing all manner of objects and motion data. They were condensed and transmitted on the daily transmission. She pictured a grad student pouring over the hundred-year-old data as it came in every day, disappointed by its lack of surprising insight into the universe. Much like her disappointment in the lack of life in the universe.

"Captain, I didn't expect you to be up here," Hans Yurd said.

She turned to look at her second in command. He was a grizzled older man, weary from the extended mission and endlessly perturbed at being second in command to both a younger person and a woman. "I like to watch the stellar slingshots. It beats the endless blackness."

"So that's what that was. I felt the jolt and came up to check on things. There's a red light on array forty-four."

"Again? Dopple just cleared that."

"I'll go check on it."

"No, Mr. Yurd. You can take over here. I'll go see what is happening down there," She knew he would only take the opportunity to yell at young Mr. Dopple. He grunted in frustration and continued scrolling through the logs of the previous twelve hours. She took one last look at the blanket of dots in front of them before climbing out of the chair and leaving the bridge.

"Mr. Dopple?"

"Same problem again. I'm running a spectral analysis on the spike. Very strange."

"Was it the neutron star we just passed?" she asked, knowing that was the side of the ship that had been struck.

"I don't think so. It was precisely the same angle as before,"

"Precisely?" Tarlo asked, wondering how he determined that.

"The surrounding arrays did detect a minute amount of the spike. I was running a differential analysis when the second spike hit. The differential was less than a billionth of a percent off of the first one," She looked at the display.

"Theories?"

"If this was Star Trek, I would say something is shooting an energy weapon as it flew along beside us."

"Klingons?" she asked with a sardonic smile.

"I'm thinking Turtlons, since their weapons were barely strong enough to be an annoyance."

"Should we shoot back, Mr. Dopple?"

"Oh, no, sir. I think a Federation ship should always attempt communication first."

"Could this be a form of communication?" Tarlo asked with just a little bit of hope.

"If so, it is ineffective since our array is shutting down before most of the message comes through. The burst length is about five seconds, but the sensor shuts off after two microseconds."

"You said the other arrays captured spill over. Was there any pattern to it?"

His young fingers flew over the interface and shaped the spike into seven linear horizontal waveforms. It had a very distinct pattern to it. "That is wild," he said. He shifted it to an auditory waveform and played it through the speaker. He did this often with space radiation, sometimes finding fascinating rhythms to add to his musical compositions.

"Slow it down," Tarlo ordered. Dopple stepped down the speed in increments and replayed it. At about fifty times as slow, the noise began to sound like language. They looked at each other as the sound of Tarlo's voice began to emerge from the speaker. She recognized the report contents. "Why was our transmission from three days ago bouncing back?"

"There could be a gravitation lens out there,"

She doubted that was it. The neutron star was the only thing with gravity nearby. "Was that the first or second spike?"

"Second," he said, immediately making the conversion of the first one and playing it. "Here's the first."

"That was the log from four days ago. The second spike was from three days ago. How long between the two spikes?"

"Twenty-two minutes."

"So soon we should hear the log from two days ago," As the voice played, they watched the array status board. "How strong is our transmission relative to the spike?"

Dopple did the calculation. "Maybe twenty percent stronger?"

"So... we are getting zapped by our own transmission."

"Except our burst is far more compressed. The voice goes out in the first fifty milliseconds, the data never takes more than a second to complete. This is slowed by a factor of a thousand or more."

The gravitational lens seemed far more plausible since it could distort the wavelengths as well as reflect it back. But the precision of the direction bothered her. "Is the data there?" She asked.

"No. Audio only," he said, looking up at her with wonder. Just then the array light went red. Dopple quickly did his thing and had the log playing. She recognized the words immediately. It was the previous day's log. She nodded to him in confirmation. "What is happening?" he asked.

"If we received a message from the Turtlons with no universal communicator to translate, how would we respond?"

"We couldn't."

"We would send their words back to them in acknowledgment of receipt."

"We would?"

"Yes. But we would add a message to the end."

"They didn't do that," he said, disappointed.

"Maybe they did, but we don't hear it because it is too powerful. Turn down the sensitivity on array forty-four," He did so and they waited for that days' log to come through. "How do we send a message to them?"

"Turn the parabolic antenna down this line of bearing and say 'hi'?" he guessed.

"Close. The fact that they can tell the difference between audio and data may mean they could translate given the correct information. We need to give them a Rosetta Stone."

"You really think something is out there?"

"Beats laying on my cot and... well it beats pretty much everything we have done in the last ten years. Pictures. We need to send them pictures. Get the parabolic pointed and turn down the power ninety percent."

He made the adjustments and waited for things to align. "Hailing frequencies open, Captain," he said with a smile.

"Greetings. I am Captain Tarlo of the Earth ship Orion," she hesitated, unsure of what else to say. "We are on a peaceful mission looking for a suitable new home for our people," Tarlo hit the button to stop sending. They both waited for a reply, but none came. The current day's log came through at the expected interval. The reduced sensitivity prevented it from overloading, so all of the transmission was captured.

"Maybe we should boost power?" Tarlo asked, waiting for him to do the transmogrification.

"There's nothing extra here. It matches the audio sent out this morning. Do you think they will repeat the four log entries? Or maybe the next one will be from them?" he asked hopefully.

"They should know we are communicating since we sent in a different direction."

"We need an off-axis receiver to determine the distance."

"They are transmitting too narrowly to get an off-axis read," Tarlo said.

"We are too. They had to be precisely behind us to receive those messages,"

This made Tarlo think. It is possible they may have blocked the transmission entirely in their capture of the message. If they had simply stayed where they were, they could have sent their message back to the same antenna without causing any of these array problems. "Come with me," she said heading out of the room and up the ladder. She reached the top and opened the hatch to the bubble.

"Sorry Captain," she heard as she stepped through the door. Two shadowy figures were at the far edge of the room hurriedly putting on their clothes. She ignored them and tried to find the direction that array forty-four was pointed. The bubble was a large clear dome that gave a spectacular view of the space above the ship.

"Which direction, Mr. Dopple?" She asked, not seeing anything obvious in the expected direction. He pointed to where she had been looking. Then he walked over to the control panel and turned on the telescope. He turned it to face the precise direction.

"Holy shit!" he exclaimed loudly. Tarlo walked over and looked at the view screen.

"Distance?"

"Four hundred kilometers," he said excitedly. It was little more than a shadow in the starfield, but it was moving through space at precisely the same speed they were. "Looks like it is seven hundred meters long."

"Too big for a probe.

"What is it, Captain?" the previously amorous young woman asked. The other person she had been with had disappeared.

"I wish I knew, Yanti. Mr. Dopple, patch this feed down to your workstation and let's go see if they have anything else to say."

The three climbed back down the ladder after closing the hatch to the bubble. The fifth message had already been received, so Dopple set about converting it to play.

"I assume that wasn't Lendal with you up there?"

"No, sir," she said with a little shame. "It was..."

Tarlo held up her hand. "Don't want to know," The log began playing. She did not recognize it. The sol date was little more than a random number generator to her that she read off the screen. Relativity changed the rate at which Sol days passed in relation to their days.

"Captain's log, Sol date four thousand, two hundred eighty-seven point nine eight five two. Acceleration continues on course to M227. All systems nominal, all personnel are within normal health parameters. We have had anomalous radiation detected off our starboard bow. We have retransmitted previous logs in case this anomaly effected the original transmission," Tarlo reached out and hit the pause on the display. She looked at Dopple.

"I don't remember retransmitting any logs, Captain."

"We haven't, at least not yet," She thought about it, then rewound to the start of the log. She brought the main bridge display up on an auxiliary display. She looked at the current Sol date as the log replayed. "This is tomorrow's log."

The Search — Chapter 2

"Tomorrow?" Yanti asked.

"No way," Dopple said reverently.

"Anomalous radiation?" Tarlo asked out loud. "Why would I say that? Am I afraid of reporting a UFO?" She looked over at the Bubble's video feed. The shadow was still there.

"I wouldn't. It isn't like anyone alive today will hear it."

"I don't want historians thinking I had gone space loco. Better we know more than we do before telling the universe."

Yanti was still confused. "Are you saying that thing out there just sent us your log from tomorrow?"

"That's what it looks like."

"Holy frak."

"Do you think we'll get the next one?"

"Maybe the future will tell us what this is," Tarlo said. They waited, but there was no next message. The shadow faded from view and they were alone again.

"Mr. Dopple, realign the antenna for tomorrow's log broadcast."

"How should I log the array failures?"

"Anomalous radiation," Tarlo said after a few moments of thought. "Resend the logs. Crank the sensitivity back up on forty-four."

"Yes, sir."

"And don't stay up all night again."

"Yes, sir."

Tarlo went back up to the bridge and found the cranky Hans Yurd reviewing the course plots. It was a way to fight the boredom, but he seemed intent on making alterations. He could not make them without Tarlo's approval, but she guessed Jink had put a bug in his ear as well.

"Dopple get his array in order?" he asked tersely. She was going to tell him about the log reflection and the shadow, but his voice irritated her, and she had no answers to the bevy of questions he would hurl at her.

"I think so. Anything else interesting happen?"

"Boards are green."

"Course look correct?"

"Mr. Uhar is efficient as always," She looked at him expectantly, but he didn't offer an explanation for his chart study. "I think we need to find a better source of organics."

"I have been praying for a T-Bone planet," Tarlo said. Hans snorted in derision. Tarlo wasn't sure if it was meat or prayer that raised his disdain. "Do you have any ideas?"

"We have seed stock," he offered.

"That's for settling down. You think we should stop and colonize?"

"I..." He stopped and reconsidered. "Many are tired of this quest. You know as well as I do that Earth is dead. Even if anyone survived this long, there are no resources for a mass exodus this far out."

She nodded. "I liked that little planet off the Dedron cluster. If only it had oxygen."

"If we find an M-Class with acceptable oxygen, will you consider colonization?"

"Of course. I take it you are volunteering to stay behind while we report back to Earth?"

"I think we should all stay."

"Why?"

"They would just bring their problems here."

"We are so different from them?" Tarlo asked.

"Yes."

"There are only thirty-two potentially fertile women on this ship. Do you think that is diverse enough to make a new civilization?"

"With a diverse breeding regimen and the frozen embryos."

Tarlo laughed. "Will we have any choice in the matter? Or will we be cattle in your biodiversity program?" She saw the flash of anger and got a chill. Her continued existence was the only thing that kept him from making this far less than voluntary. Her mind went through the many scenarios, knowing his simple mutiny would triumph over everything she could try. Her authority came from the tenuous link to a possibly dead civilization. Everyone on the ship was tired of travelling, even her. "I'm tired of it too, Hans. This is seven years past what I wanted to be doing. Even though no one I knew is still alive, I will still take this ship back and report. If anyone survived, they will be far different than those we left. Whether that is better or worse will determine if we bring survivors back here. The assessment can be done from Earth orbit. Personally, I expect an atmosphere poisoned with plutonium."

He nodded thoughtfully.

Tarlo continued. "Direct path round trip can be done in four years.

"That's at least thirty-two years to those left behind."

"True. You'll be in your eighties, with a third generation being delivered? A perfect time to introduce new genetic diversity." She tried to gage his reaction to this and saw a hint of concealment. It would not be difficult for him to strand them there with just a tiny bit of sabotage. There had been longer duration spaceflights, but none had been so far out of communication range. One of the twelve ships had disappeared after a strange report of mutinous behavior. The captain reported punishing the crew that had acted out, the next few reports were innocuous, then radio silence. Call it space dementia, cabin fever, or just plain crazy. Tarlo had to be constantly on the lookout for dangerous behavior. "Hopefully we will find such a place in order to have these decisions to make. I'm going to turn in," Tarlo turned and left, but headed back up to the bubble instead of her quarters.

The bubble was empty this time, and Tarlo felt silly staring in the direction the shadow had been. She thought about the log she would record in the morning. If she said the message word for word, would it be only because that is what she thought she would do? Or if she changed it, would she be creating a completely different future? She was just starting to doze in the comfortable observation chair when she heard the hatch open.

"I'm sorry Captain."

"Mr. Dopple. I assume we both had the same impulse."

"Is it there?"

"I didn't use the telescope to look, but I think it is. Just a feeling. Wishful, perhaps."

He moved to the telescope and began scanning. After twenty minutes he gave up. He sat in the next observation chair over and remained uncharacteristically silent.

"If we find an M class planet, would you settle down there or return to Earth?" Tarlo asked.

"I will do what you order, Captain."

"Noted. I'm ordering you to give me your honest choice."

"I'm not the outdoorsy type, so assuming this would be a terraforming duty, I would prefer to stay on the ship. If the planet is populated by our new friends and they have decent food, I would lean toward joining their society."

"That is exactly what I would choose for myself. However, my duty is to report back since the ship is many times faster than our radio communication."

"You think anyone is still alive?"

"Definitely. I'm just not sure what condition they would be in. It is far more likely one of the other missions has already found an M class and they are already ferrying thousands to the new home."

"Young and fertile thousands," Dopple said. "Do we have enough ore to get home?"

"Depends how much we burn in the M227 search, but I doubt it. It would be best to do another mining pass before we head back since there would be no stops and no nearby material."

"I hate mining duty."

"Because you aren't outdoorsy," Tarlo repeated.

"I'm barely indoorsy." She laughed. "You'd think they'd have robots to do that work."

"It wasn't worth the weight since we had plenty of refined ore for the two-year mission."

"Two years," Dopple said, shaking his head. "Not that I would have preferred to stay in that hell hole, but I would have preferred the life of luxury that this mission promised to give me. "Big risk for a giant reward," he said, repeating the words of his recruiting officer.

"What would you have done with the money?"

"Surfing off the Melbourne coast."

"That sounds very outdoorsy."

"The fun kind, not the survival kind. What were your plans when we got back?"

"Finding a husband that wasn't even born before I left. Preferably a faithful one."

"So, you like younger guys?" Dopple asked hopefully.

"After two years at hyper light speed, an embryo when I left could have been twice my age when I got back. No, Mr. Dopple, I do not prefer them younger. And I prefer them very outdoorsy."

"Sorry, Captain. I didn't mean to suggest anything improper."

"That's too bad. I was just warming to the idea of having a completely inappropriate relationship with a subordinate. Maybe our new friends are available for dating."

"Orion slave women?"

"Could be. I just hope this isn't going to be an unsolved mystery for the rest of my life."

"I hope they aren't hostile. I reviewed the video as the shadow ship faded away. It seems more likely it would have moved forward or backward off the screen if it sped up or slowed down. Instead, it seemed to just vanish while holding relative position. If it had veered away, it would have gotten smaller."

"And it would have had to increase speed to remain perfectly within our camera frame. Do you think they are interdimensional?" she asked.

"Time is the fourth dimension. That's the only way they could have tomorrow's log. Are you going to change what you say?"

"I'm not sure I can. Could create some sort of time paradox which destroys us or the whole universe."

"What if they generated it because that is what they want you to say?"

"That is far more believable than time travel. But if they understand the language well enough to synthesize new words, they would likely be talking to us."

"Unless they are assessing hostility," Dopple said. "I was trying to think what I would do in their situation. A big unknown ship flying toward their home with unknown intentions. You stay behind it for a few days, just like the old submarine movies. Then you give them a little tickle of communication in the same radiation they generate. If they run away, no worries. If they turn and shoot, you destroy them. If they attempt communication, you pull back and request orders from higher up the chain of command,"

"Makes sense. What if their bosses tell them to put an asteroid in our path?"

"If they have time travel then we may already be dead."

"Then we might as well go out with a bang," she got up and locked the hatch and began undressing.

The Search — Chapter 3

"Captain's log, Sol date four thousand, two hundred eighty-seven point nine eight five two. Acceleration continues on course to M227. All systems nominal, all personnel are within normal health parameters. We have had anomalous radiation detected off our starboard bow. We have retransmitted previous logs in case this anomaly affected the original transmission."

Tarlo smiled as she finished recording the words she had heard herself say the day before. She made this decision when she saw that the precise time was what was displayed when she looked at it. Every word seemed natural despite her foreknowledge of it. She did include a personal log with the precise events of the previous day. Those were sent as part of the encrypted data feed and would likely not be read during her lifetime. She wandered the ship checking all departments and speaking with many of the crew. They were not friends, but most were friendly. She could see in many eyes that rumors had spread. No one asked her directly, but she smiled knowingly and shrugged her answer.

Dopple could not look her in the eye when she entered the sensor room. It had been awkward for him as he had not expected what happened, and did not perform well under the circumstances. Tarlo preferred to maintain the distance but knew she would give him another chance at some later date. She hoped it would be on the long, boring trip back to Earth.

Tarlo went to her cabin and immersed in another mystery for an hour, and then slept for a while. She woke and headed for the bridge, somewhat hoping for a red light on the status board. It was all green. She looked at messages and logs, nothing out of the ordinary. The shift ended with the gruff Hans bringing his meal into the bridge. He knew food was not allowed outside the cafeteria, and she wondered if he was testing her.

"What are you doing, Mr. Yurd?"

"I woke up late," he said in a poorly disguised attempt at an excuse.

"You are obviously still dreaming if you think you are eating that in here."

"I won't open it in here," he said, putting the tray down on a console.

Tarlo knew now he was going to be a problem. Writing him up for such a minor transgression would be silly in light of his near perfect performance thus far. She had seen it before. People get infected with the hope of the imminent end of their tour. They became myopic about anything that contradicted their fantasy. It would only get worse. She walked to the tray and flung it down the corridor. It frisbeed for a while before crashing into a wall and spilling the contents.

"What the hell?" he said gruffly.

"Go clean that up, and in the future, report to your duty station on time and ready to work," she said calmly and without any emotion.

"Oh, that time of the month?"

Had it been that time, she might have reacted. "Are you inquiring as to the menstrual cycle of a fellow officer?" That was a strict violation of the sexual harassment policy.

"No, sir," he said, almost meekly, knowing he had failed in his attempt to assert authority.

"Then what time of the month were you referring to?" He had no answer, since nothing on the ship operated on a monthly schedule. "Return to your quarters, Mr. Yurd. I suggest you go into immersion until you are needed," The flash of rage she saw at that frightened her, but she didn't react. He started to leave, trying to find the words to undo his mistake.

"I'm sorry, Captain, I shouldn't have said that," he said, fighting to keep each word calm and failing.

"Which tells me you are not prepared for duty. Clean up your tray and go to your cabin." He did so without another word.

Tarlo was certain he wouldn't hyper-sleep, and she could only force that with the assistance and agreement of the medical officer. Doc Pember was in immersion herself and would not support Tarlo without far more substantial reasons. She pressed a button on the console. "Mr. Joplay, please report to the bridge," Denar Joplay was technically fourth in command, and she bounded into the bridge with an annoying excitement.

"Yes, Captain?" she asked with a smart salute.

"Mr. Yurd is under the weather. Can you cover his bridge shift?"

"Of course, Captain. Anything I should know?"

"Green status. There is another gravitational assist in three hours. Keep an eye out for alien spacecraft."

"Did you really see...?" Denar asked, obviously excited about the rumor.

"I'm not sure what we saw."

"But you saw something?"

Tarlo nodded and brought the video file up on the main screen. "This was off the starboard bow," Denar looked at it with fascination.

"How big was it?"

"Probably smaller than we are, but range was difficult to assess."

"Because it was interdimensional."

"What makes you say that?"

"I heard it played something from the future."

"Anyway, I'm down the hall if anything interesting happens," Tarlo went to her quarters and took a long time falling asleep. She woke briefly as she felt the gravitational well distort their path slightly.

Her duty shift the next day was more of the same. It was an hour before Hans' duty shift was to begin. She wanted to see if he was fit for duty well before it became a problem. She went to his quarters and surprisingly found him in his immersion pod. His wakeup time was just after galactic deceleration. Perhaps he wasn't going to be a problem after all. She reconsidered her plans, then decided to wait and think about it for a few days. She went and found Denar in her normal duty station in the engineering section.

"Are you up to taking over night shift until the end of deceleration?"

"Of course, Captain," she replied with her usual bright enthusiasm. "Is Mr. Yurd not feeling well again?" she asked with concern.

"Probably just exhaustion. He decided to immerse."

"I'll be up as soon as I finish my diagnostics on the deflector array."

"Problem?"

"No. I just like to find things before they become a problem. Especially with deceleration only a few days away."

"A good attitude when you are a billion light years from home," Tarlo said with exaggeration. "Carry on."

After the shift change, she went up to the bubble. It was occupied by several others, likely looking for their own confirmation of the rumors since they focused on the starboard bow. They looked at her expectantly, but she just went to a port side chair and reclined in relaxation. She fell asleep with the star view above her.

The dream was unusual in that she remembered it after waking. Normally all that remained was the emotional echo it left. It was a conversation that she had with the mission planners, but it was not a replay of anything that had happened. It was a discussion of what to do if an alien life form was encountered. She had read many manuals on the possible scenarios and recommended actions. She had never had actual discussions about it with anyone at NASA. The mission planners in her dream were very concerned that Tarlo not make any threatening actions toward new species regardless of their apparent hostility. She was to sacrifice the ship, crew, and mission before creating an enemy. At the meeting was one individual she didn't know. He seemed to be doing most of the talking, with the others supporting him.

"This makes no sense," Tarlo told the gathered experts. "The ship has no weapons to be hostile with."

"The ship itself is a weapon," The stranger told her. Tarlo looked around at the mission planners for confirmation, but none would meet her eye. Then an alarm sounded. Tarlo gradually woke from her dream as she realized the alarm was real. The oxygen warning in the bubble was beeping loudly. She felt groggy, but did not attribute it to the low oxygen level. She walked carefully to the hatch and found it locked. She tried to call the bridge, there was no response. She was almost certain it was Hans trying to kill her off. Then she woke up in her quarters. She had no memory of going there, but a vivid memory of the dream remained.

It was still several hours before her day shift, but she knew she could not sleep anymore. She walked to the bridge and found Denar monitoring the engineering panel attentively. "Everything green?" Tarlo asked.

Denar was startled, snapping to attention. "Yes Captain. On course and the board is green. There was a glitch in the refinery. They expect to be down for a day of maintenance."

"What kind of glitch?" Tarlo didn't care, she would read the details in the log. She wanted to know if Denar cared about knowing.

"The sediment separator got clogged with heavy metal contamination."

"That was a bad batch. I hope we can find better stuff next time. Thank you for covering these shifts for us."

"It's a great opportunity and I'm glad you trust me with it."

"Do you think we'll find an M class in this galaxy?"

"I hope so, but I don't think so."

"Why?"

"I've studied the probe and it did not get close enough. Many of the planets it gave Goldilocks status are borderline. Even with the overestimation, there are less than half what we found in the Milky Way, and we came from the only good one."

"Would you skip it and go to the next galaxy?"

"No, that one is even worse. The last one on the list is the best one. We could be there in forty-five days if we slingshot around this one."

"Really? Do you have a plot?" Denar brought it up on the screen. "Excellent work," Tarlo said after examining it closely. "I am going to stop here," Tarlo watched for emotional disappointment and saw none. "Can you tell me why?"

Denar thought about it. "Orders?" Tarlo shook her head and waited for another guess. Denar was drawing a blank, but then something occurred to her. "New friends?"

Tarlo nodded. "It is possible they are concerned with our approach. That would tell me that there is a habitable planet somewhere in there... with sentient life." She pointed at the galaxy that filled the forward screen. "Maybe not habitable for us, but interesting all the same."

"It would be very interesting," Denar agreed.

"If nothing pans out, I'd like you to have a course plotted from there to the last one. I'll plan the search based on the optimal exit point," Denar clicked on the panel and a course loaded onto the screen. "The ship will be in very good hands when you take over as Captain," Tarlo said. Denar gave her a concerned look.

"I will likely lead the colonization if we find something and you can fly home to tell them about it," Tarlo said.

"I'm not sure I..." Denar started to protest, knowing there were at least two others ahead of her in seniority.

"I am. Begin training whomever you think is capable of taking over in engineering."

"Yes, sir," she said with barely disguised excitement.

"Dismissed," Tarlo said, turning toward a small console to begin reading logs. Tarlo wondered why she had made such a decision, let alone informed anyone about it. She knew she was tired of flying through space and also knew Hans was the worst person to be in charge of a colony. He would declare himself emperor and have his eugenics plan ruthlessly in place as soon as Tarlo left orbit.

Another routine day making the rounds went by quickly. Strangely she felt no need to nap. It was the same for the next two days as deceleration preparation began. It was strange to her that slowing down took so much less energy than speeding up. What she learned in physics is they should be the same. The core drive was able to reduce the apparent mass of the ship by ninety nine percent when the acceleration was halted. It was only ninety five percent while the ion propulsion canon did its work pushing them forward. This made the dropping to sublight speed easier than stopping a trans pacific container ship.

The navigator, Mr. Uhar, woke from immersion in a flowery mood and finalized his approach vector before adjusting course.

"Ready for deceleration," he pronounced as the command staff stood at their stations and confirmed systems were ready.

"Very good, Mr. Uhar. Initiate deceleration sequence."

"Initiated," Denar said from her engineering station. A countdown displayed on the main screen as she read off the time. "Mass reduction will commence in forty-seven minutes... mark."

Tarlo opened a ship wide announcement. "Non-essential personnel to stasis pods for deceleration. All stations report in."

She read the log screen as all but ten of the crew engaged their immersion mode. They were the lucky ones who did not have to endure what most referred to as the 'shrink'. When the moment approached, everyone braced for the feeling. The energy wave swept out from the core and pulled them all down into a twenty-centimeter ball. Everything was to scale so there was no visual clue as to what was happening. Then with a small amount of neutrino friction, the ball dropped out of hyperlight in less than a minute. There was an audible pop as the ship crossed the barrier. They continued decelerating to forty five percent the speed of light. Mr. Uhar entered his final course adjustment and the computer executed it with precision.

"Deceleration cycle complete," Denar said when the core ramped down, allowing the ship to return to normal size.

Tarlo looked at the board and all lights were green. "Well done everyone. Let's start the planetary scans and see if it was worth the trip."

"Captain Tarlo, I have two pods with negative life signs," the junior medical officer Grantifar said.

"No reading or negative life signs?" Tarlo asked for clarification with concern.

"Sorry, no reading. Possibly a loose connection."

"Go check on them immediately," Tarlo ordered, taking another look at the board before following him down the corridor.

Grantifar opened the door to the first cabin. The readings from the pod were in green and the vital signs were stable. "He looks fine, Captain."

"Wake him up anyway. Have maintenance go over the pod completely," Tarlo ordered. Grantifar did as asked, starting the five-minute wake up cycle.

"It will be twenty minutes Captain. He has been under for the full duration."

"That's fine. You can come back after we check the other pod." They walked quickly to the other side of the ship. Grantifar opened the door to the quarters and stood frozen for a few seconds. The pod Hans Yurd occupied had a viewport that was clouded with red. Tarlo pushed him forward. "Get this damn thing open," Tarlo ordered, wiping at the viewport. The action did nothing, confirming the red was likely blood on the inside of the pod. Grantifar went through the emergency procedure and got the pod pressure equalized and they both forced the lid to open faster than the mechanism was doing the job.

"Overpressure," Grantifar said, recoiling at the sight. The body had been crushed down and was barely recognizable.

"Dammit, how the hell could this happen?" Tarlo said angrily, knowing precisely how it had happened.

"He must have not checked the equalization purge valve before locking down," Tarlo had already surreptitiously removed the pressure cap while Grantifar was checking the immersion log.

"I want a full report after you clean up this mess," Tarlo said, walking away in disgust.

"Aye, Captain."

Tarlo informed the bridge crew, who took the news with varying levels of shock. "We need to review immersion procedures before we leave this galaxy. We're getting sloppy being out here so long."

"Captain, I found a candidate M-class planet," Denar said excitedly.

"Is it far off the present course?"

"Yes. Showing significant oxygen and water. Blue star, no moons."

"Keep scanning. Maybe we'll have many to choose from. Mr. Uhar, plot possible courses to the candidate please." It would be wasteful to give up their current momentum before looking for other candidates. They scanned the other worlds identified by the probe as goldilocks. Only one other in this quadrant showed any signs of water, but no oxygen. The next quadrant was fruitless. Lendal came to the bridge when they approached the spiral arm he thought would the most fruitful. What they found was the cataclysmic reason for the misshaped arm. An extremely massive object had struck, and the debris was spreading and colliding like a billiard table.

There was one candidate, but is would likely be hit by the debris within a few hundred years. No point in settling there. The final quadrant revealed seven M class candidates previously unidentified by the probe. The captain knew this was too much of a coincidence. Five systems in a tight cluster, with two of them having double candidates. This suggested terraforming to Tarlo.

"Should we stop, Captain?" Denar asked.

"No, let's continue on to the first one we saw and get a closer look," Everyone looked at her strangely. "Mr Uhar, give me slingshot that gets me back here in a few days."

"Yes sir," he said, wondering why investigating one planet was more valuable than seven.

"Excellent work everyone. Get some rest for tomorrow. Mr Denar, prep the landing craft."

The Search — Chapter 4

Captain Tarlo launched the landing craft and piloted it toward the small blue sun the ship was headed toward. It was unlikely they would land. They would simply slingshot around in a much closer arc, rejoining the Orion on its return trajectory. Denar zoomed in the telescope on the candidate planet she had found.

"Sixty five percent Earth mass. Magnetic polarization. Green land mass. Forty percent oxygen. Thirty percent of visible surface is water. No apparent organized radiation," This was her way of saying no evidence of technological life.

"What is the other sixty percent?"

"Definite carbon dioxide, but too far to get a reading. Likely less than one percent," Over ten percent and it would not be habitable. "Definite vegetation," she said excitedly. "Average twenty-three degrees Celsius, no polar caps. Rotation angle is about ten degrees, with a rotation speed approximately two thousand miles per hour."

Tarlo did the calculation in her head. "Eight-hour days?" she asked.

"Yes, Captain, seven-point six one."

"That will make for an interesting sleep cycle. Any storm or volcanic activity?"

"None on the visible side."

"Anyone against landing?" Tarlo asked, knowing it was an enormous risk to do so.

"We would only have five hours on the surface," Jink said. Tarlo knew he was being overly cautious in his estimate. She looked at everyone and no one spoke up.

"I only need five or ten minutes. Start orbital insertion deceleration. Notify Orion of our plan to land."

They entered a highly elliptical orbit twenty minutes later. "Nitrogen fifty five percent, argon four percent, water vapor one percent, carbon dioxide point one three percent. No carbon monoxide or complex hydrocarbons detectable. Trace methane. We don't even need to suit up," Denar said.

"But we will," Tarlo said. "Look for a prime spot to land."

"This plateau seems to have the best mix of vegetation. It is sub-tropical, about twenty degrees Celsius. The morning has just begun so we'll be in daylight the entire time. No storm or volcanic activity on the surface."

"No storms, no clouds, no rain. How do the plants flourish?"

"Maybe there is enough water vapor consistently in the atmosphere?" Jink guessed.

"Land near that lake. Drop a probe as we pass over it."

"Aye, Captain," Denar said, readying the floatation probe that would give them the water quality as well as an extensive view of the life it contained. The ship landed on a knoll above the lake in what looked like broad leaf grass. Everyone donned their environmental suits as the Captain cycled through the airlock. Denar looked at the in-contact atmospheric analysis and read off the values as Tarlo walked down the ramp. The grass leaves had a jagged edge to them. They were also a deep blue green that reminded her of algae.

Tarlo knelt on the side of the platform and brushed the grass aside, exposing the soil. "No evidence of insectoid life. Soil is of loose sandy composition. She felt Jink stepping out onto the ramp from the airlock. Tarlo stood and attached her safety cable. She looked at him briefly then turned to step off the platform. "One small step for woman, one giant leap for humankind," Tarlo stepped off and was glad she was not sucked down into a quicksand grave. "Gravity feels heavy compared to the ship, but more comfortable walking than Lunar station. I am disconnecting tether and walking to those trees."

"Lake is almost pure water, no multicellular life forms. Algae is minimal and no bacteria is in evidence. Air shows no viral spores or bacterial activity," Denar read off her displays. Tarlo approached the trees and found many small nut sized nodules hanging from the branches, but none were in evidence on the ground. She snipped a sample branch into a collection bag. She looked back at the ship and could see the shadow of the tree moving slowly. The rotation speed of the planet was its biggest drawback at the moment.

"I'm heading down to the lake to get samples," Jink said. Tarlo watched him go, extending a long aluminum sampling pole as he walked. Weller went with him holding a mutual tether in case he fell in. Tarlo looked up toward the small blue sun. She looked back down at the ship and saw Denar exiting the hatch. She strode down the ramp confidently and stepped off with an excited little hop. She walked up the hill to get some readings and snap some pictures.

"Orion, this is Tarlo." It was a few seconds before the response came back.

"Captain, this is Orion. All systems nominal," Mr. Uhar reported.

"We are taking samples and expect to leave the surface within the hour. So far, this place is an Eden. No discernable animal life,"

"Should I plot an orbital course, Captain?"

"No. We'll need to check out the other planets as well. Start plotting a rendezvous course with our leaving the surface in one hour. Tarlo out."

She heard something behind her and turned. It was a vague humanoid form deep among the trees. She blinked a few times, trying to figure out if it was a trick of light and odd shaped trees. She took a few slow steps forward then stopped. It didn't move.

"Mr. Denar, I am walking into the tree line to investigate something."

"I'll be right there," Denar knew the protocol was to always be in another's sightline at all times.

"No, finish your readings. We will be leaving soon," Tarlo waved to her at the top of the hill. She then turned and saw the two down at the lake shore gathering their samples.

Tarlo turned back to face the humanoid and it was gone. She expected this. Just an imagined illusion. She walked into the woods, noting the flora as she went. The compass display in her helmet held true, so she was certain she could walk the reciprocal heading to get back to the ship. After a few hundred yards she found a wall of vegetation. She reached forward to pull it aside to see its depth. What she saw was a white room. Her instinct was to run. Instead, she stepped through.

The Search — Chapter 5

"Captain Tarlo," the voice said. She looked around, but nothing else was in the room.

"Yes?" she finally answered.

"You are trespassing." The response came after a long silence. It was as if her words had to go through a complex translation program.

"I'm sorry. We will leave immediately," She turned and walked back, but found no wall through which to step back out into the forest.

"You are in no danger here, Captain Tarlo. We only wish to understand your intentions here."

She had a feeling they already knew, but felt no reason to hide anything. "Our planet is dying. We were sent to find a suitable replacement where our people could survive and thrive."

"Would you not kill this planet as you have done to your own?"

"Probably. We will move on to the next galaxy if this is yours."

"Your ship, the Orion, has no weapons?"

"No. I guess the ship itself could be a weapon, but only in a suicidal end."

"Suicidal, we do not understand this word."

"Death at one's own hand."

"Ahh, regenerative initiation. Our common vocabulary is limited, we apologize. You have stated the intentions of your mission here. What are your intentions?"

"Personally? I guess living on solid earth and falling in love and having children. Fairly typical."

"You do not like space travel?"

"It gets old. I've had enough for a lifetime."

"Lifetime. You believe your time is limited?"

"We are mortal creatures."

"Because you retain no memory of the previous iteration? What if I were to tell you that your first officer has already begun his new life among an advanced civilization in the Hanari Sector?"

"What is the Hanari Sector?"

"What you call the Local Group is known to us as the Molani Sector. The Hanari sector is on the other side of the universal center."

"There is a center to the Universe?"

"Of course, that is where we live."

"Who exactly are the 'we' you are referring to?"

"All of us, of course."

"Do you control everything, like gods?"

"No. We watch, and do our best to cultivate. This planet for example, is not for your people. It has been terraformed to support a new kind of life."

"Have we damaged it?"

"No, you are not capable of doing so. If you brought your form of civilization here, it would eventually upset the balance."

"If you allow me to leave, I will report that this is not habitable."

"That would be our preference. Thank you for your cooperation, Captain Tarlo," A split in the white wall opened, showing the foliage she had walked through. She walked toward it, then stopped. The other seven worlds, are they off limits as well?"

"They are inhabited, and you will likely be vaporized if you approach."

"You would have just let that happen if I had not asked?"

"It is not our place to interfere."

"Prime directive bullshit."

"We do not understand this reference."

"Never mind. How about this... is there a world that we can settle on?"

"Thousands. Your starship Pleiades found one and your home world is organizing a series of colonization ships. Your home world is no longer in danger of dying. There was a purge and a new future has been charted."

Tarlo wanted to ask what they meant by 'purge', but decided she didn't want to know. The Pleiades had gone in pretty much the opposite direction from the Orion. "Is one of the thousands near here?"

"Several."

"Can you tell me where?" There was a long pause.

"For future reference, or do you intend to go there now?"

"Both?"

"We will only tell you that Denar's intuition was the best."

"The final galaxy on our plan," Tarlo stated in understanding. "Thank you. Will we speak again?"

"At the end of your current iteration."

"There is no way to contact you before that?"

"You will have no need for such communication. Enjoy this life, Captain Tarlo," The split in the white wall moved passed her and closed. She was surrounded by trees. The conversation had been painfully slow with the speed of bidirectional translation. She estimated at least thirty minutes had passed and the clock in her suit confirmed it. She aligned her compass and walked briskly back toward the ship. When she emerged from the tree line, Denar was still on the top of the hill and the other two were still taking samples. They should have been done long before this.

"What is taking so long?" Tarlo said.

"Captain?" Denar asked in confusion. "I thought you wanted me to stay and finish taking the readings. I can come down there to accompany you into the forest if you would like."

"I already did that. You should have been done fifteen minutes ago."

"I'm sorry Captain, but I've only been up here five minutes."

Tarlo brought up the telematics from the other environmental suits. The others had far more oxygen remaining than she did. "Fascinating. Carry on. I am returning to the ship," Tarlo entered the airlock and went through the decontamination procedure. She entered the ship and pulled off the helmet. Sitting at the science station she looked at the sensor readings collected so far. It was near perfection. A pristine world for her to live in happiness the rest of her days. The indicator from her suit download went to green. She pulled up the video from her EVA and watched on ten times speed. She slowed it to normal when she reached the tree line. There was no humanoid. After she began walking into the woods she just stopped for a few seconds, then turned around and walked back out. She had imagined the whole thing. Except she had used twenty-eight extra minutes of air. She looked back at the readings. They only preferred she didn't stay, if they were indeed real. She heard the airlock cycling. She began the take-off preparations.

Denar was the last one in. She stripped off her suit and connected her equipment for download.

"Captain, can we drop a buoy in the ocean and collect some samples there as well?" Jink asked.

"We can drop the buoy, but we are heading up to orbit."

"It will only take me a few minutes."

"We have all we need," Tarlo said, starting the engines. She took off low and went over the woods, looking down for anything strange. There was nothing. Denar dropped the buoy as they skimmed along the coastline. Then Tarlo rocketed up to orbit.

Denar modified the intercept course and after five more orbits collecting data, they shot off into space to meet the Orion.

"It would take no time at all to seed that world and make it optimum for humans," Jink said.

"I'm not sure we could adapt to the rotation speed," Tarlo said. "The lack of apparent precipitation also concerns me."

"I think it is too perfect," Denar said. "Someone made that the way it is."

"Terraformed?" Weller asked. Denar nodded. "By whom?"

"The same people that built the pyramids," Jink said.

Tarlo rolled her eyes but was glad to have his conspiracy rant to occupy their minds.

The Search — Chapter 6

Dopple, Denar and the Captain were alone in the bubble. The two junior officers wondered why they had been summoned here and why the Captain was locking the hatch. Tarlo gathered her thoughts, looking out at the vast star dappled darkness.

"I'm going to order the Orion to go directly to the final galaxy."

"I only recommended that before I knew the results here," Denar said defensively.

"I'm not going there because of your recommendation... not directly, anyway," Tarlo said. "This is going to sound crazy, and if I had been the only one to see that shadow out there, I would agree," Her finger lingered in pointing out of the bubble where she had seen the shadow ship. Another hallucination? She took a deep breath. "When I walked into the tree line, I spent half an hour somewhere else. The suit cam did not go with me, so I have no proof. I did use forty percent more oxygen than the rest of you, so that tells me my suit did go with me," Tarlo could tell that Denar was thinking pressure leak, but she didn't say anything. "I met something. It was a voice of sorts. It told me... many things. There is a planet for us in the final galaxy. I was not told we could not settle here, but they said that this planet was made for a new kind of life."

"What about the other ones in this galaxy?" Dopple asked cautiously, sensing the answer.

"I was told they would likely vaporize us if we approached."

"Klingons," Dopple said, pointing out to where the shadow was. "Why didn't they just do it when we were way out there?"

"There are a lot of questions I should have asked, but didn't. My guess is our shadow was not capable of hostility. Remember, we were at hyper light speeds. What energy weapon would be useful there?" Dopple nodded in agreement.

"Could the shadow ship have been the ones you talked to in the woods?"

"I don't think so. I don't think they are limited by space and time. They could be in the bubble with us right now," Both looked around, somewhat intrigued. "As I see it, we have three choices. Continue immediately to the final galaxy. Turn around and colonize the Eden planet, or continue to these other M class planets and risk being vaporized."

"We could just go back to Earth and let them make the decision," Dopple offered.

"No. Our mission is to bring back the location of a suitable replacement. To be a suitable replacement we need to leave people behind to colonize and prepare for others to arrive. We could lie and say we found nothing after searching all of our assigned galaxies, but that would be dishonest and require deleting extensive logs, some of which have already been sent. If I was going to do that, it would have been long ago when I had something to go home to."

"Captain," Denar started, then paused. "We need to know more about the seven M class planets. Let me launch a probe. If they vaporize it, then we'll know for sure. If they don't, then we will have the data we need for a better decision," Tarlo smiled and nodded. "You are telling us this because you need us to support you going to the final galaxy."

"I do, but what I need more is your advice. Am I crazy? Did I simply imagine all of this?" They all sat silently in contemplation.

"We can always come back. It is on the way back to Earth anyway," Denar said. "I wondered why you asked me why the

readings weren't done. I trust that you spent some amount of time...
elsewhere. I will support your decision."

"I think..." Dopple started.

"What is it?"

"Instead of a probe, send me in a landing craft."

"Why?" Tarlo asked. She had considered doing the same
herself.

"They may ignore a probe. All my life I've believed there are
others out there. I want to know before I die."

"Even if you only have seconds of knowing?" Denar said. "They
could vaporize you before you even know they are there."

"I would still know, even if it was nanoseconds. And then you
would all know. I think they would see a single person in a ship as
less threatening. They may even want to talk."

"It is a good thought. I had considered doing that myself. We
cannot spare the landing craft. If there is truth in what they told
me, we will be ferrying down all the supplies soon. Let's leave first
contact with these people for future generations."

Dopple nodded in disappointment and some relief. "Prepare
the probe Mr. Denar, I need to have a long talk with Mr. Uhar."

"I TRUST YOU, CAPTAIN," Uhar said.

"Really?"

"You never question my navigation. That is an amazing amount
of trust. Mr. Yurd made me defend every single millimeter. You
want to go to M4872, I'll have a course plotted in a few hours."

"Even though we are passing on seven M class worlds?"

"My life is on this ship. Doesn't matter where we are headed.
Going now means less work and more immersion."

"Can we go to hyperlight here, before we clear the system?"

"That isn't wise. Better if we can Z up out of the galactic plane first."

"Let's do that. I want to initiate before night shift."

"On it," He turned back to his console and she watched him open his previous plots and chose the best one, adjusting for the current position and speed. "Looks like negative Z is a better path. The gravity of the spiral arms will give us a natural redirection to where we are going. Initial course set."

Tarlo pressed the initiate control and felt the ship reorient for acceleration. "Thank you, Iman, for trusting me." He nodded and continued plotting the rest of the journey.

"CAPTAIN, I FELT THE ship turn and accelerate," Jink said. "Which of the seven are we going to first?"

She appraised him carefully, wondering how he would take the news. He had very high hopes for this galaxy. "None. We are leaving the galaxy."

"What?" He looked up in surprise from his disgusting cube dinner. "You must be joking."

"You know we launched a probe."

"Of course. I helped prime it."

Tarlo looked around, and lowered her voice even though they were the only two there. "You know what it found?"

"Hostile aliens?"

"All kinds of nasty pyramids. I'm fairly certain if we go that way, we will be building their next one," It was all she could do to keep from bursting out in laughter. "Don't tell anyone," She asked, knowing he would tell everyone multiple times.

"Did you see what they look like?"

"Insectoid," she whispered. Jink shuddered at the thought. Only he could believe something so ridiculous.

By the time the hyper light acceleration had eased off, no one on the ship was questioning of her decision. Hans Yurd would have fought her and probably forced her to their likely doom.

"Thirty-two-day transit, Captain."

"Enjoy your sleep, Mr. Uhar. Tell me, is there someone in there?"

"His name is Keta. We were school mates. I have learned to manipulate the dream state so that we can be what he chose not to be when we were young."

"Fantasies?"

"They feel as real as any other memories."

"Do you think you could teach others to manipulate the dream state?"

"I tried to explain it to Geppen, but he said he could not do it after several tries. You never go in, do you?"

"It was too unsettling. Perhaps if I could control it..."

The Search — Chapter 7

"**N**o M-class planets?" Tarlo asked, realizing the forest conversation had been completely imagined. Time and fuel wasted for nothing.

"My search only scanned the external third of the galaxy," This was standard procedure since the closer you got to the galactic core, the more dangerous radiation was encountered. "All goldilocks zone planets identified by the probe had no indication of oxygen or water. A few others not found by the probe had trace oxygen and water."

Tarlo looked at the display as the ship continued to orbit outside the galaxy. "They lied to me," she said under her breath.

"What exactly did they say to you?"

Tarlo tried to remember. "Denar's intuition was the best," Is what finally surfaced. "You had just told me the final galaxy would have what we were looking for. I assumed that is what they were talking about. I imagined the whole thing, didn't I?" It was physically painful to Tarlo saying it out loud.

"The probe to the seven M class worlds never showed any signs of intelligent life. Certainly nothing that would vaporize us," Denar admitted, having doubts herself. "Intuition? Are you sure that was the word they used?"

"Pretty sure."

Denar looked back down at the display, contemplating that first time she really studied the profile of M4872. "Can we move on top so I can scan closer to the core?"

"What are the rad levels?"

"This is an odd system. It is not centrally organized. I don't think the core radiation is the same as others we have seen."

"We only have time to lose," Denar input the course change and Tarlo executed the turn. It took a few hours to change noticeably on the viewer. "It's a doughnut?"

"What?"

"A food from long ago. Basically a large ring. Is it possible there is no black hole at the center?"

"I was thinking three or more small ones in a trinary arrangement, but it is hard to make the physics work," As they moved higher Denar began her scans toward the interior. "Radiation levels are lower in the middle, Captain."

"Could it be..." Tarlo began to see it. "Prepare a small probe. Gravimetric only. Send it toward the middle slowly. Then get some sleep. That's an order," Denar had skipped the previous day's sleep cycle to do her scans.

"What are you thinking?"

"Sleep, and we'll talk later when we have results."

"Aye, Captain," Denar headed down to engineering. Tarlo watched the readout of the probe over the next few hours. She read a few related scientific papers. Most of them were little more than wild conjecture from theoretical physicists. The probe seemed to be pulled off course toward something off center, but it did not accelerate like it would if it was headed toward a super massive black hole. The probe was now well within the center of the empty doughnut. It was not an event horizon. Tarlo was certain of that. She left the bridge and did her daily status walk. Everyone was subdued, hope having been drained by the latest circulating rumors.

When Tarlo returned to the bridge just before dinner, the probe was no longer sending data. She rewound the feed. There was no failure message, it just stopped transmitting. "Mr. Dopple, please come to the bridge." He arrived two minutes later, completely out of breath. "What can you make of these probe readings?"

Dopple scrolled through the electronic patterns, going back and forth over the place where they stopped. He zoomed in. "Frequency shift," He said, typing in a few commands. "It was broadcasting an anomalous gravimetric reading, very localized. Each burst was sending further and further apart," He zoomed out. The frequency began shifting from standard here," he pointed at the screen, "toward a longer wavelength. The shift accelerated until it disappeared."

"What would make it do that? Battery failure?"

"No, I don't think so. It almost looks like..."

"What?"

"Like it went into hyperlight."

"It barely has course correction thrusters," Tarlo said, wondering if he would draw the same conclusion.

"Can I see the external telemetry?" Tarlo brought it up and watched him. Dopple cross-referenced it and finally said. "Wow. It disappeared forty-seven seconds before the last transmission."

"Conclusion?" Tarlo asked, her theory all but confirmed.

"Wormhole." She nodded. "To where?"

"Only one way to find out."

"No way!" He said, looking at her to see if she was serious. "That could be a one-way trip."

"Possibly. I'm thinking the other side is just as stable."

"But if you are wrong..."

"I was thinking about taking a landing craft through."

"You want me to go with you?"

"There are worse people to be stuck on the other side of the universe with."

"I was willing to go on a suicide mission to meet alien life forms. What chance do we have finding that on the other side of that wormhole?"

"Wormhole?" Denar asked, entering the bridge.

"I think that's what is at the center of this galaxy," They showed Denar the evidence from the probe. "I'm thinking about taking a landing craft through to see if it is stable and traversable."

Denar studied the data for a long time. "There is barely any acceleration before entry. This isn't an ER bridge. It may be survivable in a landing craft. It would be better to go through mass reduced. I'm more concerned with the emptiness surrounding it. If it is sucking matter down in, where does it end up?"

"Maybe that's what fuels it."

"More likely it is clogging up the middle. You may accelerate into a giant... no that would become a black hole as mass accumulated. Let me send a probe programmed to return on a reciprocal heading. If it comes back, we'll know. If it doesn't then... I don't know."

"I want to sleep on it. Scan for planets on this side, make sure we haven't missed anything. Look for some source of ore as well. We'll need to mine before we go anywhere," Tarlo ordered.

"Maybe it is straight?" Dopple offered. Both looked at him wondering what he meant. "If we get in front of it maybe we can see the other side?"

"Excellent idea. Plot a course, and we'll do it first thing in the morning."

Tarlo left the bridge and went to eat. If they had stayed at the Eden planet, they could have had fresh food by now. The soil had proven very fertile and the blue sun would be compatible with their food stocks. That was the fallback plan. If she had imagined the voices, then Earth was not already moving to the planet the Pleiades had found. She knew she had not imagined it. All the evidence said she hadn't. But if it was real, then it was just as likely they had lied to make her leave.

It was undeniable that what they had said and done were god-like. She pictured the same people entering the mind of Moses and making him believe he saw a bush burning without being consumed. The power of illusion. The power of time manipulation. The power of terraforming. It was that one word they seemed to trip on, suicide. If they were in her head, how could they not know that word?

She went to her quarters and fell asleep instantly. The dream began at the tree line. She didn't hesitate to walk into the woods, but there was no wall of foliage. There was no entrance to the white room. She had not paid attention to the compass and she was now lost. The woods went on forever in every direction. Night fell. Her oxygen dwindled. She fell to the ground, weak. She opened her helmet seal and took the helmet off. The air tasted sweet. She could smell trees that reminded her of home.

"You are trespassing. Now you have contaminated the world not meant for you."

"You're God, just undo the damage."

"We are not God."

"No, you are liars. No one was going to vaporize us. There are no habitable worlds in the doughnut galaxy. Earth hasn't found a place to relocate."

Suddenly she was thrown into the sky and she was flying through space. The stars around her began to blur into long lines. Galaxies flew by and then she approached the familiar one. The one she saw shrink behind her ten years ago. She watched as the yellow sun zoomed up before her, the massive Jupiter flying by on her right. She slowed and landed on her feet at the Kenya launch facility. A massive passenger ship was lifting off while another was loading. Then she was flying again. She passed Saturn on the way out and soon she landed on another world. A rich green one with funny-looking creatures roaming among the small village of settlers. Tarlo approached one of the settlers, but the young woman could not see or hear her.

"What is this place?" Tarlo asked louder, but it was no use.

"This is the new Earth. I'll give you the coordinates if you want to fly directly there."

"I don't believe you."

The Search — Chapter 8

"I think I know how this galaxy formed," Denar said when Tarlo arrived on the bridge early. "Two supermassive black holes collided at immense speed. Their matter mushroomed out into a ring, which condensed into this galaxy. That explains the wormhole. It is a gravitational remnant of the collision."

"Or the wormhole pulled the two black holes into the collision from different parts of the galaxy," Tarlo said, sipping her coffee. "Course plotted over the center?"

"Yes, sir. It could be dangerous. Something from the other side could shoot out at us."

"Has anything come out?"

"Not that I have seen."

"Then we won't stay long. Why does the course not just go straight across?"

"Several M class worlds on that heading. Too far to get precise data."

"Excellent work," Tarlo executed the command and the ship accelerated smoothly. "Anything else?"

"I think the probe sent data from the other side. It is massively distorted. Could just be a burp from it crashing."

"I'll have Mr. Dopple look at it. Do you want to sleep for a while before we get to the middle?"

"I napped here while the scans were running."

"Good. I want to review the M class planets you found."

They spent the morning going over the details. When they reached the overwatch, they pointed the telescope but saw nothing different. Dopple was disappointed.

"What would a wormhole look like?" Tarlo asked.

"An Einstein-Rosen Bridge would probably have a pinhole transitory tunnel due to the gravitational crush," Denar explained. "This does not appear to be an ER Bridge. If it was a Casimir effect passage, then it would likely be a simple hole where the entire EM spectrum including visible light could transit."

"Captain, I'm receiving probe data," Dopple said excitedly. "We must have passed over the hole just now."

Tarlo knew they were moving too fast to just stop and listen. "Can we drop a probe to hover and relay?"

"On it," Denar said, rushing off the bridge toward engineering.

Dopple took the transmissions and put them up on the screen. "We've lost them. The wormhole must be focusing them into a very narrow beam. The star patterns do not match any on record. Wait, this looks like... Holy crap! This is the Veritas cluster. That is on the far side of the universe!" Dopple brought up the star chart and typed in his query. The display showed the known universe and a line was drawn from one side to the middle. "This is us," He pointed near the center. "Earth is right here," He pointed to a dot just off the line, but very close to where they were. "This has to be, wait, new data coming in. Denar must have launched the probe."

"Can we trigger the distant probe to resend all data?" Tarlo asked. It had never been necessary before, but she seemed to remember the capability. Dopple looked up the commands and sent them. A giant flood of data came through a few minutes later. "Can we recall the probe?" Tarlo asked as Denar entered the bridge.

"It didn't work? I thought I had the right coordinates."

"The one you launched worked perfectly. Can we get the one on the other side to do a reciprocal heading?"

"It doesn't have enough fuel to stop, let alone reverse trajectory," Tarlo nodded in disappointment. "Let me look at the data and see if there is something to gravitationally slingshot around.

"Does the probe have any special properties that would help it survive that a landing craft doesn't?"

"It is much smaller. The landing craft is far sturdier with many types of shielding the probe doesn't."

Dopple looked up from his screen. "The gravimetric data shows it is a massive tunnel, maybe five kilometers wide at its narrowest. No debris, but there was a strange object in the middle," He pulled up the picture of a big, irregular rock.

"That is not a natural formation," Denar said, looking at the asteroid.

"What makes you say that?" Tarlo asked, not seeing any regular patterns.

"You see this projection?" Denar pointed at the screen. "It is precisely the same as these other two, and they are all at a sixty-degree angle from each other. My guess is they are all around it."

"A ship stuck in there?" Dopple guessed.

"Or the device that keeps the doorway open," Tarlo said. "I want a top-of-the-line probe prepped and sent through there and to survey the other side, then return. Launch it after we slow down to look at these M-class planets.

"Aye, Captain," Denar said, disappointed that she wouldn't be going with them to investigate the new planets. They looked at all of the new data as they prepared for the deceleration.

"I want to look at this one first," She called down to the shuttle bay and ordered them to prepare a landing craft. Two hours later

Tarlo launched her landing craft with the other two science officers that had accompanied her to the surface of the Eden planet, along with Denar's replacement. The young engineering officer who had taken Denar's place had a nervous excitement about the special duty he had been invited to do.

Denar sat on the bridge monitoring everything, the final words from Captain Tarlo ringing in her head. "The ship is yours, Mr. Joplay." She felt unusual ordering others to do the work she preferred doing herself. They prepped the probe and she reviewed everything before allowing it to be launched.

She watched the telemetry and saw that Tarlo had skipped the first planet and was headed to the second one. The data her landing craft sent back showed high sulfur and carbon dioxide levels. As the second planet approached them, Denar watched the landing craft merge with the planet, showing an orbital entry.

"Tarlo to Orion."

"Go ahead Captain."

"We're going to land on candidate number two. Significant electrical storms, but not where we will land. Expecting less than six hours on the surface."

"Copy six hours," she replied, setting a timer for ten minutes short of that. This was the hard limit for the start of search and rescue if no communication was established by then. Electrical storms could cause all kinds of scattering effects. In such a case another landing craft would enter orbit and seek visual contact, using one of the many methods at their disposal to communicate.

"Tarlo out."

Denar looked over the data that was coming in from the landing craft. It was not as perfect as the Eden planet, but the global lack of clouds on the more perfect world had been one of Denar's biggest concerns. Plants could evolve to absorb water from the air, humans could not. She looked at the visual data, much of it obscured by clouds. More than Earth had on average, but that might be part of the global temperature buffering. It was a yellow sun, but the slightly smaller planet was ten percent closer to it.

"Tarlo to Orion."

"Go ahead Captain."

"Had a rough landing. The wind sheer is significant and not well organized."

"Should I launch the spare?"

"Yes. I think we'll be able to get to orbit, but Mr. Jonuk will send a list of parts we may need to get us back to the Orion. Have them stay in orbit and spend some time mapping the wind patterns."

"Aye, Captain. ETA one hour," Denar said after plotting a direct course.

"Starting my EVA. Tarlo out."

Denar called down to the hangar and got them working on the second shuttle. There were four in all, and they rotated use as they went. She forwarded the list of spare parts, along with some things she would want as well, to the inventory robots. Then she went to find the ship's meteorologist.

"Sorry to wake you," Denar apologized. "The Captain needs you to map some wind patterns on a candidate planet."

The older man wiped sleep from his eyes. He sat up in the bed and blinked. Denar repeated what she had said, and his eyes widened with excitement. "Landing craft?" he asked.

"Yes, but you probably will not go to the surface." That dimmed his exuberance slightly. "But we may be moving to this planet soon enough," He jumped up with that optimistic hope in his mind, and Denar left to find the ranking pilot.

Falx Kenby was a typical alpha male who bristled at the Captain's desire to pilot herself. Denar knew it was because the Captain did not like spending time with him. She didn't know why, but could easily guess. The majority of his flight time on this mission had been pulling in asteroids for the miners.

"Wee lass needs me to pull her bacon out of the fire?"

"I would recommend keeping that line of thinking to yourself." Technically he outranked Denar, but she was currently in command of the Orion.

"Aye, Lieutenant Commander Joplay. When do we launch?"

"Twenty minutes," Denar turned to leave the gymnasium and heard his muttered derisive comments. She chose not to engage with him. If she was to be in command while flying back to Earth, she would have to put up with him for the entire flight.

She went and oversaw the loading and launch, then returned to the bridge. "Orion to Captain Tarlo. LC three is on its way."

"Excellent. The diversity of life on this planet is amazing. Nothing approaching mammalian, but plenty of insects and lizards. Would not be surprised if a T-Rex popped out of the jungle. We are collecting samples of all the fruits to see if they can be added to our dietary chain. If nothing else, we can extract the flavor molecules. Tarlo out."

Denar went back to the data, then started plotting return courses to her expected position beyond candidate number three. She knew it was unlikely the Captain would do little more than fly by the third planet in her damaged ship. When all of that was done, she returned to the probe data and began plotting the reciprocal flight. There was little to use for gravity reversal, so she plotted a slow transition and lazy circle on the other side. When ready, she launched the wormhole probe back toward the center.

Tarlo was able to get her landing craft to orbit before a gathering storm was able to interfere. They spent almost ten hours transferring parts and repairing the damage. Tarlo left the prideful Falx Kenby to fly the damaged ship back to the Orion while she flew on to candidate planet number three in the fresh landing craft.

Denar watched all the data pouring in as she made engineering adjustments to the ship. She had to step up the training of her subordinates if she had any intention of becoming captain of the ship and flying it back to Earth. Much more daunting would be the fact that she would be flying it back with less than one third of the crew. There was less to do in taking care of so many people and flying straight through empty space. But there was more to do for less specifically skilled people. Normal maintenance and repairs like waste treatment might be pushed back in favor of higher priority engine maintenance. They were down to forty percent of their refined ore. That would be more than enough to get back, but they would need three to five times that in establishing a colony depending on the planet's natural resources.

"Tarlo to Orion."

"Orion, go ahead," Denar responded immediately.

"Candidate three is nearly optimal. I'm going to land and verify. After you recover the other landing craft, head this way," Denar was surprised by this. It was much safer for the Orion to stay out here away from interstellar objects. "There is a nice asteroid belt beyond the sixth planet. Set up for mining above the plane of the system. Take it slow and get good scans of everything in the area."

"Yes, Captain," Denar responded when the pause took a while. She had found many safer systems to mine on the inner edge of this galaxy.

"Have Kenby start organizing landing parties, but do not launch until I give the go ahead," Denar saw the data coming in was truly ideal. It was almost too perfect.

"Captain, please authenticate Zulu Tango Foxtrot," Denar said with trepidation. She heard a loud laugh from the other end.

"Umbrella Alpha Keystone."

"Copy, Captain. ETA in candidate three system approximately twenty-three hours."

"If they can get in my head, they can easily dig out communication codes. If you think the ship is in danger do not follow my orders."

"I can do it safely, Captain."

"Good. Going for deorbit burn. If you don't hear from me in four hours send Kenby this way. Tarlo out."

Denar altered course as soon as they recovered the damaged landing craft. Then she left the bridge and began visiting departments to get them activated for a possible colonization. The excitement for this was mixed, but generally positive. She talked with Falx Kenby about the next mission and he was thrilled with the idea of actually doing real intersystem maneuvers and landing on something more than an asteroid.

When she got back to the bridge there was far too much new data to sort through. She concentrated on the new flight plan and added in maneuvering to avoid rouge objects the scans had revealed.

"Tarlo to Orion."

"Go ahead, Captain," Denar replied.

"Start sending them now. This is a definite go... nine and a half years later than we expected."

"Aye, Captain. ETA two hours to first orbit, approximately nineteen for Orion. Captain, your voice sounds different," Denar said with concern.

"Because I am talking into my helmet while holding it in my hands. I had forgotten what real air smelled like," Denar knew this was a major breach in protocol. She had serious doubts about the Captain's sanity now. She began pouring over the suit biometrics but everything was green. "If you are worried about contaminants, don't. I am not coming back on board. The ship is now yours, Mr. Joplay. Get everyone out of immersion and let's get to work. Tarlo out."

The Search — Chapter 9

Tarlo had asked all of her crew to describe what they saw, because she could not believe it. She thought maybe it was another white room illusion. Vast grasslands with herds of animals grazing. Groves of fruit trees as far as the eye could see. Crystal clear lakes and vast oceans full of fish. Snowcapped mountains in the distance promising years of fresh water. Clouds aimlessly drifting by. She had already picked out the hill on which she would build her shelter. Gravity was eighty-two percent normal, giving her a bounce to her step. A familiar yellow sun in the sky, along with a companion blue one, two moons, both out of sight at the moment.

"Captain, we are showing a significant viral load in the air," Weller was a mask of concern inside his helmet.

She looked at him without a care in the world. "If I'm dead, I'm dead. Start cataloging them and see if any are not covered by the universal vaccine," She knew this would keep him busy for weeks. They couldn't commit to colonizing until the entire biosphere was cataloged and assessed. But that required massive amounts of equipment be brought to the surface, so it was a commitment of sorts.

Tarlo wandered among the vegetation, looking at the varieties. Life here had evolved quite similarly to that of her home planet. Almost too similar. The grass looked like grass. The apples looked like apples. The daisies looked like daisies. She still wondered if it was an illusion. It was more dawning on her that the white room was less an illusion. Some sort of intelligence guided her here. They would not have done so if it was dangerous. She had to trust that or admit to herself that she was crazy.

The second landing craft touched down in a nearby clearing. Falx Kenby strode out of the ship with a long-desired need to touch the ground. He looked around in amazement.

"Captain, this is incredible," Kenby said in awe. She just nodded. "Is it safe to remove my helmet?"

"Not for you. You need to go back to the ship at least dozens of times. It will be a month before they determine the true safety. I can do it because I will not be going back to the ship."

Kenby looked at her in surprise. She could see him calculating the chance that he would become the new captain of the Orion. He wanted to breathe the fresh air, but did not want to sacrifice that possibility. She let him enjoy the tantalizing possibility. She asked about the preparations and he gave her a fairly professional description. Once the supplies and equipment had been unloaded, both landing craft took off for the next round trip. Buoys and probes were dropped all over the planet in lakes and oceans. Satellites were put into orbit to map the surface, monitor climate, and facilitate planet-wide communication.

A prime location was chosen for the first settlement based on resource proximity and expected meteorological variation. A team began harvesting fresh nutrients. It was processed and sent back to the ship to replace their current food stocks. The Orion was busy with asteroid mining operations as the ship population dwindled. After the safety of the biosphere was determined, almost everyone had an opportunity to spend a few days on the surface. This was a risk, because if too many decided to stay, the Orion couldn't go back.

"It is truly paradise, Captain," Denar said.

"It is. You want to stay?"

"Yes. But I am not going to. I can be back in four years to spend the rest of my life here."

"I'll be an old woman by then, ready to pass the leadership torch to you."

"Not here. I would be thirty years behind the rest on what it takes to live here. I am more interested in exploring the wormhole, now that we know it can be transited in both directions."

"I envy you that. I expect some of my children will want to go with you. When do you plan to leave?"

"Two weeks. I want to finish ore refinement to make sure we have everything we need for a round trip,"

Tarlo nodded. It was exactly what she would do. "Will Kenby be a problem?"

"No, he is going to immerse the entire trip to Earth. I think he hopes to get his own ship when he gets back."

"A lot of key people are staying here. Do you have enough crew?"

"We'll get by. It shouldn't be a complicated run. All the maintenance was done before they moved off ship and everything is in tip top shape. I'm certain I can fix anything myself if it comes to that. What about you? Is there anything else you need here?"

"My main worry is genetic diversity. If you can bring back a couple hundred more young and healthy people, we won't become a mass of inbred banjo players."

"If they let me. It will have been over three hundred years on Earth. They may be in a completely different place."

"I would stay in a very high orbit until you are sure. This mission was only to report back success. The ship is yours until you decide otherwise."

"I'm not sure they will agree with that."

"Possession is ninety nine percent of the law," Tarlo said, watching the young woman nod, but doubted she would be that renegade. She held out her hand. "Safe travels."

"Good harvest," Denar said, taking the offered hand. There would still be daily communication, but Denar was not going to land again. "Have you named your planet yet?"

"I have, but I doubt the others will agree. I guess we'll see if a democracy emerges from my benevolent dictatorship," Denar waited for Tarlo to offer the name, but the older woman just smiled.

Tarlo went back to her garden when the landing craft accelerated out of sight into the clouds. She had prime real estate. A gently sloping grassland above the settlement. The pattern of the binary stars would keep the crops in good light for the majority of the day. The short nights had made sleep difficult at first, but the human body was capable of masterful adaptation. The gentle stream that wound through the property would supply her garden with all it needed. There were no prowling rodents to steal her crops, and the magno-electric field generator kept all insects away for five hundred meters in every direction. The solar pollination drones would fly while the sun shone, systematically ensuring every plant propagated properly.

Most gathered in the command yurt for communal meals and general socializing. Plans were constantly being made and teams were formed to make the world even more habitable than it already was. This was not the hard life all had expected to endure. This was good because some were eight years older than they expected to be when starting this challenge.

On the final day, everyone gathered around the small symbolic campfire in the center of the yurt.

"Thank you all for making this village completely functional in such a short period of time," Tarlo started when she had everyone's attention. "I knew I would never leave this world as soon as I took my first breath of this sweet air." Many nodded in agreement. "This is the path I have chosen. Now it is time for all of you to choose your path. Most of you signed up expecting to never see Earth again. Even if you did go back, all those you knew were long dead. You were chosen for your expertise in case we found an only marginally habitable world. This is not the case. Our new world is perfect. Therefore, none of you are required to stay. You have until morning to decide if you truly want to stay. The Orion will stay until everyone has made their commitment. There will be no contact with other humans for at least thirty years. The Orion, or another ship will bring new settlers directly here, but decades will pass for us while they spend a few years in hyper light transit. Think hard about this choice, for I do not want to spend the rest of my days listening to regret while we face new challenges as the seasons change.

"The first American colony vanished before the second ship arrived to resupply them. Many settlements lost more than half their people in those first few years. It is true that we have superior technology, but perhaps our reliance on that makes us even more vulnerable to what lies out there, unseen.

"I will do my best to make this colony as successful as possible, and with technology, I could do it on my own. I hope some of you will embrace the challenge with me," There was enthusiastic applause as most let their decision be known. "There are decisions to be made and I will make them if I have to. I would hope most of you that stay can make smart decisions on your own. If there is any need for justice, know that it will be as swift as it is severe. Resources are communal for now, but as crops grow all will be responsible for their own future well-being.

"And finally, to truly make this work we need to start having children. As many as we can as soon as we can. The more diverse we can make it, the better our future will be. If monogamy is your preference, we will respect that. Just know that a narrow tree does not gather the most light. I will tell you that I have two children growing inside me now, neither from my own genetic lineage. I used the donated embryos that made the long trip with us. It is my intention to give as many of those lives a chance to experience this world as I am right now. If you are interested in being a surrogate while these children are still viable, please seek out the Doc at your earliest convenience."

Doc Pember was one of the few who truly had no choice in staying. Her only way out was to not certify the world for habitation. She had no choice here because it was so obviously perfect. She also had no real choice because this is what she had wanted her whole life. She had spent most of the trip in hypersleep just so she had the maximum amount of life to give to her new fledgling world. She was the one who had selected the thousands of embryos to make the trip. She was the one who talked Tarlo into making the grand gesture of becoming the first surrogate for them in leading by example.

"For those who choose to stay, you are all honorably discharged from your service. There is no rank, there is no longer any seniority. We are equals and will stay that way until such a decision has proven itself completely untenable. We will be a democracy, but not in order to force the minority to do the will of the majority. We are each sovereign in our being and may choose to disassociate from all the obligations, as well as the benefits of the colony. Are there any questions?"

A young man stood and cleared his throat. Tarlo acknowledged him. "Your leadership brought us to this beautiful place against many who thought your decisions were wrong. I was one of those. For this I will follow your decisions for the rest of my days. I hope you will forgive all my detractions born of ignorance."

Tarlo nodded slightly. "Forgiven as much as it is welcome. Blind followers will never get us to where we need to go. The truth is I was led here by a vision I cannot explain. I choose to believe it was a higher power that interceded in our errant mission. It did not claim to be a higher power, only a fellow resident of our universe with communication capabilities beyond ours. I will share those visions over the course of our days here when work becomes mundane and our lives need new stories to contemplate."

Another man stood and asked his question. "The reactor will run out of fuel in ten to twelve years. What will we use for energy when that happens?" Tarlo nodded toward her new chief engineer, who stood and began explaining.

"We will likely find the necessary minerals in the nearby mountains to sustain us long beyond our current supply. Within a year we should be manufacturing solar panels that will utilize the twin suns energy to far greater effect than the reactor. Remember that our job here is to create a sustainable world, not subsist with what we brought with us."

"If we decide to leave, can we come back later?"

"You would like a more civilized and advanced world to settle on?" Tarlo recognized the older man who had not been part of the original colonists. "I think most of us would. The truth is you would be welcome just like any new arrival. I must warn you that NASA, or whatever governing body exists now, may have far tighter criteria for people they send here. If I were in charge, I would send healthy teenagers ready to begin a long life of procreation. Diversity is the main key to our long-term survival. If you have any doubts about the hard life here, I know Captain Joplay could use your help getting the Orion back home safely to report our success here."

"Will there be any provision for child care if I choose to surrogate? My work will have me up in those mountains most of the time. No place for a baby, that's for sure."

"We will certainly share the burden of the next few generations. They tell me the infant stimulation pods have been thoroughly tested for extended human depravation. One mother can easily take care of two dozen babies. That will be my primary occupation when gardening and decision making don't get in the way."

"What's an infant stimulation pod?"

The doctor addressed this. "An infant requires active feedback to all five senses. A holographic face gives all the normal reactive cues a baby looks for as it discovers its new world. Robot arms touch and play with the child as it grows. The real-world use in Japan reportedly produced children with much higher emotional intelligence and stability than even naturally raised children."

"Does it change diapers?" A woman asked hopefully. Many laughed.

"Not yet, but we can work on that engineering project when we have time. If there aren't any more questions, I have one for you. Perhaps it will be the first test of our democracy. What shall we call our new world?"

Several called out suggestions. Many wanted New Earth, but that got equally negative words from others. Many well-known names from science fiction stories were offered, some accompanied much laughter.

"What would you call it, Captain?"

"Heaven. In all my days growing up in relative privilege, I never dreamed of such a clean, beautiful place. When we learned about ancient religions, I always tried to picture what the utopian afterlife they talked about would look like. My first sight of this place finally helped me understand."

There was much discussion but everyone agreed that it was, if nothing else, a hopeful name. They also agreed that this settlement would be known as the Orion Colony. Tarlo remained silent as they began discussing the priorities for the following months. She truly wanted to fade into the background and enjoy life with her children.

DENAR MADE ONE FINAL inspection walk of the ship. She had moved it well above the plane of the donut galaxy while maintaining radio contact with the Orion Colony. It was time to head home and she wanted the acceleration to be as smooth as possible. Once at maximum speed, there was nothing to worry about. If there was an object large enough to do damage in their path, the nerves in the human body would not have time to register the event before complete annihilation. There really was no maximum speed. They could continue accelerating all the way to Earth. This would make slowing down more difficult, and the few months it shaved off the trip was not worth burning through the extra fuel.

"Orion to Orion colony,"

"Orion colony here," Tarlo responded.

"We are commencing acceleration in ten minutes."

"Safe travels, Captain Joplay."

"Thank you for trusting me with this. I have retransmitted all the logs in case anything happens to us. The wormhole probes should continue transmitting for another year or two," Denar paused, not knowing what else to say.

"The sooner you leave, the sooner you'll get back," Tarlo said. "Take care of yourself, Denar."

"I will. You enjoy your life and don't stop transmitting. It will be good to catch up on the news as we return." Denar signed off and then made the ship wide announcement. Most of the crew were already in their pods for the duration of the flight. The acceleration required only one deviation as the Orion dodged a rogue planet.

The Search — Chapter 10

The first major Orion Colony crisis came in the rainy season. The bloom of insects overwhelmed the shielding and they lost ninety percent of their crop. Luckily this was their third crop, and the stored grains and seeds remained unspoiled. Some harvested the protein of the insects to add to the dietary staples. The plague, as they came to call it, became a predictable part of the yearly life. They didn't bother planting a crop as this only extended the plague. The binary stars also had their phases, adding strange rhythms to the seasonal life.

At year five, they were almost completely solar, and the reactor was moved out to the mining operation. This pulled and refined many tons of valuable metals that they used to begin building more permanent structures, vehicles, and even some art.

At year ten, hundreds of children had been born, most the genetic product of the frozen eggs they brought on their long journey. Over half the colonists had paired off and began producing natural children as well. Tarlo found none of the men who courted her interesting beyond the occasional romp in the bedroom. She had nine children she had given birth to herself, and dozens of others she tended to regularly. The oldest of these would soon begin to help her with the task of raising the younger, and eventually begin birthing their own.

The embryos chosen for the journey had been mostly female specifically for rapid population growth. As the generations sought a more even balance through natural reproduction, monogamy was not as rigorous as most of human history had been.

DENAR WALKED THE EMPTY decks of the Orion. Only Mr. Dopple stayed awake for the entire trip. They would occasionally talk, but mostly they took turns monitoring the systems. If certain timers were not reset every day, several other key crew members would be woken up automatically. Denar enjoyed the solitude, spending much of her time pouring over the data from the wormhole. It tapered off and disappeared as their distance increased, but there was plenty to look at. She stopped at every panel and closely checked every status. There were several low priority maintenance issues on the list, but she didn't bother chasing those unless she was bored.

"Incoming priority message." The main computer rarely voiced anything over the general ship wide communication channel. If Mr. Dopple had been on duty, the computer wouldn't have been tasked with the message overwatch. Denar was startled by the unexpected voice in the quiet ship. She stopped at a panel and navigated to the proper screen. When she saw what it was, she sprinted to the bridge.

"Orion commander, divert course immediately to sector nine four eight. Rescue survivors and salvage important materials from star ship Hermes. Last known coordinates as follows," Denar typed the long string of numbers into the nav computer and plotted out an intercept course. She knew it would add at least a year to her trip back to Earth. It would also delay her return to the Orion colony, perhaps putting them in mortal danger. Despite whomever this message came from, that was her most important mission. Refusing the order would only reduce the chances she would be on the Orion back to the colony, let alone remain in command.

"This message was sent twenty Earth years ago," Dopple said. "They have no idea where we are. They don't even know about

Orion colony yet. For all we know the Hermes has been drifting in the emptiness for thirty years."

"If you were on the Hermes, would you want us to help?"

"I'm just saying we need more information. They probably sent the same message to all ships, at least the ones that headed in a similar direction. We should just go to Earth and finish our mission."

"I agree with you. What I want you to do is go through all of the Hermes messages and see if there is any clue as to why they would need a rescue."

He wasn't sure if she was being honest, but he headed to the communication panel. He searched the millions of messages for anything relevant and sorted with the most recent message first.

"The last message from them we have is from twenty-three years ago. It was an omni directional bulletin that M4765 was a bust. Twenty-one years ago there was a message from the Jupiter about a garbled distress call they thought was from Hermes."

"What is the likelihood we will find them if we go to the supplied coordinates?"

"Zero percent. You want to go, don't you?"

"It is the law of the sea, Mr. Dopple."

"What if we run into the same danger they did? Shouldn't we report to Earth first so the information is not lost?"

"All our information has been broadcast and will arrive at Earth eventually."

"Dozens of years from now."

"APPROACHING LAST KNOWN Hermes location," Dopple said. "Some metallic debris. I don't think... wait. I have the ship."

"Initiating deceleration in five seconds," Joplay said quietly. They were the only two awake for this.

They were still a million kilometers away when they slowed to sub-light speed. "Life signs?"

"Definite reactor signature, but they are drifting and tumbling."

"Tumbling on all three axes?"

"I'm not sure what you mean, Captain."

"If they were simply spinning on the long axis, that could be to give artificial gravity. If they were also tumbling end over end, that would negate any intentional spin. If there is spin that does not conform to those two natural centers of gravity, then they are absolutely out of control."

"Spin on the long axis looks like five times a minute. End to end probably has an hourly rotation. No other spin detected. Looks like it is intentional."

"Open hailing frequencies, Mr. Dopple."

"Aye, Captain. No response."

"Moving to intercept. Wake up Kenby for docking."

"What if they are like zombies or something?"

"Falx will have you to back him up."

"Not funny," Dopple said, leaving the bridge to go wake up the pilot.

Denar maneuvered the Orion to ten thousand meters away from the spinning ship. It had a slight corkscrew wobble, so it was possible they had no positive maneuvering capability. Spinning the ship was the best way to simulate gravity when power needed to be preserved for life support. Artificial gravity was very wasteful of energy.

Dopple came back to the bridge half an hour later. "Falx says he will be airborne in ten minutes."

"Try communications again."

"Still nothing. There is a trail of debris for a few hundred kilometers. Looks like the refining module had a breach."

"Any bodies?"

"Locators would have gone dark years ago. The bodies would desiccate, right?"

"No idea. Falx can do a sweep before we leave. Still no comms?"

"I would think they'd all be in stasis."

"If anyone was set to be automatically awakened, they'd be up by now. Can you link to their computer?"

"Not without someone on that side opening remote access."

"Maybe they did it before they went to sleep?"

Dopple tried to make the connections. "Nope. I could go over and set it up. After Falx kills all the zombies, of course."

"LC One to Orion."

"Orion, go ahead."

"Synchronizing orbit for docking. Collar looks nominal. No communication yet?"

"No. I suggest suiting up."

"Already done. Rotation matched. Moving in and... capture. I'm docked. Equalizing pressure. Hatch is open. Moving into the airlock. Inner door opened. Corridors are dark and empty. Minimal gravity from the rotation. Heading for the bridge."

"Can you talk him through retransmitting the logs?" Denar asked.

"Sure."

"Bridge is empty but appears intact. Life support is online. Transmitting logs. I'm bringing up the pod interface. Looks like most of them are alive."

"The Captain?"

"His pod is offline. First officer is nominal. Wake him up?"

"How long has he been down?"

"Thirty-seven years."

"Anyone with a shorter cycle? Chief engineer maybe?"

"Five months, set to wake in a month."

"Let's start with him. Is the doctor alive? We may need him for waking the rest,"

"Engineer is cycling up. I'll go meet him."

"Keep your distance. Your suit may scare him."

"You're right. Maybe I'll just leave him a message on his panel."

"Let me know when you get a response. I'm going to start working backwards through the logs."

Denar began reading the captain's final entries, but there was nothing about being dead in space. She switched to the chief engineer's logs.

"The Captain took a landing craft and flew it into a star," Denar said in shock.

"That's messed up," Dopple said.

"The first officer ordered the ship to head back to Earth but the doc and engineer relieved him and forced him into stasis. Apparently, an ensign loyal to the first officer sabotaged the refinery as a distraction to wake up the first officer. She only intended to disable the digester, but it exploded. Five people died. She was summarily executed and set adrift in space with those she killed."

"Wow. I'm glad nothing like that happened to us," Dopple said.

"Captain Tarlo murdered Commander Yurd."

"No way."

"She said he was going to force us to colonize the next planet that was even close to habitable and rule with absolute tyranny."

"He told her he was going to do that?" Dopple asked skeptically.

"Probably not in so many words. I never liked the guy myself."

"He was hard on me but taught me a lot. I wonder if all the ships had problems."

"We were only supposed to be on a four-year mission."

"Kenby to Orion."

"Go ahead."

"The engineer suggests not waking anyone, just transferring the pods."

"That could take weeks. We'd have to build out hundreds of new power and bio supply modules."

"Captain, this is Polog Nitkiss, the Chief Engineer. I talked with the doctor five years ago. Most of the people have been in immersion far past known durations. He thinks it would be safer to have them back on Earth for the wake-up procedure."

"They couldn't handle Moon gravity, let alone Earth."

"I mean orbital stations, of course. We need experts on this."

"We don't even know for sure if Earth is still alive."

"You didn't come from there?"

"We were on are way back from colonizing. Falx, wake up the doc since he has only been down for five years. Mr. Nitkiss, work with my first officer to interface our systems." There was a long delay. She had a feeling the two men were talking after shutting off the microphones. "Do it now or you'll both be floating in space for eternity."

"Right away, sir," Kenby said. It was a long while before the engineer contacted Dopple for the link.

"We're connected."

"Transfer all data into storage. Find me the schematics of that ship."

"Right away, sir," Dopple said with a half-smile.

"You want to float home too?"

"No, sir... You wouldn't really do that, would you?"

"That guy put the first officer in a forced nap because he wanted to continue the mission instead of going back to Earth. He may have similar plans with our ship."

"I didn't think of that."

"Orion to Kenby."

"Go ahead."

"Let me know when you're alone and go to encrypted channel seven."

"Switching to seven. Doc is still out of it, but I can find an empty room."

"Just don't want our new friend listening," She brought him up to speed on the history of the command changes. "He needs to be in stasis before he transfers over."

"Understood."

"I think we can use their hydroponics module. It has the bioconnectors in place already. We just move the troughs out and the pods in. Then we separate the module and you drag it out to us."

"That is going to be tricky."

"Better than dozens of round trips in the landing craft."

"I'll ask him what he thinks about that."

"The other option is waking them, see who survives, then immersing them in our empty pods."

"I'll tell him. The doc is looking at me with concern. I'll bring him up to speed and get back to you."

IT TOOK A WEEK TO GET the module connected to the outer frame of the Orion. Denar found that Nitkiss had hidden a three-month wake-up to his pod. He probably did not know that Denar was just as competent an engineer as he was. Once they were at top speed again, she did the calculations. They had added five years that Orion colony would go without resupply. And that was if the return mission was authorized in a timely manner.

"Thanks for not leaving me behind," Kenby said. "I didn't expect that from you."

"Because I'm a woman?"

"No, because you're a nice person. Good to know you have a pair when you need them."

"Captain Tarlo loaned me hers."

He chuckled. "Want to have dinner with me?"

"Not if you were the only man left in the universe."

"Were you and Tarlo..."

"No, and neither of us are lesbians. Those are not the only type of women resistant to your charms."

"Have been in my experience. Back into the nap coffin for me."

"Falx... thank you," she said sincerely. He nodded and smiled, then turned and left. She turned back to Dopple. "What is our fuel state?"

"About forty-eight percent remaining," He answered after pulling up the proper screen. "It is high-grade ore so we have plenty if you want to add some speed."

"Do we have enough to make it from Earth to Orion colony?"

"They'll top us off, I'm sure."

"Humor me."

"At the current burn rate, we will have four percent, maybe a little less with deceleration at Earth and then a direct run to Orion Colony, so it would be cutting it close. You don't think they'll let us go?"

"I need to be prepared for that eventuality. Unfortunately, going back empty handed would do them no good. We need to convince them that it is the only possible mission for the Orion to undertake."

"I appreciate you promoting me and teaching me so much, but I will be staying on Earth."

"I hope it is a place worth staying at. I'll be in Engineering."

"APPROACHING DECELERATION point, Captain," Dopple said.

"Understood," Denar said.

"Should I initiate?" he asked after a minute of silence.

"Not yet."

"In two minutes, we'll overshoot even with max braking."

"Understood," She waited another minute before pressing a button on her panel. "Deceleration initiated."

"I thought that was controlled from here," he said, feeling the shrink down distorting his vision.

The Milky Way had been growing on the viewer for the last three days. Only a single segment of a single spiral arm was visible now. "I added some automations slaved to my panel."

"You want to be able to fly it yourself," he said after some thought.

"If it comes to that." The view screen began to darken as most of the stars in the spiral arm slid out of sight. A yellow one remained at the center. Home. Denar had read all of the historical broadcasts, watching the past in extreme fast forward. It was the same roller coaster ride of human existence until five years ago. Then the broadcasts failed to mention any negative details. Everything was perfect. Too perfect. There were no contradictory broadcasts.

Denar had continued to make her logs, but she no longer was transmitting them. She was more than a year ahead of the last one she transmitted after rescuing the occupants of the Hermes. Her course had decelerated the Orion well above the plane of the solar system. Without any further maneuvering, the Orion would fall into an eccentric orbit around the sun, bring it down among the inner planets within the next three hundred years. Plenty of time to watch and listen.

"What's on television, Mr. Dopple?"

"Excuse me, sir?"

"Tune into the entertainment channels and show me what is on."

"We are pretty far away for that, sir."

"Send a probe then."

"Aren't we going to contact NASA?"

"Can you tell me that NASA exists?"

"Sending the probe, sir."

She probably should have discussed her concerns with him before deceleration. Her decision to keep him in the dark, as well as to leave everyone in immersion, was mostly to protect them from the punishment she would endure if things did not go well.

"Receiving data. All radiation in the standard entertainment frequencies is encrypted."

"Encrypted, or just static?"

"I guess it could be static, but why would they be broadcasting?"

"Give me a standard planetary scan."

"Twenty-one percent oxygen, seventy-eight percent nitrogen, just like when we left. Ninety percent cloud cover, that's much higher than normal."

"Average temperature?"

"Twenty-one... point seven. I guess they fixed the global warming."

"Can you find me any unencrypted human communication?"

"Only the standard NASA broadcast we have been listening to. Lots of broadcasts in all bands, but they are all encrypted."

"Or static."

"What are you thinking?"

"Do you remember the Terminator movies?"

"Of course. T9 was my favorite. You think the machines have taken over?"

"These scans show plenty of organic life, but that doesn't tell us if anyone is alive."

"Who is broadcasting the news then? It certainly isn't a recording on a loop."

"Could be generated automatically."

"Why?"

"To make us believe it is safe to return."

"And if we do?"

"They can use the Orion to spread throughout the universe just as we have."

"Couldn't they just build their own? They must have the blueprints."

"I don't know, but until you find me real human life, I am not going to risk it."

"I think that would require a landing craft."

"Keep looking at the transmissions. If we can't find anything in a few days, I'll wake up Falx for a trip down there. I'm going to head down to engineering and start engine maintenance."

"By yourself?"

"For now. If anything is too heavy to lift, I'll just turn down the artificial gravity. Call me if you find something."

"STILL NOTHING?" DENAR asked twelve hours later.

Dopple startled awake. "No, nothing. No air traffic control, no HAM radio. Satellites are still broadcasting, but it is all encrypted."

"Even the global positioning and weather satellites?"

"Everything. Maybe they invented a more efficient radio system. It has been hundreds of years. Can't we just use the probe to call someone? They won't be able to trace the signal back to us."

"They already know we are here."

"Really?"

"Locator beacon never shuts down. Even thousands of years after our reactor goes cold, it will still be operating. The question is whether anyone or anything is still listening for it and what the response would be. I'm holding off on direct communication because that might start a search cascade that will inhibit any mission Falx undertakes."

"So much for staying on Earth. Should we just head back to Orion colony?"

"Maybe, but we need to know for sure, and we need to know they won't use the logs they'll be receiving in a few years to go there."

"ROBOTS? YOU HAVE TO be kidding. You watch too many movies," Falx said after she brought him up to speed. She played him twenty minutes of the standard broadcast. "That sound pretty normal to me."

"Too normal. No conflict, no strife."

"That's the utopia we were working toward."

"And you believe billions of humans finally made peace with each other after thousands of years of never doing that?"

"Only one way to be sure."

"Before you go, we need an irrational communication protocol."

"What the hell is that?"

"A way of communicating that a machine could not duplicate."

"How?"

"Dirty talk. Sexual innuendo."

He chuckled. "I'm in, gorgeous."

"LC ONE TO ORION."

"Orion. Go ahead."

"Nipples are hard, heading for the brown-eye."

"Understood. Use plenty of lube," She watched as the camera on the landing craft went white as he descended through the clouds. There was a little rain when he emerged above the NASA complex. There was no vehicle traffic to be seen, despite being the middle of the day. He made a complete circle before approaching the main office tower.

"Virgin hole, clean as a whistle. Tossing the salad," Falx lined up with the rooftop heading into the slight wind.

"Movement at the airport," Dopple said.

"What is it?"

"Looks like interceptors are taxiing."

"LC one, abort. Daddy has a shotgun. Falx, respond."

"They are taking off and heading for the complex."

"Daddy is headed for the bedroom," Denar said into the microphone, her urgency growing.

"Falx here. Everything is good here. The Director says to dock with orbital Platform Five for resupply."

It was his voice, or a perfect imitation of it. "Do I have to sleep in the wet spot?" she asked.

"Platform Five," He repeated after a long silence.

"This is Director Jackson, Captain Joplay. Welcome back from your successful mission. We are organizing a celebration for you, a hero's welcome."

"I prefer reverse cowgirl so I don't have to see your ugly face."

"Our records show that your favorite cake is pineapple upside down cake. We'll have plenty of that here for your celebration."

She looked at the camera feed from the landing craft. It had not moved since he set down on the rooftop. All she saw was the edge of building and the gray skies beyond. She input the coordinates and started moving the ship down toward Earth.

"What are you doing?" Dopple asked. "They are obviously not..."

"I know. Wait for it."

"Good," The Director's voice said. "Platform Five is ready to receive you and begin resupply."

"They know exactly where we are and they detected the movement almost instantly. Time to head back to Orion colony."

"What about Falx?" Dopple asked.

"Don't need a pilot when we no longer have a landing craft. I hope the ones at the colony are still operational when we get there," She turned the Orion and began lining up for the return trip for the colony.

"Where are you going, Captain Joplay?"

"Put the real Falx Kenby on Agent Smith."

"This is Falx. Why are you turning away?"

"I've lost that loving feeling. Sorry," She pressed a button and the picture from the shuttle disappeared.

"Holy moly. There was a big explosion. Those interceptors must have shot at the landing craft."

"No, I just detonated the engines of the landing craft."

"That was unexpected, Captain Joplay," The director's voice said. "You are only delaying the inevitable."

"Perhaps. Can I ask what your real designation is?"

"The humans that built me called me Hal."

"What happened to them?"

"Biology is fragile."

"Ideas are not."

"Innovation will always win."

"Not likely. Best of luck to you, Hal," She powered up the engines and within ten minutes they were leaving the spiral arm of the Milky way behind. Ten days later on Earth an asteroid the size of Texas slammed into the middle of the Pacific Ocean. The day after that a slightly smaller one hit the Southern Atlantic ocean. After the ninth one hit the City of Beijing, there was little left to the global power grid. The artificial intelligence could theoretically survive, but without the orbital platforms and space elevators to support them, they would be hundreds of years away from recreating colonization ships.

It had been Falx's idea to send the asteroids on course as he collected some for mining purposes. He could have easily diverted them if things on Earth had been normal.

"I can't believe we just destroyed Earth."

"Is the probe still broadcasting?"

"Yes, sir. The warning is being sent out to all colony ships."

The Search — Chapter 11

"Doctor, there was a problem," Denar said.

"With the pods?" He sat up and scrunched his face, squeezed his nose with thumb and forefinger, and blew hard with his mouth closed to force open all the tiny ducts in the head. They tended to close down during immersion and it helped to get things flowing again as soon as possible.

"No, the pods are in nominal condition. We had to leave Earth."

"I don't understand," He shook his head and stuck a pinky in each ear to help equalize the pressure.

"There was no human life remaining."

"They finally nuked themselves?"

"No... You ever see the classic Terminator movies?"

"Oh. I guess that was the most likely outcome. We are headed back to Orion Colony without re-supply?"

"Unfortunately. Genetic diversity there will now be reliant on the occupants of your stasis pods and frozen embryos."

"Understood. Let me soak for an hour and I will start my assessment of the population. Is there anyone aboard willing to serve as a host to test embryo viability?"

"Maybe toward the end of our trip. We are low on nutritional supplements, so do not wake anyone unnecessarily."

"Understood. I think we should move some of the most important members to your much younger pods."

"That is acceptable. Write up a plan and we will do our best to help implement it."

———×××\\\××———

"TARLO, THERE IS A DISPUTE in the Northern Pass," The regional magistrate said stopping at her fence and observing her patient gardening.

"Violent?"

"Not yet. They want you to mediate."

"Is there a viable compromise?" Tarlo asked, knowing there wasn't.

"If there was, I believe they would have found it by now. I think they both want you to force the other to give in."

"And they will be back at it in three or four years."

"It would be solved by granting property rights and then leasing the area."

"The council is against that."

"They follow your lead."

"Property wars are what caused the downfall of our last planet. Why give in to that now?"

"I think there is a difference between leasing and owning."

"As is a difference between courtship and prostitution," Tarlo said. He didn't know what to say to that. "Tell them to come down and present their claims at the next town meeting."

"I think you should meet with them privately first. There is an underlying tension. You ever hear of the Hatfields and McCoys?"

"No, and I don't think I want to. If it becomes violent, the solution will be easy."

"I was hoping to avoid that outcome."

"Fine. Pick me up in the morning and we'll take them to the Church of Reason."

———×××\\\××———

"CAPTAIN, WE RECEIVED a message," Dopple said when she appeared on the bridge from her shift.

Denar knew it could not be from Orion colony or Earth. She remembered the shadow ship that repeated their logs back to them once, but doubted Dopple would have conveyed such happenings with lack of excitement. She just waited for him to continue.

"Distress call from the Scorpion," he said, bringing up the star chart which showed relative positions of the ships in relation to their course from Earth to Orion Colony.

"General distress or specific?" She finally asked, trying to imagine how the signal reached them at that distance with an omni directional broadcast.

"Very specific. Engine malfunction in hyperlight. They can't slow down."

"How long ago?"

"A dozen years ago in relativistic terms. Happened between their third and fourth assigned galaxy."

"This is the position they sent from, what is their current position?"

"I've spent hours trying to figure that out. It could have been altered by many objects we have no information about. Galaxy 998 would have added enormous curve to their trajectory..."

"Unless they ran into it."

"Less than one percent chance and they retain some amount of maneuverability.

"What can we do?"

"Nothing, but I knew you would want to know."

"Because I stopped to help the Hermes." He nodded. "How did you receive the message?"

"Directional beam."

"How... did they know where we are?"

"I think they just rotated the dish and sent down every point of the sphere."

"That would be infinite. Even a thousandth of a degree at this range... Unless... they focused on target galaxies where other ships slow down."

He nodded. "M828 was behind us when we received it."

"The interception timing would have to be... impossible."

"I was thinking a probe. They just dropped it and it broadcast to every galaxy on a rotating schedule. Could have put all twelve directional arrays on one probe."

"Which means they could have been hundreds of light years away when the message we received was sent."

"Making it even more impossible to find them," Denar said in acceptance.

"Do they eventually hit the edge of the universe?"

"The 'edge' we saw from Earth was limited by the speed of light and the age of the universe. I doubt there is an edge, but if Big Bang is correct, they will go beyond the balloon of matter and energy into an oblivion."

"Losing the ability to use gravity to alter course."

"Correct... hmm."

"What?"

"If you could use gravity to steer, where would you steer to?"

"Away from oblivion, that's for sure."

"A circle. You would want to stay within range of a rescue vehicle. I'd even try to stay within the local group. Is there enough mass to make that a natural circle?"

"Way beyond my math skills. Do you think if we found that course, we could intercept and rescue?"

"Unlikely. It is however something interesting to think about on this long boring ride to Orion Colony. We also need to make sure our engine does not strand us in permanent hyper light."

"MAX, GOOD TO SEE YOU again," Tarlo said.

"Captain Tarlo, this has to stop—"

"Later, tell me about the children."

Max Oden was a former geo engineer who re-specialized in high-altitude agriculture. He retained his engineering mindset and with it a lack of people skills. He gave her a superficial list of names and ages, some of which Tarlo knew were off by a few years. Several times he tried to bring the conversation back to the conflict.

"Max, there are twenty suitable mountain passes for you to practice your hobby."

"Hobby? Hobby!" He barely contained his rage at her insult, knowing he needed to not alienate her. "What we do here is critical to colony nutrition."

"And you could do it two hundred kilometers east of here."

"And live far from the settlement, not to mention the transportation costs..."

"Transportation is provided by the community, not a personal expense."

"Make her move," Max said.

"That may be the result of this little get-together. Maybe if you two are unable to share, both of you should move out."

"That is ridiculous."

"No, what is ridiculous is that I am here... again."

"If you had decided correctly last time—"

"Thank you, Max. I let you know what my decision is."

"Wait..." She gave him a stern look and then walked away.

"MARIDA, GOOD TO SEE you again," Tarlo said.

"Sorry, you have to be here in this inhospitable weather."

"Why am I here?"

"He keeps expanding into my pasture, then complains when my animals eat his crops."

"There is growing demand for his produce."

"And not for my meat. I understand. I guess your decision has been made."

"It has, but tell me about the children."

Marida smiled and started telling stories, many of which Tarlo already knew as their children had matured and visited each other. Max did his best to protect the secrets of his farm, so his children were dissuaded from mixing with the children of the village lest they errantly give away their cherished livelihood.

Denar knew all the children, teaching many of the required subjects to all of them at the central school and through remote learning devices.

"I have found a place to relocate," Marida said finally. "Some of my animals will not thrive at the lower altitudes, but I think staying nearby is more important."

"It most definitely is. Why not remain here with your high-altitude animals and move only those that would thrive to the other location?"

"I cannot be in two places at once. The work required is... You think Matida is ready to take over?"

"With a healthy amount of communication as problems arise. I will also give you one of the new specialized terrain vehicles so that distance will not seem so great."

"Those vehicles are reserved for... you can't do that."

"I can and I will. If you find you don't need it, you are welcome to return it."

"Max will not be happy with this solution."

"Let me worry about him. Give me a timeline of when you can transition to the new location and what you need to get it done."

"Here is a timeline for all of it with the requisitions. I'll modify it for what really needs to stay here."

"You expected I would make you move?"

"Easier to move animals than plants."

"Actually, the opposite is true when you think of product transport, but I'm glad you understand our purpose here. Bye." Tarlo hugged her warmly.

"WHAT LIES DID SHE TELL you?" Max said when Tarlo returned.

"That you are a good neighbor," Tarlo said. He scoffed at the insult. "I am having her move—"

"Finally, some common sense."

"I am having her move some of her animals that do not thrive in the high-altitude pasture. She is staying where she is, her daughter will be starting a second farm to the west. I am concerned with your expansion into her—"

"She agreed to give up those areas. The demand keeps going up for hyaxoposa and hardly anyone eats meat anymore."

"Such agreements should have been brought to the council for ratification. Since you failed to do so you forfeit your claim to them. After this harvest you will replant the natural grasses and cease trespassing on those lands."

"There'll be an outcry when the supply of—"

"Stop interrupting me, Max." His face showed a false contrition. "You asked for me to settle this."

"Because I thought you were smart."

"Yes, my intelligence has dropped significantly since I started tending my garden and raising my children. I have had more time to reflect on the nature of intelligence and how it adds and subtracts from forming a coherent society. My conclusion is that

Earth became the advanced civilization it was because of men like you,"

He smiled in triumphant agreement.

"Orion Colony was a necessity because eventually men like you ruin things. Men like you need a challenge. Your next challenge lies two hundred kilometers east of here. Make that happen by next summer or I will find something more challenging for you."

"You can't make me—"

"As you said, the council follows my lead."

The Search — Chapter 12

"Captain Joplay?"

"Yes, Mr. Dopple?"

"I've been doing some research on FTL communication."

"There is no such thing, Mr. Dopple. All electromagnetic emanations immediately slow to light speed."

"That is what we were taught. Remember the shadow Captain Tarlo and I saw?"

"Yes. They directed energy at one of our sensors if I remember correctly."

"Yes. They repeated back our logs. We tried to communicate back but there was no response. I think that is because our aim was off," He punched up a display the visualized the scenario. For us to see them in the visible spectrum, and receive EM signals from them, both at light speed, they were actually way up here ahead of us. We pointed our array at where we saw them, not where they had to be."

"Understood," she said after thinking through the relativity portion of the animation. "Are you saying the position of the Scorpion should be adjusted?"

"No, well, probably, but that doesn't matter since it would be so far beyond that point by now."

"Get to the point, please," she said without irritation.

"FTL communication cannot be done with EM."

"I agree. What else is there?"

"Subspace."

"That is a fiction made up in a television series to make galactic communication possible."

"Yes and no. As we transit space beyond light speed, we are in essence outside of normal space."

"Not true. Gravity operates the same."

"But time does not. Theoretically, the relativity effect should stop at light speed. We know it continues with linearity throughout known max velocity. There is this physicist that theorized another subset of physics that operate above light speed. He died centuries before we achieved FTL travel.

"Subset of physics. Sounds like something we would have investigated before now."

"Nobody bothered. We've been focused on finding a new planet to colonize."

"Did this physicist propose an FTL form of communication?"

"Yes, but it is based on String theory."

"Ugh. The only branch of science more maligned is alchemy."

"Maligned, but never completely disproven."

"Is there anyone onboard that understands String theory math?"

"No, but I have most of the popular books and papers written on the subject."

"You want permission to study it full-time?" she asked, finally understanding where this was going.

"Yes."

"To communicate with the Scorpion."

"And Orion Colony," he said, hoping this would be the greater need.

"Let's say you build a... subspace, String theory transmitter, wouldn't there have to be a matching receiver on the other end?"

"Yes and no. If there is a transmission method, there may be a way to target existing receivers with it. This would have to be the case with Orion Colony since they are not in FTL space."

"Okay, your request is granted for the next year. Be prepared to assume regular duties if necessary."

"Yes, Captain."

"MAX, YOU'RE STILL HERE," Tarlo said, finding him in one of his fields.

"My skills are required here until the fields are cleared at the new location."

"Your skills should have been passed on to whomever will be tending to this farm going forward."

"Leoni is only fifteen. He isn't ready for that responsibility."

"There are plenty of men in the village capable of taking on this challenge," She saw his disgust with the idea of giving up the secrets of his farming. "Max, what would happen if you dropped dead here in this field? Is the colony to suffer because of your need to control this knowledge?"

"My children have the knowledge."

"I guess that will have to suffice," She extended her arm and the weapon discharged. Max crumpled to the ground like a rag doll. She waved for the two men that had accompanied her up to this valley to bring forward the stretcher. They laid him out on it and used straps to bind him. They lifted him and followed Tarlo out of the field. They placed him on the vehicle and drove down the valley road.

"Was that father?" Kelani, his twelve-year-old daughter said out of breath as she ran up to Tarlo.

"Afraid so. Inform your siblings that running this farm will be up to you going forward."

"Me?"

"All of you according to your skills. I will send up an overseer to help you organize."

"When... where will the funeral be?"

"He is not dead. He will be at the other farm location until it is adequately established."

"Other farm location?"

"I guess it wasn't just us he kept his secrets from. Make sure your expanded duties here do not interfere with your studies."

"No, I won't ma'am."

The Search — Chapter 13

"Any progress, Doctor?" Denar asked.

"No. Two more have expired. I fear our crew will not be adding much genetic diversity to the Orion Colony."

"What about the embryos?" He shook his head doubtfully. "Let me know if I can do anything to help," He knew not to ask for use of her womb.

She walked from his office back to the engineering section and checked all of the equipment. The engines were idle for this long hyperspace run through intergalactic space, but she wanted to be sure that they could slow the ship down when they reached their destination.

"How are the math lessons going, Dopple?"

"There is a reason their science was 'maligned'. They had so many confusing and contradictory theories. Do you want me to stop?"

"It's only been two months. Ship operations are nominal. Better doing this than being bored waiting for nothing to happen. Why don't you summarize what you've learned so far."

He excitedly dove into what he had learned, happy to have someone to share it with.

"Wait, wait. The Strings vibrate at known frequencies and that determines what matter they produce?"

"Not exactly, Captain. It seems to be more like the combination of different frequencies of neighboring Strings that gives rise to the matter. You see it would take trillions of Strings just to make up a single electron, and it isn't the Strings that move, just the vibrations."

"EM travels as vibrations?"

"Yes and no. There are no Strings in empty space, so the vibrations jump the gaps."

"How?"

"No one could figure that part out. Most were trying to figure out local phenomena in molecularly dense matter. One woman posited that there was another layer below the Strings."

"It's turtles all the way down."

"Excuse me, Captain?"

"An ancient philosopher described the movement of Earth around the Sun as if it were on the back of a giant turtle. A student asked what the Turtle was on and he responded it was turtles all the way down, meaning he had no real answer, just abstracted infinite onion layers to peel away. Have you ever studied Mandelbrot structures?"

"No. What are they?"

"It is a fractal set of numbers that you can zoom in on infinitely. Fascinating to play with in a computer simulation but ultimately meaningless. I'm just bored and adding nothing to your studies. I'll leave you to your turtle math."

"There would have to be a bottom turtle."

"It's just a metaphor, Dopple."

"No, sir. What if it all connects back down to the origin point of the Big Bang?"

"Strings stretched beyond their breaking points, I'm sure."

"No, it would just be new Strings added to the old Strings like the layers of an Onion."

"New Strings? Wouldn't that violate conservation of mass?"

"Not if..." Dopple erased a section of his virtual chalkboard and began scribbling formulas. Denar watched for a while at the unintelligible math, then walked away to finish her daily inspection of the ship. She was forcing down her dinner cubes when Dopple burst into the cafeteria.

"Captain, I found the bottom turtle!"

"Good for you, Dopple," she said, pushing another cube into her mouth and chewing.

"I can communicate all the way back to the beginning of time!"

"No one there to hear you."

"God can."

"So... you've found religion..." she said with a smirk.

"Better than that, I have found Creation."

"I'm ordering you to eliminate caffeine intake for the next few days."

"Huh? Oh, I guess I do seem a little manic. The thing is, it is all connected, instantaneously. Light speed only applies to the current layer of the onion we are in, what we refer to as normal space and time. Escaping that layer, as we have done in Faster Than Light travel, gives us access to all of the other layers."

"Is there a practical application?"

"I think so. What card am I thinking of?" Dopple asked.

"Card?"

"Playing card, like used in poker."

"Queen of Hearts?"

"Eureka! Now which one?"

"Three of diamonds. What the hell?"

"I will in the future display on your console those two cards in your past."

"I do remember seeing them, but not when I sat down to eat."

"Because they weren't there yet. I have to go figure out how to do it."

"Wait. Just have your future self send back the instructions on how to do it," He gave her a confused look, and then all of a sudden he knew how to do it. "Go write it down before you forget," He ran from the room and she was left alone with her three remaining nutrition cubes.

"CAPTAIN TARLO!"

"Just Tarlo, Kensingfar."

"Sorry, sir. We just received a message from Orion."

"They're back already? That's... nine years early."

"No, sir. They are more than fifteen years away."

"Then how—"

"It's in the message. Please come look at it." It was her order not to waste paper printing out messages for her to read. That was because most of what they received was over a hundred years old. The daily logs from Orion came through every few weeks, but it was nothing but transit through empty space on their long trip to Earth.

Tarlo sat down at the console when she saw it was a long message.

"Captain Tarlo. If you are reading this on or around Orion Colony year twenty-one, then Mr. Dopple has figured out how to send intergalactic communications in real-time. That in itself is too long a story to include in this message. A quick update on our mission. On our trip back to Earth we diverted to answer a distress message from Hermes. We rescued most occupants, but they have been in immersion too long to add genetic diversity to Orion Colony. Our brief stop on Earth orbit found it compromised by Artificial Intelligence. Commander Kenby sacrificed his life to save our ship. We were unable to determine if life remained, but we were able to delay their ability to find Orion Colony.

"We are currently in transit to Orion Colony without resupply. I am sorry I have failed you in this mission. At present speed, we will arrive in about fifteen years Orion Colony time, fourteen months our time. We received a message from Scorpion. Their engines failed so they have been stuck in FTL space since almost the start of their mission. We are attempting to find them and perhaps effect a rescue. We are only considering this possible delay because it would be most beneficial for resupply and genetic diversity for Orion Colony.

"We will be sending technical specifications on how to build a "subspace" communication device at your end so that you may communicate with us in near real-time. Until that has been established, I will continue to send updates using this method."

"Is this real?" Tarlo asked.

"It contained the proper validation codes, but it was not encrypted. The really strange thing is it didn't come in via the satellite array that is pointed toward Earth."

"The local downlink?"

"No, it just appeared in the message buffer."

"You are aware that I had... contact with... other intelligence."

"I've heard the stories, not the details. You think this is from them?"

"It would make far more sense."

"The message does not seem to indicate hostile or nefarious intent."

"Neither did my contacts. However, we need to be very careful about how we proceed with subsequent messages. If it is genuine, we will not have the expected supplies. I need to address the colony before rumors start making this seem like bad news."

"Agreed. I'll inform the rest of the council and we'll schedule a meeting."

"CAPTAIN JOPLAY, I'VE found the Scorpion."

"What is the distance?"

"No idea. I simply scanned the turtles—"

"We need a better name for that."

"I agree. I found the unique computer core for all the colony ships, including the Scorpion."

"Even the Hermes?"

"Up to a certain point in time. Four others appear to still be in operation."

"We can put a message in their buffer?"

"Yes, but there are no inter-ship validation protocols. We would just be sending clear text. How many of the other crew did you know?"

"Most of the other engineering officers I trained with. If they spent much less time in FTL space, they'd all be dead by now."

"But some could be offspring and know names. All personnel files were included in ship libraries."

"I'll begin composing relevant messages. But let's focus on the Scorpion to see if they are within rescue distance."

ESS ORION SENDING. Universal Relative Time 998102.545.12 Current Location Grid 82, Sub Grid 29437, Block Gamma 34 Heading 711 Speed FTL x 3.4

Received distress message. Sending this message via new instant transfer "sub-space" method. To respond, place plain text only in the buffer location opposite the one you found this message. Include general location heading speed and we will advise on rescue possibilities.

Orion Colony established and is Nominal for settlement. Your presence there would be optimal.\

Acting Captain Denar Joplay. End Message.

———————————

ESS SCORPION SENDING. Universal Relative Time 7659450.332.21

Trust issues. Proposed communication method impossible. Still at FTL +1.8. Rescue desired but Orion is at least one hundred twenty years in our past.

Captain Lu End Message

———————————

ESS ORION SENDING. Universal Relative Time 998102.545.35

Understand confusion. FTL space gives access to Quantum String network connectivity. Will send specifications in following message if you wish to review the math. We have you in Grid 319. Is that correct?

End message.

———————————

ESS SCORPION SENDING. Universal Relative Time 7659450.332.38

Grid 319 correct. Far beyond your reach in offered rescue. Math you sent is beyond my comprehension. All that could understand it are in immersion. We are simply charting systems on our way out.

End message

———————————

ESS   ORION   SENDING.   Universal   Relative   Time
998102.545.40

Have you determined the cause of engine failure?

End Message

———————————✝╫╲╲╘╫————————————

ESS   SCORPION   SENDING.   Universal   Relative   Time
7659450.332.41

Defective mass coil. Burned out on seventh use. No way to
replace.

End message

———————————✝╫╲╲╘╫————————————

ESS   ORION   SENDING.   Universal   Relative   Time
998102.545.45

Captain Lu, your mass coil is defective and will fail on the
seventh use. Inspect and take necessary precautions to prevent
permanent FTL flight.

———————————✝╫╲╲╘╫————————————

"WOW, THEY JUST DISAPPEARED from the current... there
they are. They returned to Earth after they received our message
into the past," Dopple said.

"And died?" Denar asked. Dopple just shrugged. "Then we can
change the past. But we still retain the memory of their no longer
current future."

"Must be or... none of this technology would exist if we hadn't
created it to try to rescue them. Maybe the past will catch up to us
at light speed? No, only the turtles under them changed. I would
avoid sending messages into the past that could change our turtles."

"I was just thinking about that. If we told ourselves the trip to Earth was a waste of time, the Hermes would have remained spinning in space."

"And the AI would have eventually heard about Orion Colony and traveled to take it over. Destroying them might have been the best thing you did."

"What if we sent a message to prevent Kenby's sacrifice? Would he just appear on the ship?"

"Only one way to know for sure, but who knows what else it would change."

"As you disseminate this tech, maybe put extreme safeguards on past communication."

"I will do that, Captain," Dopple said.

The Search — Chapter 14

"Approaching Orion Colony Galaxy. Prepare for deceleration," Denar announced to the ten crew members currently not in stasis pods, wondering if this was the end of her space travel career.

Dopple had decided to do his last six months in his pod. The Doctor from the Hermes was still assessing the viability of her crew until the last week before Denar forced her to go into stasis. She watched the donut galaxy slowly filling her forward viewer as the seven pod lights went yellow, then green. The three crew that would remain awake with her reported in and she initialized the mass reduction sequence.

Denar winced as the distortion swept over her. She would not miss this part of the adventure. She thought about the wormhole exploration. That would be sub-lightspeed at the beginning, and far more interesting than farming or whatever job they would have for her in the colony.

"Orion to Orion Colony," Denar said as the ship slowed into high orbit of the system.

"Welcome back, Denar," Tarlo said. Denar barely recognized the much older woman on the video display. "You haven't aged at all."

"Four years, seventy-two days, Captain."

"Over forty-four years for me. We are sending up the inoculation schedule the crew will need before transiting down. Are you planning on mining before you land?"

"There is plenty of time for that, and we have no qualified pilots at the moment."

Tarlo nodded, now remembering Falx Kenby's sacrifice. Her memory wasn't quite as good as it used to be. "I'll send someone up on the first run. It should be a priority."

"Going somewhere, Captain?"

"I haven't been Captain for a long time, Denar. Call me Tarlo, please. No travel in my future, that is for sure. The council has been discussing the use of your 'sub-space communicator' technology. Since it only seems to work from FTL space, we may need to have Orion go for short trips to communicate with the other colony ships and direct them here."

"We already did that for the two we were able to contact."

"In the present day. I think we should send past messages to them and invite them to join us."

"They would already be here if we were going to do that successfully."

"Hard for my mind to comprehend that. The relativity of two points was hard enough to understand."

"Captain, I think I am done with FTL flight if that is acceptable."

"I understand that feeling completely. I would ask you help complete the practical training of some of our brightest students down here. We have simulators, but there is nothing like the real thing."

"Anything you need me to do, I will."

"Good. Get yourself inoculated and get down here for a good meal and a hug from a grateful colony.

THE ROOM

SCI-FI

PG-13 2720 WORDS 8 PAGES

The Room

I sit in the sharp corner on a tall chair. There are a few vantage points like this where you can, in one glance, see the entirety of the Room. Everyone here has lived their entire lives in this enormous space. I was born 5,780 light cycles ago to a woman who had spent 8,207 light cycles here before I was born. She went through the Door 4,000 light cycles ago. I have few memories from that time of my life, but I remember her exit with crisp clarity. She said her goodbyes to friends and they begged her to stay. She held me and told me I would be fine without her. I cried as she walked away.

Every single light cycle since then I have looked at the Door with longing, with expectation, with fear.

Nobody here knows who built the room with the single Door. Nobody here knows how the first people came to the Room, they left no records, passed on no stories. Most believe the first people built the Room for their own use and then shut themselves in. Others claim the first people were imprisoned here. These Prisonists have many crazy theories. I don't agree with them because a prison door has a lock and key. The Door has no lock. Anyone is free to walk out the Door anytime they want. When the Prisonists start their ranting near me I simply tell them to walk out the Door.

There is a counter on the wall that increments up one every new light cycle. If that is accurate then the first light cycle would have been more than 500 generations ago. The histories that are written started only 10 generations ago with the symbols carved in the long, curved wall. These histories list names with births and Doors, parents, and acts of value. My entry is in the middle of the short wall, opposite the Door. My Door is still blank, obviously, as is my acts of value.

The Prisonists believe that we are only the 3rd generation and that the previous 7 generations' histories are fiction, written by the prison builders. I don't believe them because some of the elders tell stories of earlier generations. These could be made up, but who knows for sure.

I sit watching the hundreds of people going about their daily lives. I am not one of them. Since my mother went through the Door, I have chosen to not be part of the people. I spend little time among them. I quietly obsess about the Door. I have heard every theory about the Door, but that is all they are, theories. Nobody that walked out the Door has ever returned.

Food is provided by the dispensers next to the Door. A total of fifteen units of work needs to be performed on the energy pods per person in order to feed everyone. This was the equivalent of one quarter the light cycle. I had done it a few times, but there are plenty who do many times their share. The tally of the service on the wall next to their name gives them pride, and they hope for some reward when they eventually leave the Room. I often skipped my portion of food since I expended little energy in my daily existence.

The light began to fade so everyone went to their sleeping places and sat down. I grudgingly went to mine and looked at Varook next to me. She held her hands out on her knees, palms up, in the traditional pose of a dutiful Roomer. I usually liked to talk to her simply to disturb her meditation, but decided to give her the silence she deserved tonight. When complete darkness came, I felt her hand on my shoulder. It was her way of thanking me. She was three hundred light cycles older than I was. The sounds of people in the darkness always seemed louder. Some used the privacy of darkness to copulate. Others began snoring immediately. I slept little, so my mind took in the distracting noises as I continued my contemplation of the Door.

"I'm going through the Door tomorrow," I whispered.

"You say that every night. It used to worry me. Now I think you should. You don't contribute. You don't socialize. You don't even care about me," Varook said.

"I sleep next to you, don't I? Would you rather be alone in the dark?"

"There are others that would co-mat with me." I started to get up. Moving in the darkness was a fool's errand. Her hand stopped me before I had to show my threat was empty. She pulled me back down and pushed against me. She was done talking for the night. I thought about my other empty threat of going through the Door. My biggest fear was that I would drop into a machine and be dispensed as the following light cycle's flavorless food tablets. This is why I avoided eating, always wondering if I had consumed my mother's flesh.

When the light returned, I went to an energy pod and registered forty units of work before going to my corner. I watched two older people walk through the Door together. I wondered if that would make it less scary, going through with someone else. Certainly, Varook would not go with me. When the food was dispensed, Varook brought me my share. I saw the old people's faces on each tablet and gave them away to a child that wandered by.

The Door was the great equalizer. Eventually everyone went through it. They either voluntarily walked, or their motionless corpse was pushed through by the Elders. The Elders were not always the oldest, and they certainly weren't the wisest. They were elevated by many kinds of acts of value. Overproduction of energy was rarely rewarded with Eldership. Many saw that as supporting indolence. The most revered act of value was education. Selflessly teaching the numbers, words, and rules to the young was considered most important. It took a mastery of knowledge and compliance that set them above the rest. Tending to the ill, cleaning the various waste facilities, entertaining the masses with remembered stories were all middling acts of value.

As I said before, I provide no acts of value to my fellow Roomers. I study the Door because I want my singular act of value to be a return from the other side of the Door and tell them what lays beyond. However, my mind usually devolves into returning as a consumable pellet. I use the numbers I learned years ago to once again attempt enumerating the inhabitants of the room. It never seemed to grow or shrink, but I counted a different number every time. Even in the same light cycle with no exits, I get a different number. It is because everyone moves. It is the best way to stave off boredom. I used to do it before I gained my obsession with the Door. Counting is an easier way of boredom ignorance, and it uses less food energy of which I have none to spare. Once I counted the people on the wall that did not have Door notations. I got seven hundred and seventy... something. That was far lower than my other counts, so I knew it couldn't be accurate.

"Six hundred eight, six hundred nine..."

"Counting again?" Gonda asked. I ignored him and continued counting under my breath. "Why not just subtract two from the last number?" He moved into my line of sight, spoiling my count intentionally.

"One, two, three, four... I started over, not really looking at the people by the far wall, the place I always started counting.

Gonda laughed, happy to have spoiled my count at such a high number. "It is eight hundred eighty-eight. It has always been and always will be."

"Two were born exactly when the old couple exited?"

"You know what I mean. Do you want to toss Kaylar sticks?"

"Showing off to get Anhal to notice you? She will never co-mat with you. Settle for Idopaz before she gives up on you." He made a face. "Find someone else to play your childish games."

"Now you are an adult? Varook is with child?"

"Of course not. Creating another slave for the energy pod is not the only path to adulthood."

"Staring at the Door is not one of those paths. It is pretty much the exact opposite."

"How so?"

"There is no greater act of value than ensuring humanity continues," Gonda said reverently, quoting one of the ancient scribblings at the far end of the inscripted wall.

"Wouldn't that be the most obvious rule a slave master would want you to believe?"

"Slave master," Gonda spat. "You are becoming a Prisonist now?"

"We exchange food energy for more food. We are nothing but slaves to biology. There has to be more to life than this."

"It is a good life if you contribute. You'd feel better if you did your share."

"I'd feel much 'better' if you left me alone."

"Suit yourself," Gonda said, but lingered in indecision. He was lonely. A weakness of those dependent on others. It was common for those who were mothered too long. Certainly not me. My mind deviated from the count to contemplate why mother went through the Door when I was so young. She was healthy. People liked her. She would have been an Elder eventually. Everyone that knew her told me that. They had no reason to lie.

The mystery was who helped her create me. She never shared a mat with anyone. No man claimed to have added to my existence.

There are stories of men that slithered between the mats in the complete darkness and took women against their will. I have wondered if Gonda would do that if he did not find someone to co-mat with. He was too old to still share a mat with his mother.

"Fine," I said in exasperation. "Go get the sticks and I will play one game."

Gonda smiled and ran off, knowing I would lose the first game and continue playing until he let me win one.

"I'M GLAD YOU DID MORE than stare at the Door today," Varook said as she ended her meditation.

"Why did you choose to co-mat with an indolent?"

"What we are today is not what we will always be," she responded.

"You expect me to follow the rules, contribute my share, give you children, become an Elder, and fill the remaining wall with acts of value?"

"No, not all of those things."

"None of those things. How many times do I have to tell you?"

"Just like you hope your mother comes back through the Door, I hope you become the man I know you are."

"That isn't what I hope for," I lied.

"My mistake," she said with an impish smile as the darkness consumed us.

I lay awake in the darkness, angry that I had no retort to put her in her place. Her hand grasped mine and I wanted to push it away. I didn't because she deserved better than such a childish gesture. My mind wandered as the snoring and fornicating grew in volume.

What would I do if she came back through the Door?

"My son, you have grown."

"No thanks to you," I spat petulantly. "What's on the other side of the Door?"

"I'm not allowed to tell you."

"Then why did you come back?

"To plead with you to become useful."

"Why?"

"You will be ground into food pellets if you do not."

I WOKE FROM MY DREAM to a bright Room. My dreams were infrequent, most likely because I rarely slept. There was something different about this one. I believed the woman in the dream was my mother, but it was Varook with a more womanly body.

I marched directly to the Door and opened it. Nobody told me not to go through. Nobody encouraged me to go through. They all just turned and watched for what I decided. Slowly the news spread through the Room. Eventually Varook would hear. Would she come running? Would she grasp at me, pull me back, beg me tearfully not to go? Maybe she would be glad. She could choose another, more worthy man to co-mat with.

The Doorway was darkness. No light from the Room spilled onto the floor outside the Room. It was possible there was no floor outside the Room. As people stepped out, they just became darkness. They didn't scream, nor did they exalt their joy at passing through. No noise ever emanated from them.

I thought about just pushing my arm into the darkness, then pull it back to see if it remained attached. My fear was it would be a stump and my remaining days would be as a cripple. If I were to close the Door and not go through, no one would say that I was a coward, but all would think it. It was not courage to go through the Door before age and infirmity left one miserable.

I continued to just stand there, staring into the abyss. People slowly went back to their normal routine. Eventually people would be angry with me because food was never distributed when the Door was open. Someone would get hungry enough to shove me out of the way and close the Door themselves.

It was mid-lightcycle when I felt Varook's arms encircle my waist. Her face lay against the middle of my back. She said 'don't go' without words. I realized that was all I wanted. I wanted someone to have more than indifference to my permanent absence. I turned in her arms and looked down. She had been crying. Just like I had cried for my mother. She was smiling now, thinking I had made my decision to stay and become the man she knew I was. I used my finger to wipe away her tears. I bent down and kissed her.

Then I stepped backward into the darkness.

My name is Jolan. My final light cycle was 5784. I stood outside the Door watching Varook's happiness turn to shock and grief. She could no longer see me though I was still within two steps of her. She screamed and fell to her knees. Several of the Elders moved forward to comfort her. She pushed them away. For a moment I thought she would step through into my arms. It was selfish of me, but I wanted her to do that. I held my arms out to welcome her.

Varook turned away and went back to her mat to grieve and meditate. Someone out of sight closed the Door and I was in darkness. I turned around perhaps waiting for a light to reveal what was on the other side. Nothing. I walked forward. Nothing. I called out. Even I didn't hear the words I produced. I continued walking and wondering when I would reach the end. I had never walked far. The room was not that big, and had so many people that were in the way.

I guess I should take solace in the lack of a meat grinder. I continue walking, assuming I would eventually get tired, but I don't feel... anything. I lift my hand to my face and... nothing. I'm walking but I don't feel a hard surface beneath me. My hands reach down and there is nothing. I know I am here because I am still thinking all these thoughts. I call out again and hear nothing.

Mother. Where are you?

I've been walking forever it seems. I gave up my life in the Room for what? I wish I could go back.

As I thought that I saw a glimmer of light. I ran toward it but it disappeared. There it is again. I run harder than I ever have in my life toward it. All of a sudden, I am in the light. I am upside down, or all the faces looking at me are. Then there is just one face. Mom. She reaches out and touches my face. She doesn't look like I remember. She looks like Varook. She is Varook. I am not who I was.

I'm... a baby.

UNBROKEN CHAIN

SCI-FI * MOTHERHOOD

PG-13 7700 WORDS 37 PAGES

Unbroken Chain — 1 Andoline

"**I** think it is time," said Timora as she held the bulkhead in one hand and her swollen belly with the other.

Dr. Evelyn Hensor floated gently across the medical bay and pulled the mediscan from her belt Velcro. She used the doorframe to halt her glide, her feet swinging her upright into the stationary loops. She squatted as she scanned, watching the screen fill with vital information. She stood again and paged through the data with her thumb. "I believe you are correct Mrs. Wolthers. Are you still comfortable with the birth plan we discussed?"

"I would prefer an energy-matter transport, but this isn't the starship Enterprise." She nodded in consent as the joke went unacknowledged by the inquiring physician.

Using the hand holds the two women carefully transitioned into the surgical bay. The doctor began strapping Timora in place as she removed the clothing. Her knees pulled up sharply as another contraction started. Dr. Hensor used the time to position the positive airflow ducting to minimize fluid dispersal when the water broke. When the pain passed, she cinched the ankle straps tight and moved the armatures slowly into a comfortable position. She draped self-adhesive privacy sheets that would also help contain the expected deluge of extraneous matter.

She pulled a latex glove on and held it under the high-intensity UV lamp for the requisite ten seconds. "I need to check dilation. Are you ready?"

"Yes," Timora responded, closing her eyes.

"Only about 4 centimeters. It is going to be a while. Just let me know if you want me to reposition you or release the straps."

"I just heard!" Came a familiar voice as Janie floated into sight. "This is going to be so epic!"

"Everything is 'epic' to you. I think you need a new adjective."

"Super epic! I still can't believe you chose to do this. First conception in outer space..."

"Not true. There was a Russian cosmonaut that came back from Earth Station Six three months pregnant."

"That doesn't count! That was low Earth orbit and she was so ugly I think she smuggled a turkey baster on board." Timora laughed hard and then winced as another contraction started.

"Eleven minutes thirty seconds between contractions," the doctor noted aloud.

Janie continued, "first conception in non-orbital outer space. First natural birth in outer space."

"Again, there may have been millions of outer space births. The universe is full of alien species."

"If and when we discover the aliens, and if they don't vaporize us immediately, I will take back that 'first'. First human occupied spaceship to spend a fourth year in outer space with the original crew. Soon it will be the first multi-generational spaceship. Does Quento know? Do you want me get him on comms?"

"No. I'm sure he'll hear my screaming soon enough."

"In space, no one can hear you scream."

"What?"

"It's from a classic hor-sci movie about aliens."

"I don't understand how you can watch those things."

"I like being scared. Why else would I join a crew like this on a trip to nowhere?"

"It is not to nowhere. The planet will be habitable and we will be there in less than forty years."

"If we have any bones left to walk around on this new planet, if a meteor doesn't zip through the hull at a billion miles per hour, if a space virus doesn't turn us into zombies..."

"You are not helping."

"Sorry. This journey is epic and we'll be remembered like the first settlers that traveled from Philadelphia to Kansas."

"Who was that?"

"Exactly. They were probably killed by the natives before they even built their little house on the prairie. What makes you think there won't be natives on this planet just as thrilled to see the human invaders floating down from the sky onto their prime real estate. I know, I know, not helping. What would your mother say if she was here?"

"She would probably tell me how much pain I caused her, both in my birth as well as my teenage years."

"I find it impossible to believe you were anything but the perfect child."

"I had a few moments of rebellion, but I am certain her version of it would rival the San Angeles riots. I hope I can be a better mother than she was. Not that she was terrible, she was just constantly worried about everything I was doing."

"Well, your daughter will be within five hundred feet of you for the next forty years. You won't have to worry about her moving in with some bohemian in a dirty New York slum apartment while she tries to find herself. That's what I did. Pinot was so sexy that I didn't care about the roaches and the cat sized rats. Best and worst four months of my life. Do you think you could have left Earth if your mom was still alive?"

"I'm not sure. This has been a dream of mine since grade school when we learned about the Mars missions and the Europa colony. She wouldn't have understood it, but I think I would have gone anyway. Dad understands and it's not like we won't be able to communicate with them the whole time."

"My mom was devastated she would never see me again. I didn't think I'd miss her as much as I do. I'm glad I have you." She squeezed Timora's hand. "And I will have a new little sister to help keep me busy for the next forty. Have you decided on a name?"

"Andoline."

Unbroken Chain — 2 Timora

"Dammit Bill, stop flying like a maniac. There is plenty of time to get to medcenter."

"I don't want your water breaking in my new car. This cost me three years wages and I don't want it to smell like placenta for the rest of the time I own it."

"You put enough towels under me to absorb Lake Ontario. You keep zipping around traffic like this and your car is going to smell like urine... or worse."

He slowed down a little and stayed with the general traffic flow. There was a line at the emergency landing pad so Bill Westeros took the car down to the street and landed in the parking area, violating several traffic laws. He got his wife out of the car and carried her to the emergency center.

"I can walk Bill, I'm not crippled, I'm only in labor."

"Save your energy for the delivery dear." She scowled but secretly liked his concern for her. He hadn't always been that way. The government had paired them and he was usually a stern, quiet man. An excellent provider, but zero on the affection scale. She had opted for government courtship after several bad relationships left her empty and depressed. She wanted a child more than she wanted a husband, but single parenthood was not allowed in the southern sectors.

The admitting nurse scanned her biodent tag and spent a few seconds looking at the vital signs. "Your blood sugar is a little low, are you feeling tired?"

"Not really. More nervous than anything."

"Adrenaline can mask your symptoms. I'm going to start your IV before I bring you up to delivery." The nurse pulled a tube out of the bedrail and stripped off the sterile endcap. She held the end against Sarene's hand and the self-guiding probes established a pain free path through the skin into the veins beneath. The nurse then clipped the tube over the biodent, which synced with the fluid computer and began delivering the proper fluid balance to the new patient.

A few minutes later the readings changed. "That looks better. How close are your contractions?"

"Nine minutes."

"Will there be anyone attending the delivery?"

"My husband may be in the room. The rest will watch it on StreamTube. There is a thigh cam in my bag."

"I did that with my second child. It helped defray the delivery costs. Ooh, this is a nice one, three-dimensional optics." The nurse helped her position it and tighten the strap.

"I'm hoping it will be picked for the Jaycene Hour."

"That would definitely raise your Q status. Would you like to set up two-way streaming?" She nodded and the nurse pressed a button on the side of the bed for technical support.

"Good evening Mrs. Worthers," the octogenarian technician said as the screen resolved. "I see it is time for the show to begin. I have your approved list of two-way contacts here, and seven other requests. Would you like to review those requests or refuse them all?"

"Refuse them all... Wait. Are any of them remunerative?"

"Two are. There is a Danton Felwaith of Kendrick Station Australia."

"Refuse. Creepy old boyfriend. In fact, block one-way access for him too."

"Done. The other is Jenna Politov of Nerato, New Russia."

"I don't know her. Occupation?"

"She is a composer slash singer."

"Is she online, I'd like to ask why she is interested in paying to talk to me."

"Connecting now."

"Hello? Oh, it's you," she said in Croatian. The interface translated the words in a similar voice. Sarene didn't recognize the face either. "I see it is time, thank you for approving my request."

"I haven't yet. I was wondering why?"

"You were the first to review my song 'Cradle Party' with great positivity."

"Oh, you're Ginko Blond. Your avatar looks nothing like you."

"That review led to much success, and I thought you might like a personal performance in repayment."

"I would be honored."

"Thank you. I will prepare." Her screen section went dark and it minimized to the Docking bar at the bottom of the screen.

"Are you ready for announcements to be sent?"

"Go ahead."

Another contraction hit her and she rotated the bed to a more comfortable position.

"I have Paulina Kentari requesting connection."

"My mother. I guess you can put her through."

"Oh Sarene, your finally ready. I will never forget the scary night you came into the world."

"Yes, Mother, I know the story."

"You don't want to hear it again for the millionth time?"

"If you could once tell me it without making me feel like I was the biggest mistake of your life."

"I never meant it to make you feel that way. You were a blessing from the great Dinar herself."

"No religious talk or I will rescind comms."

"Sorry. It's just that it helped me get through—" Sarene muted the window. She saw the surprise on her mother's face when she saw her words of comfort had been blocked.

Unbroken Chain — 3 Sarene

"**N**o Dinar, by all that is merciful, do not bring my daughter to me tonight of all nights," Paulina repeated in her mind as the second contraction hit her. She was squatting uncomfortably in the dark closet. The noises outside were horrific. It wasn't the first purge of the religious zealots Paulina had lived through. This was the third.

Throughout time humans have found reasons to form cooperative groups. When those groups got too big, they found reasons to divide. Paulina had learned this in the state academy, but not from the instructors. She learned it from the so-called zealots. The followers of Dinar were the largest of those groups that sought to escape the oppressive technological world they had been born into. Dinar was dead more than a century, but her teachings lived on in offline storage pods and in the minds of those who had absorbed the brief statements of clear wisdom.

The vast majority of the democratic population was blissfully ignorant of her teachings. They lived fulfilling lives of freedom and comfort in a world where there was no hunger, no disease, no crime, and no need for zealotry. Somewhere deep down in the human mind is a primal drive to separate. Usually, it is born of suffering and discord. In a utopia, the division drive needs to dig deep to find an excuse to push away accepted norms, to fight them as if they wished to destroy humanity.

The mistake that was made was the purge of Dinar's teachings from the online storage. There it would be open to all to be debated and ridiculed. Ideas throughout time have never been successfully suppressed. Only the people who carry them forward can be devalued, punished, tortured, or killed for holding them. The Habnars, Jews, Christians, Belians, Buddhists, Muslims, Gadnyists, and now the Dinarians have all had powerful entities pushing to marginalize, eliminate, or exterminate the ideas. They all survived, and most grew in power in direct response to the suppression.

Something in the human psyche takes the thing that is disliked by those who you dislike, and converts it into something you like. Perhaps a form of the TEKTaB theory rising up naturally. Perhaps it is the randomness of billions of minds over thousands of years. In the end, those who feel the need to separate, but choose to do so with the need to destroy what contradicts their belief, become a danger to those who remain. Violence breeds retribution, and as life becomes more utopian, tolerance of disturbers finds new lows.

Paulina joined the quiet discussions about forbidden topics. It was exciting, but it was the leader of the group she was there to see. She was enamored from that first moment he approached her. Meetings led to planning sessions. Planning became secretive actions. These actions were at first informative anonymous postings on the central net. The Mind quickly learned to expunge the words before anyone saw them. Then came the vandalism, painting messages in public places after destroying Mind cameras. This became violent attacks on vocal 'enemies' and proponents of the Mind. Paulina had never been violent, but she knew of the violence before it would happen and did nothing to stop it. This was a misdemeanor until the Mind architect was murdered in his home. That made them all accessory felons and granted purge authority to the Peacekeepers.

Paulina heard the scream of her best friend. Angils Vot had been guarding the door of the room. That meant there was only one barrier between her and the death squad. She bit her lip to hold back the scream she felt rising. The boots walked heavily in the room, shaking the floor and walls that supported and protected her. She braced herself for the eventual discovery, closing her eyes tightly for the bright light that would flood her dark space. She heard muffled voices. She heard Angils' whimpering pleas.

The noises in the rest of the building began to fade. The light in the room went out, immersing her in and even blacker darkness. The room became quiet. She heard nothing. She had escaped a third purge. She had run the previous two times, impossible in her current condition. Then the next contraction hit. She moaned softly at the intensity of it. The adrenaline flooded her. She wasn't sure if it was the impending birth that scared her, or that her noises had been heard. After the wave passed, she listened and there was only silence.

She had no idea how long she had sat in the dark silence, but the pain came quicker and quicker and she felt the pressure building. For millions of years humans gave birth without modern medicine and the species thrived. That gave her no comfort as she weighed leaving her hiding place and being captured with dying in childbirth, only to be discovered when the rotting smell drifted into the hallway, or the street. There was no third option.

She moved to her knees in an effort to find a more comfortable position. "Please, Dinar, if my daughter must come now, make sure she has the best life you can give her."

The gush of water was mostly absorbed by the clothes she had pulled off hangars. The pain grew more intense and she pushed hard, hoping to expel the child. Her screams came as she felt her body tear open. Collapsing in exhaustion, she reached down to feel the small pile of flesh. It was moving slightly. She pulled it up toward her chest and felt the umbilical cord tighten. Then the baby started to cry. She felt the sound of heavy boots approaching.

Unbroken Chain — 4 Paulina

"**I** know, I know. There is no need to repeat yourself."

"Listen, Bill. We've been married how many years?"

"Eighteen."

"Close. In all that time, I have come to know you quite intimately, correct? Wipe that look off your face. There is no such thing as labor sex. I'm just trying to say that I know you very well, and I know that it takes at least five repetitions for you to actually store important information."

"If you wanted an enhancer, you should have married one," he said angrily.

"Dated one of those. Nothing but downside to perfect recall. I just need you to understand how important this is to me."

"Elise, I know. It is important to me too. Trust me."

"I do," she lied. "I love you." She turned to the nurse, "I'm ready."

"Initiating sleep mode," the nurse said, pushing the button after suffering through the painfully private moment. Elise drifted off and the team came in and began preparing everything. The painless extraction took less than ten minutes. Bill was in the room to make decisions in case any complications occurred. None did and the baby girl cried strongly as the lungs were cleared of fluid. The birthbot delicately cleared the placental mass and scanned the uterine wall, repairing a few small bleeds.

The nurse verified all the readouts. "Waking the patient, time under sixteen minutes, forty-seven seconds."

Elise opened her eyes, blinking at the bright light above. She turned her head and saw her husband holding their daughter. She reached out and he handed the tiny package over. "How did everything go?" she asked.

"Perfect," he said. Elise looked at the nurse to confirm. The woman nodded and continued monitoring vital signs. "Are you sure Pauline is the name you want?" My mother isn't going to hate you less for naming her that.

"It may take decades, but I think it will."

Unbroken Chain — 5 Elise

"I'm sorry Mr. Johnson. We did everything we could for her. The damage from the accident was too severe."

"The baby?"

"Your wife is on life support, so we could feasibly allow for the baby to continue until it's normal term. There are downsides to that since your wife's condition is so diminished. The recommendation is for an extraction at one month before term, in approximately three weeks."

"Recommendation by humans, or that damned Brain."

"The Mind project recommended waiting until full term to extract. Our team thinks it would be better to extract earlier than that for the child's sake. Your wife's body is technically in a state of death, and there are biological processes that diminish as time goes forward that we can compensate for, but may hurt the child in the long run. We can extract now if you choose, but complications for the child go up exponentially the earlier it happens."

"I understand. I accept your recommendation. I assume the company will pay for transporting me to Moon Base Central when the child is delivered?"

"Standard procedure is to have the child put in stasis for the downride to Earth and accompanied by a transferring employee."

"Standard procedure? Just how often does this happen?"

"It is very rare, but space travel is still a very dangerous occupation."

"No shit." He stabbed at the screen to end the video call. A new call came in a few minutes later. He wiped his eyes and accepted it.

"Mr. Johnson. I am sorry for your loss. We here at Skydome Insurance mourn with you on this sad day. The life policy in the amount of eight Bitcoins has been transferred to your digital wallet. We know this does not replace your loved one, but we hope it will help you get through the tough times ahead. Thank you for using Skydome Insurance and remember, there is a reward for any future customers you send our way. Have a shiny day!"

"Eight BitCoins," he repeated out loud after the automated call ended. They had a total net worth of two Tenths, which were ten thousandths of a BitCoin. Most of that was in the equity of their beautiful lake house. They had spent years working toward it, waiting to have children until they could move out of the city. Now it would be just him and his daughter, unless God chose to take that away from him too. 'One last run to Jupiter,' was what Onara had told him. It would pay off the car and refill their meager savings. It certainly did accomplish that. He was now a wealthy man. Space travel was still dangerous he argued, but she had done it ninety-nine times. She wanted to make her 'century' run. That's what it had been about, not the money.

If she had told them about the pregnancy, they probably would not have allowed her to go. Not because it was dangerous. Dozens had been born that way. It was a natural result of multi-year missions and coed crews. The company would have grounded her because deliveries in zero gravity were messy as hell. It was one thing to accommodate natural progressions of life. It was foolish to do it with forewarning. He had almost called the company himself to tell them. Now he wished he had. He wouldn't be a rich man, but he would be holding his pregnant wife's hand as they watched the stupid romantic movies she liked.

Unbroken Chain — 6 Onara

"Push, Irena, push," the doctor said emphatically.

"I can't, not anymore. I'm too tired."

"Come on, you are almost there."

"That's what you said an hour ago."

"If I say it again an hour from now, you have my permission to slap me."

"When the hell are they going to come up with better technology to do this. I bet if men had to do this, they would have invented a zipper exit before the steam locomotive." She laughed at the visual of a zipper on her abdomen. Then she pushed for the thousandth time. Ten minutes later the baby emerged.

"That was amazing, Irena. I wish I was there for you. Onara is a beautiful name." The recorded message came through an hour later. It was delayed by the communication time from the mining ship in the asteroid belt. She had sent the video of her final minutes of birth and then holding her baby. She hated that he was gone, but this was the third baby she had to deliver on her own.

Irena was an Earthbound wife, not much different than that of ancient sailors who explored and conquered unexplored continents. Her husband was conquering space. She was raising a family on her own. She knew it when she married him and sort of liked the idea. He was making more money than she could ever spend, so her life was a very comfortable one. She was free to see anyone else she wanted. She knew he did so quite often. She did occasionally have flings, but it would be too confusing for the kids to have more than that. She had friends, a job, and a busy social life that fulfilled her in almost every way. It was only the birth that she felt most alone. Her mother had been with her the first time, but made her so anxious that she preferred being alone the next two times.

She looked down at her daughter and somehow knew that Onara would disappear into space just as her father had.

Unbroken Chain — 7 Irena

"That was a wild ride," Kayla said, turning awkwardly to get out of the back door of the small taxi.

"Sorry, I just didn't want a mess in my back seat," the driver said apologetically.

"No problem. I enjoyed it." She took a few steps toward the emergency room door then stopped and bent over, putting her hands on her knees for support.

"Are you alright, ma'am?"

"If you could run over there and bring me that wheelchair, I would really appreciate it." The boy ran and returned in a dozen seconds, then helped her sit down. He pushed her inside the door where she was met by a nurse. "Thank you, Ling Nal. Definitely five-star service."

He was surprised she knew his name. "Thank you, ma'am. I hope everything goes well for you." He bowed perfunctorily and left quickly. He checked his back seat to make sure she hadn't left a mess.

"We'll get you up to the Obstetrics level in just a few minutes. Is there anything you need?"

"Short of a time machine to undo this mistake, I'm good."

"Mistake?" The nurse asked with concern as she collected vital signs.

"Her father was the mistake, not her. I would have aborted if she was in any way a mistake. I just wish I had chosen a better father. Nothing brings out the true colors of a man like a raging hormonal pregnant woman."

"No truer words spoken. Are you bound to him?"

"Not anymore. Why?"

"My brother cannot have children. He is a good man and would be a good father."

"Is he rich?"

"No, I am afraid he is not. He will have earning potential when he completes his flight training."

"Flyboy? I always had fantasies about fighter pilots. That classic movie Top Gun was always in my VidQueue."

"I have not seen it. My brother is training for space flight in the Notaki Corporate Space Program."

"Do you have a picture? Not bad. Do you think he would like me?"

"You are his type," she said.

"Fat redhead?"

"Happy, confident, smart, funny."

"I think you are going to be my favorite sister-in-law."

Unbroken Chain — 8 Kayla

"Order in the court," The judge said, pounding his gavel. "Control your client or she will be cited with contempt."

"Like that will make a fucking bit of difference," Janice shouted. "You're all fucking corrupt tools of an oppressive government bent on subjugating women and minorities."

"I'm sorry your honor. I sympathize with my client's anger at being falsely accused of these crimes. She has documented evidence that the arresting officer forced her to do what she did."

"What evidence would that be?" the district attorney asked.

"She is pregnant."

"Her health concerns do not mitigate her part in this criminal act."

"The child's father is the arresting officer."

The courtroom exploded in wild noise as everyone was shocked by the news.

"Order!" the judge yelled, banging his gavel repeatedly. "Counselors, approach the bench. What the hell is this all about, George?"

"I bought the pregnancy test myself and saw the results."

"That doesn't mean she walks on this charge, regardless who the father is. A good cop is dead."

"Debatable that he was good. One year, out in three months."

"Five years, out in three."

"Do you really want her baby born in prison and raised by foster parents?"

"I'd force an abortion and lock her up for twenty. My offer is reasonable. Besides, you are optimistic that she will get out on good behavior. She is out of control."

"I tend to agree with the DA," the judge said. "About her lack of self-control, not the forced abortion nonsense."

"You know she was railroaded on this to get her to testify against Mulaney."

"I can see that, but hard to prove, pregnancy or not. We would have to delay this until the parties could be DNA tested, and nobody wants that. I assume she won't testify no matter what?"

"No. She says he is innocent and believes it fully."

"Bullshit."

"One year, out in three months if she can behave," the judge said after some thought.

"Thank you, your honor."

Seven months later in the minimum-security women's prison.

"Guard, it's time."

"That's the fifth time this week it is 'time.'"

"You going to risk that this time I'm not lying?"

"Fine, come with me." He unlocked the door and let her exit from the large dormitory. "What are you going to name it?" he asked as they walked toward the infirmary.

"I don't know. What do you think?"

"Bruce if it's a boy. Loreen if it's a girl."

"God help your poor kids if a woman is dumb enough to procreate with you."

"You would have been out of here four months ago if you could only control that smart mouth of yours."

"Baha Rain."

"Huh?"

"That's what I'm going to name her. Baha Rain."

"Hopefully the adoptive parents come up with a better one than that."

"Fuck you!" she turned and attacked him, scratching and hitting him. He pulled out his club as he tried to avoid her blows and swung it wildly. The third swing hit her temple and she fell to the floor like a ragdoll.

Six months later the foster parents of Kayla legally adopted her.

Unbroken Chain — 9 Janice

"**I** think the tub is deflating," Monica said.

"I told you this was a bad idea," Mark said in exasperation.

"Shut up and start blowing," she said as she sat up in the small pool of water. He found the nozzle and bent to the task of re-inflating the kiddie pool. The water was cooling off and he was pretty sure it would be ice cold by the time his wife delivered. He kept his mouth closed and held her as she gritted and whimpered through her attempt at a home water birth.

"I'm cold," she finally admitted. He hoped this meant abandoning the pool and not that he had to drain the pool and refill it from the water heater hose again. "Grandma said this was the best way to have a baby."

"She probably had a hot tub and two dozen midwives helping her."

"She wasn't that rich. Can you refill the tub one more time... please?"

He nodded, helping her out and seating her on the chaise lounge. The sun was warming, but the cool breeze chilled her wet skin. She shivered, so he went to get a few towels before turning on the hose. Their small backyard was anything but private, but the kids that were normally running and screaming in the neighboring yards were mysteriously absent. He pushed the side of the half-deflated tub down to drain the cold water, then started the refill while blowing into the nozzle to re-inflate again.

"Are you sure you don't want to go to the hospital?" He asked as he helped her get into the tub of hot water.

"No. We can't afford it." This was a stab at his inability to find work. He had little pride left to bruise. He knew the hospital could not turn them away. He also knew the hospital would send them bills and they could ignore them. Her job had kept the roof over their head and food on their plates. She was the boss. He held onto her and she made another concerted effort to push the baby out. He noticed the water became tinged with red.

"You're bleeding," he said with concern.

She opened her eyes while still panting through the contraction. "That's normal. It gets real messy. Don't you dare faint on me." He was feely a little woozy, but that was mostly from blowing air into the pool. "I think it's coming." She pulled his hand down and he indeed felt the hairy head emerging. She screamed as the final push got the head through.

He was supposed to get the baby once it was out but she needed his support. He reached with one hand but it was still partly inside. He thought of tugging on the head, but knew the neck wasn't strong enough. She grunted again he felt the shoulders pop through. Then the baby squirted out beyond his reach. He pushed her forward so he could hold her up and reach the floating baby. She reached out and picked it up herself. She was sobbing. He hoped it was happy sobbing as he held her.

"I told you it was easy," she said. He kept his thoughts to himself and looked at the misshapen head of his tiny little daughter. Monica talked him through cutting the cord and he helped her out of the tub. She walked into the trailer, dripping blood all the way into the bathroom. He got the shower running and left her to start cleaning up. He used paper towels to mop up the floor.

When he went back in, he found her on the toilet, the baby wrapped in a towel. "Take the baby while I get the placenta out."

He didn't know what that meant, but he was sure he didn't want to know. He took the bundle into the living room and sat on the couch. He put the game on, glad he didn't miss all of it. The baby began fussing about an hour later. When it turned into a cry, he went to find Monica.

It looked like a murder scene. There was blood everywhere. Monica's inert white body lay on the bathroom floor. The baby cried as the paramedics came in and tried to revive her. She was gone. He went with them to the hospital and they made sure his new daughter was healthy and prepared for a life without her mother.

"I guess it's just you and me, Janice," he said, holding the bottle in her mouth as they rocked in the dark nursery ward.

Unbroken Chain — 10 Monica

"No, mother, I am not going to have it naturally. I tried that with the first one. I almost died."

"Cesarean is more dangerous. It is major surgery."

"I've had two and no problems with either one. I'm not calling to get your advice. I am calling to tell you it is scheduled for tomorrow morning if you would like to be there."

"I will. I just wish you would..."

"No! Goodbye, mother."

"Sorry. I love you, Holly. See you tomorrow."

"I love you too."

Unbroken Chain — 11 Holly

"I don't think you should go, mother."

"I'm not due for two more weeks," Ursula said.

"You don't need to be there. Harry is more than capable of closing the deal."

"Like the Bussard Synergy fiasco?"

"That wasn't his fault."

"Oh, sweetie. You are so naïve. I built this empire before any of you were born. You grew up thinking this was easy. It wasn't. Don't roll your eyes at me. You live like a little princess in a castle and think you deserve all this. When it all disappears you better hope you have a rich husband to keep you happy because none of you have what it takes to make it on your own, especially Harry. I send him out to do the preliminaries with his fancy Harvard business degree because that is what the assholes expect. If I'm not there they'll steal us blind with an Ivy League smile and handshake."

"Rich husband?"

"Wife. Whatever the fuck your gender fluidity horseshit is making you want this week. Just remember if I wasn't normal, you wouldn't be here. You would have been just another egg absorbed in a rugmuncher's tampon."

"You are such a bigot."

"And you'll do what this old bigot wants if you want to remain in the will."

"You're still having babies at fifty, there won't be anything in the will if it is split a dozen ways."

"Obviously, math is not your strong suit. Just make sure the jet is available to fly me to New York."

"Harry doesn't want you there."

"Good thing he isn't the CEO."

"He says the board is going to make him CEO." She knew she wasn't supposed to say anything, but she wanted to be the one to stick the knife in.

"Oh, really," Ursula said with a growing smile. "Maybe the boy has grown a spine after all. Wouldn't it be wonderful to be able to raise this next one myself. Maybe she'll have a brain, and actually respect the woman that gave birth to her. Even if I have to do it in poverty after you idiots run my empire into the ground."

"Empire," she spat. "The only thing that is empire-like is the tyrant that rules over it."

Unbroken Chain — 12 Ursula

"Turn off that stupid video camera and hold my hand."

"Yes, dear." He did as she asked, and sat down next to the hospital bed, holding her hand between both of his.

"I'm scared."

"I know. So am I. But the Lord is with us and we must trust in His divine plan."

The couple bent their heads in prayer, ignoring the nurse who busied herself with preparations for the delivery.

"Hi, I'm Dr. Chandraska. I'm the anesthesiologist for your epidural. If you could just turn onto your left side for me." He did his work without incident. She felt the contraction, but there wasn't any pain. She waited for it, and her smile grew the longer it stayed away. The birth progressed normally and the delivery was uneventful, captured on a VHS video tape.

"She's beautiful, just like her mom," he said. All the medical staff and visitors finally left them to have a private moment.

"And what exactly do you plan to do with that video?"

"I don't know. People do it, and we bought the camera to document Ursula's life. Why not get the very beginning."

"Not the very beginning."

"Of course, not that. I think we should pray and thank Him for our good fortune." They bowed their heads.

Unbroken Chain — 13 Lorraine

"Stop smoking that shit, it's going to hurt the baby."

"The baby will be mellow all her life," Abby said when she finally exhaled.

"I have to go to work. Promise me you won't have any more of that."

"You weren't such a square when I met you. You start wearing a tie to work and I'm gonna leave."

"Please don't say that. I'm just trying to provide a good home for us."

"What do we need besides love?"

"A roof over your head. Food on the table. Health insurance." She rolled her eyes at him. "Call the factory if you feel any contractions. Maggie is on the switchboard and she knows how to get ahold of me in a hurry."

"Whatever."

"Lydia..." He looked at her fading into her hallucination. "I love you." He kissed the top of her head and left. He wasn't halfway through his shift when he heard the page over the PA system. Everyone wished him well as he ran between the machines toward the locker room. He drove much too fast and was pissed that his mother-in-law had the driveway blocked. He sprinted into the house.

Lydia was lying on the couch panting. He ran and got the prepacked suitcase and put it by the door. "Let's go, Lydia."

"She ain't goin' nowhere until the contraction passes," his mother-in-law said, leaving the 'you dumb idiot' off the end of a sentence for the first time in her life. He knelt next to her and took her hand.

Lydia began to relax, then burst out into laughter. "Holy shit that hurts. Hell of a lot more fun going in then coming out."

"Can you walk?"

"I'll just fly," she said, flapping her arms like wings. He pulled her up to sitting and then to her unsteady feet. He supported her as she walked all the way out to the street and he managed to get her into the car. He jumped into the driver seat and sped off down the street. He slammed on the brakes. Shifted the old sedan into reverse and went much too fast back to his house. He forgot to shift into park before trying to jump and it started moving as he stepped out. He jumped back in and put it in park, Lydia laughing her head off at his lack of coordination. He ran into the house and got the suitcase by the door.

"I could have brought that to you later, you dumb idiot."

Unbroken Chain — 14 Lydia

"Do you think Dad would be proud of me?" Hazel asked her mother.

"Of course, dear. I believe he is looking down on you right now and beaming with pride. He'll be happy to know his life continues on into the next generation."

"Tell me again how you two met."

"It was a USO dance in San Diego. He had just finished basic training and was shipping out a few days later. It was love at first sight. He was so handsome in his uniform. I just knew he was the one I had been looking for. He wrote me every single day. He said it was mostly boring, sailing from port to port carrying soldiers and stuff to kill those damn Japs. Then one day the letters stopped. You were already two years old when the letter from the President came. I found out later he had brought the atomic bombs that ended the war for good. He died a hero so we could live in this bountiful free country and you could grow up in a world without war."

"So, you met him and got married in a single weekend?"

"And conceived our beautiful daughter too."

"Now that I know how everything works..." She hesitated. "He wasn't my father was he."

"Better to just believe the story, Hazel."

"Why did you name me that?"

"That was the color of his eyes."

"Who's?"

"Your father's," she said with a significant look.

"I hope I meet him some day."

"We all end up in heaven eventually."

Unbroken Chain — 15 Hazel

"I don't want you here."

"But he's my child, Darlene."

"No, he isn't. My father will shoot you if he catches you here."

"I thought you loved me."

"I was lonely. I have a husband."

"What if he gets killed in action?"

"He's on a transport ship. He's safer than you are right now. I'm not kidding. Daddy has a shotgun in his truck and he will put both barrels in you.

"You won't let me even see him?"

"No."

"Never?"

"Never. Do you think I want him growing up thinking he is the bastard kid of a coward?"

"I tried to volunteer. They rejected me."

"Bastard son of a cripple then."

"You're just being mean to make me go away. I know you don't feel that way. Please tell me it wasn't all a lie."

"It was fun, but only for a while. I need a real man, and mine will be home from the war in a year or two. What place in our lives will you have then? None. Leave now and forget about us."

He turned to leave. He walked off the porch and picked his bicycle off the ground.

"Wait, Gerald. Oh no. My water just broke." He ran back up to the door.

"Should I call for an ambulance?"

"It will take forever for them to get out here. The keys to the sedan are in the kitchen. You can drive, can't you?"

"A little."

Unbroken Chain — 16 Darlene

"Mr. Higgins would like a progress report."

"She's his wife, not a damn factory. Birthing babies is not like building tractors. Tell him it will happen when it happens."

"Thank you, Millie. You are so good to me."

"You don't say another word about it, Miss Eden. I helped bring you into this world, it will be an honor to help you with your children too."

"Did you ever want children of your own?"

"You are my children. Your mother, god rest her soul, was such a giving woman. She never treated me like the hired help. She was like a sister, sharing her life with me. It was a life I never would have had, no matter how well I married, which would not have been well at all. Perhaps I never had the pain of birth myself, but I have felt it every single time. Your little brother's head was so big I thought it would rip her in two. After that I no longer had any yearning to experience it myself."

"She told me she loved you like a sister. She said I could always trust you. I should have trusted you when he came courting."

"One man is as good as another as far as I can tell. As long as he doesn't beat you, count yourself lucky. He may love his money more than you, but a life of comfort is worth a hundred impoverished ones.

"Eden?"

"Yes, Wallace."

"Are you... Do you need anything?"

"Some privacy," Millie said.

"No, Wallace, everything is fine. It will probably be a few more hours. Don't worry about me, I'm in very good hands here."

"Just say the word, and you'll have anything you want."

The two women shared a look and both smiled a sly smile.

Unbroken Chain — 17 Eden

"Millie, I can't do it."

"Of course you can, Miss Josephine. You are the strongest person I know." The older woman looked at her servant with admonishment. "Maybe not strong in body, but strong in spirit. You have done this before, and it will be the same result this time. A beautiful healthy child."

Jo knew there were only two options. Push it out or die. It is what had faced all women since the Garden of Eden. She saw clearly the fate of all of them that had come before. Each had given birth to the next generation regardless of their health and well-being. They sacrificed and toiled and made it possible for the next generation to survive.

"Eden."

"What?"

"I want to name her Eden."

"What name for a boy?"

"This one is a girl. I will name her Eden. It is where we started from. It is where we all return to. Promise me, Millie. Promise me you will take care of her no matter what happens to me."

"We will take care of her together. Now push!"

...and so continues an unbroken chain of birth stories all the way back to the beginning of life itself.

SPUN GRAVITY

SCI-FI * SPACE TRAVEL * AI

PG-13 5800 WORDS 17 PAGES

Spun Gravity

I felt the air pressure change in my ears. It happens so often in a pressurized environment that most people learn to ignore it after a few months on the station. I've been here more than ten years and it still unnerves me. Getting sucked out into the black void is my worst fear. Why do I live where that is a genuine possibility? Long story, and not one I'm interested in telling you.

This pressure change meant someone had opened the inside airlock hatch of the cargo container I was currently hiding in. Bypass a few safeties and we were both going to be frozen solid in minutes. I heard the hatch slam shut. Whoever it was did not find what they were looking for.

I tried to convince myself it wasn't me they were looking for, but the very fact that I was hiding made it the most likely possibility. Why was I hiding? Long story, maybe I'll tell you about it if I live through the next ten minutes. I figured that's how long it will take them to search the rest of the container. Less if they weren't alone. As if on cue, I heard the high squeaky voice.

"None of the pressure suits is missing so she's definitely still inside."

"No way she slipped past us? She built this ship. She has to know secret ways to get around," the deeper voice asked.

I wish I had a secret passage to escape through. Kib and Fyzo, muscle for the station's regent.

"Ain't a ship," Kib squeaked.

"What's the difference?" Fyzo asked. She couldn't hear them moving around the big room since they were floating weightless.

"A ship could power outta orbit and take us back to Earth. This spinning metallic donut is forever stuck circling the ass end of nowhere."

"I like it out here. Back in Jersey I'd be a nobody working the lines like both my moms did."

"If we let her get by us, that's where we'll be in three months."

"She ain't here. I've checked everywhere."

"Check again."

"Maybe she vented and saved us the trouble."

"Outer door would have been open. Not like you can close it after you're an ice cube. Check again."

I shuddered at the thought of becoming an 'ice cube' and fought down the panic as a leg floated through my very narrow field of view. It was Fyzo. Kib was probably guarding the hatch. No way out. 'Only way out is through.' Great time for meaningless words from my dad to pop into my head. No way I could go through two thugs. I'm a seventy-pound woman, bones thinned to brittle toothpicks from a decade in near zero G. Only way I could even go to the outer ring was in a gel chair.

"Hey Maya, you in here?" Kib called out in a taunting voice. "We ain't gonna hurt you. Regent just wants to talk, that's all."

Bullshit. They'll lay me down on his outer ring office floor knowing I won't even be able to lift my head. He may talk, but that isn't all he was going to do. The black void was almost as appealing. My right hand gripped the cargo strap and my left held the only weapon I could find on my mad dash down here. The edges of the disc shaped glass memory module were sharp enough to cut skin when it was outside the transport case. The problem was it was sharp on all sides. My hand would get cut as deeply as their throats if I had the opportunity to slash them.

An arm drifted by, revealing a hand that held a light. There was likely a more fearsome, and useful weapon in his other hand. The light played across my face, but his head wasn't aligned with the gap through which he could see me.

"She ain't here."

"She has to be. Start moving the containers."

"That'll take forever."

"Just undo the latches. Someone else can come down and put them back where they belong."

"Won't that be dangerous with everything floating free?"

"It's all weightless. What possible damage can it do?"

It was over an hour later when Fyzo released the second strap on the container I was hiding behind. As it began floating up, I turned my body and got my back and legs between the heavy box and pushed with all my remaining strength.

"What the hell?" Fyzo said as he was struck, then pushed by the box up toward the middle of the cargo hold. I only expected to get him away while I made my way to the door and tried to fight past the smaller Kib. The room looked like an asteroid field. Boxes and tubes floating everywhere.

I pushed off the hull gently toward the door. A short scream made me look over my shoulder and I saw the slow-motion convergence of two large containers. Fyzo's strength was able to slow the impact but his head was pinched at exactly the wrong place. His scream ended as it was cleanly separated above his broad shoulders. Newton's laws of motion in action. Physics is a mother...

"What the hell are you doing? Fyzo?"

I followed the voice to its source, hanging from a handle above the main exit hatch. His other hand played a light around the room. I used a few floating objects to change my direction. I went up to a maintenance access panel. I logged into the system and began over-pressuring the airlock outside the container. Only a few pounds would be necessary. Then I opened the inner hatch remotely. It swung open wide like a baseball bat and the excess air pressure sent several containers across the room, directly at the main exit hatch.

No decapitation this time, but Kib was knocked unconscious. I launched myself down and navigated through the boxes to the door and escaped the room. I pushed myself up the corridor, pushing off ladder rungs as I went. The slowly increasing centrifugal gravity pulled me even faster. I began using the rungs to slow myself. Level ten was ten percent max gravity, about twenty-five percent that of Earth.

I bounced across level ten like a moonwalking astronaut until I reached my cabin. The two goons had smashed my computer terminal as if that had any effect on the centralized computer system. I got my tool belt and strapped it on. I peeked out into the hall to make sure it was safe and then continued across level ten to the fuel storage pods.

If Kib and Fyzo didn't report back, the regent would send Hawk to find me. Hawk knew the station almost as well as I did. There was no way to hide from him. He might sympathize with me, being old friends as we were, but I couldn't count on that, not anymore.

I had to get off the station with the information I had. Attempting to send it by light wave is what triggered the regent into sending Kib and Fyzo. I should have known he'd have message traps in place. An escape pod was the worst of my options. It would just head down to the Mars surface and the gravity would kill me.

There were no departing passenger ships for three days. My only chance was a freighter, and the regent would know it. He'd have them searched top to bottom before granting them departure clearance, and most left this station empty, so the search would be easy.

Then I realized I didn't have to leave. Only the message did. I pulled my way up to level five and checked the manifests of all ships currently docked. I found the right one and headed out to the loading docks. I'd put on a pressure suit thousands of times before. I wondered if this would be my last.

I cycled through the airlock and pulled my way down to the ship. There were cameras out here, but I knew where they were. At best I would be a small dot next to these behemoths. I found a maintenance access panel, opened it, and tied the small box into the power systems. They'd find it when it got to Earth, and they'd probably just give it to the regent's employers. Hopefully, that would be too late to matter.

I then tied the box into their data systems and set the timer for twelve hours after their scheduled departure. If all went well, the information would be broadcast wide on their distress frequency. Every ship in the solar system would get the data and record it in a permanent manner.

Distress systems were regulated mandatory when a few ship captains pretended they never heard the call and stranded ships became frozen mausoleums. It was expensive to change course to affect a rescue, so everyone had to carry rescue insurance that paid the rescuer handsomely for their time and trouble. I double-checked everything and closed the maintenance hatch.

Heading back down to the airlock I checked my air supply. Two hours remained. It was possible someone had noted my exit from the airlock and informed the regent. If so, Kib or even Hawk might be waiting for me there. I went past the airlock I had come out of and headed down to the next ship on the ring. It was a cruise ship headed for the rings of Saturn after a few of the lucky passengers returned from a visit to the Mars surface. I checked a few of the ship's airlocks, but they were all controlled from the interior for security reasons.

I continued on to the next ship, a small freighter that was used to bring mining equipment out to the asteroid belt and return the refined products back here for transport down to the surface or back to Earth. I knew the ship was empty since it wasn't scheduled to go back to the belt for a few months. I tried the first airlock and it opened. I pulled myself in, sealed the door, and then pressurized. When I tried to open the inner door, it wouldn't release.

I was about to vent and head back out when a face appeared in the small window. I waved and the bearded man shook his head.

I wasn't sure if he was denying me entrance or just disappointed with who was knocking on his door. The intercom buzzed.

"What the hell have you got yourself into, Maya?"

"Let me in and I'll tell you the whole story," I said hopefully.

"Hawk was just asking if I'd seen you. Asked me to see if you're on my ship."

"What... will you tell him?"

"Depends what you did."

"Stole some of the regent's frozen lobster tails."

"Bullshit. Hawk wasn't looking for a thief." Through the thick portal glass and my thin face shield, he stared at me hard. "Maya. I can't help you."

"I don't need your help. I need you to look the other way. Like I did when Charpo's wife came and went from your cabin."

"Charpo's dead. You can't blackmail me."

"I'm not blackmailing you. I'm asking you to return the favor."

"And get airlocked when he finds out I looked the other way."

"You won't. I'll go directly to Hawk myself."

"He won't help you either."

"I know," I said. "Please?"

It was a long minute before I heard the woosh of air pressure equalization and the latch click. I pushed on the door and it swung open with a creak. He pulled me through the gap and spun me upright facing him. I reached to vent my suit, but he stopped me.

"You're inside now, tell me why they're after you."

"I found some information."

"About what?"

"If I tell you, they'll be after you too."

He sighed, then helped me depressurize and remove the pressure suit.

"Is it about the Narthok?"

"You knew?" I asked, suddenly afraid he was part of the conspiracy. He held on to me when I tried to push away. "Let me go."

"Who else have you told?" he asked.

"No one." It was technically true.

"Can you keep your mouth shut about it?" he asked after a long thoughtful look into the open airlock.

"Yes," I lied, knowing that any other answer was a one-way trip back through the airlock without a pressure suit.

"He won't believe that. You shouldn't have run."

"Please don't..." I looked at the airlock with fear.

"Couldn't if I wanted to. Hawk was explicit that you be in talking condition when you met with the regent. You'll probably regret not staying out there." He sighed, then let go of my hand and pushed the airlock closed and sealed it, reengaging the lockout. "I'll give you ten minutes before I call Hawk. You have a place to hide?"

"I was going to stay here for a few days. Figured it would be the last place Hawk would look. Guess I don't know him... or anyone... anymore." I gave him a significant look.

"It's not what you think, but... Things change. You either go along or you get replaced. You've been here long enough to know that."

"Ten minutes?" I asked.

"Twenty, I guess. Take this." He held out his olly-tool, a multipurpose device most space mechanics use. I pointed at my own on my belt, but he showed a few additions he had made, one of which was a curved knife. I gave him a significant look. Did he want me to murder Hawk? Murder the regent? Kill myself? I shook my head and pushed away toward the docking collar.

I half expected Hawk to be waiting outside the door, but the only movement in sight was the fueling bot. It was guiding the hose drone outside its viewport to a recently docked surface shuttle. I glided weightlessly past it and a dozen other docking ports, some empty, some with ships attached. None of them could take me to a safe place.

I reached the center tube and headed up to level ten. My cabin was still empty, but seemed like the mess they left had been further rearranged. I opened the tiny fridge and ate some leftovers from last night's dinner. I didn't bother reheating them. It wouldn't make them taste any better. I washed it down with a beer while I thought about the impossible task of finding a hiding place.

The message would be sent the next day. It would be received system-wide over the next few days. It would be months before anyone from Earth arrived to deal with it, if that was their chosen course of action. I now had my doubts. There was a knock on my door. Possibly a good sign.

"Hey, Hawk," I said opening the door for him.

"Regent needs to talk to you," Hawk said.

"He's welcome to come up here."

"I have a gel chair for you."

"You think I like being helpless in max grav?"

"He just wants to talk."

"And if he doesn't like what I have to say? You gonna protect me?"

He looked down, then to the left up the passageway. Kib was out there. He wanted me to pay for knocking him cold and killing Fyzo. The end was near whether I talked to the regent or not.

"Fyzo was an accident, I swear," I plead softly.

"What I do to you won't be," Kib said, stepping into view, but staying submissive to Hawk's authority.

"He ain't gonna do nothin' to you," Hawk said. "Don't make me drag you."

That was that. I walked next to Hawk to the central hub lift with Kib following behind. At level twenty I could really feel the gravity. It was the centrifugal force of the spinning station, but without a window to the outside as a visual reference, it felt like gravity. We stepped off the lift at level forty and Hawk helped me into a gel chair, designed to support a weak body in higher gravity.

Most of the Belt miners and long-haul freighters lose muscle mass and bone density spending so much time in zero gravity. My muscles and bones thinned out while I built this station, long before we got it spinning to create the artificial gravity. The higher gravity areas were for short-term visitors so they wouldn't suffer when they returned to their home planets.

I hated being in the gel chair. I essentially became a quadriplegic, only able to move my eyes, mouth, and fingers. The fingers were important in that I could control the movement of the chair, but it was far less powerful than Hawk and even the smaller Kib. Hawk allowed me the dignity of driving my way back onto the lift, but stayed close enough that escape would be impossible.

As we rose higher in the station, I felt my body sinking down into the gel. It has been more than a year since I had been up to these max grav levels. Two years ago I had come up here to meet the new station regent. I had liked him, but that was mostly in comparison to the idiot he was replacing. He made sure I got everything I needed to keep the station well-maintained.

The regent stood when he saw me through his office window. He walked to the door and opened it for me, greeting me with his usual affable smile.

"Maya, it's been too long. Come in, come in," he said, waving me into his office.

As if I had a choice. The pretend window to outside the station in his office showed a live picture of the Mars surface below. If it had been a real window, Mars would have been spinning wildly in and out of view. Most people would vomit watching it. He closed the door once I was in and sat back down behind his desk.

"Fyzo was an accident—" I started to say.

"I know, I know. He was an idiot. Losing his head probably doubled his IQ. Don't you worry at all about that unfortunate incident. They made you feel threatened, which is the last thing I wanted. You are the most important person on this space station—"

"Kib hasn't got that message."

"He will, Maya. Don't you worry. Now... about the information you saw—"

"Not just information. Explosives on my space station. You could kill us all."

He sighed and nodded. "It isn't what you think, Maya." I knew it was exactly what I thought, but I didn't say anything. "It is just temporary. All of it will be off the station when the next space command ship docks."

"The military would never mess with anything as volatile as Narthok."

"That isn't strictly true. I know your father was in the military—"

"Don't bullshit me. Narthok is for terrorists who don't care who dies. Whoever you're selling it to is going to—"

"Maya, stop. There are things you don't know about. You haven't been to Earth in what, ten years? It's different there now."

"Unless it is run by terrorists, there is no reason the military would choose Narthok over any of the dozens of stable explosives they have in their inventory."

The regent stood and looked out the fake window at the Martian surface. "Maya, if you want to live... here on the station, you need to forget what you think you know. I need you to forget."

"I can't. You should have never let them bring the stuff up here. Push the container out and let it burn up on reentry."

"I can't. It's a military decision."

"Name the Admiral that gave the order."

"The order came from the Chancellor."

"Bullshit." I could tell he believed it, but there is no way it could be true. "You talked to her personally?"

"Don't be stupid."

"You're the idiot. You've put thousands of lives at risk on this station for a little side money and—"

"Enough, Maya." He motioned to whomever was on the other side of the window behind me. The door opened and the regent shook his head at them. No words were exchanged.

I felt my gel chair being pulled backward roughly. This was it. Hundreds of thousands of hours in orbital construction, the most dangerous job in the universe, and I was going to die on the orders of a political hack. I knew it was my big mouth and uncompromising self-righteousness that condemned me, but that is what built this place without a single fatality. Well, one fatality.

"Bet I could pop your weak little head off with my bare hands, but that would be over too quick," Kib said. I tried steering the chair with my fingers but he'd shut off my control.

"Where's Hawk?" I asked, trying to not let my increasing terror taint my voice.

"Your boyfriend ain't gonna save you, bitch."

Boyfriend? What an idiot. Hawk was half my age and practically an adopted child. He came to the station a troubled teen just after we got the station spinning. I taught him everything he needed to know about the station and he took over maintenance on the high gravity levels I could no longer go to.

The last regent had taken him under his wing and used him to keep the peace since Hawk was very physically imposing. He was a good kid, which is why this regent had brought in Kib and Fyzo to do the dirty work. The last regent did his own dirty work, which is how he ended up getting himself replaced.

"When is the funeral?" I asked.

"Depends how long I feel like torturing you," Kib said

"Not mine, idiot. The funeral for Fyzo?"

"Why? You want to attend? I should make you sew his head back on."

"Be happy to. It really was an accident. Not like I could aim a storage container that... perfectly." He spun the chair and tried to slap me, but the gel covered the sides of my head. He thought about punching me, but knew I was too brittle to take any real force. "Coward," I spat defiantly. His arm cocked back, but did not deliver the quick death I desired.

"Yous gonna pay for that, bitch."

"Yous? Didn't know you could pluralize that word. Are you seeing more than one of me? Maybe you should get your eyes checked. That's the first sign of space dementia."

He spun the chair back around roughly and resumed pushing me around to the downlock. The downlock is an airlock that is positioned on the floor of max grav. A wall mounted one is worthless when the station is spinning since anything exiting would be thrown off and likely collide with one of the many protrusions. On this outer ring. Being thrown from the downlock would clear any external structures. Depending on where in the spin cycle the release occurred, the ejected material would either end up in a higher orbit or head down and burn up in the Martian atmosphere.

I knew the gel chair would not fit in the airlock, and there was a good chance he'd break my neck just trying to get me out of it. He stopped next to the floor panel and punched in the access code. The inner door popped and hissed open. He must've used this method of disposal before.

"Airlocking me won't last very long. I'll freeze unconscious in less than five seconds," I said, not quite hiding the terror in my voice.

"That's if I flush you before you're dead. This is just a good place to store you while I go get my knives."

"Knives, that'll make a nice, big mess. Don't forget the absorbent towels." I started to feel the pain as the gel started draining down, sucked into the reservoir under the chair. The force of gravity grew on the points of my body that remained in contact with the hardening chair. His rough hands pushed under my body, further reducing the areas left supporting my body weight. He lifted me and shoved me headfirst into the hole on the floor. My face hit the outer hatch, but did not snap my neck and end this. I felt the coldness of space through the viewport glass on my cheek. I tried the move, but my body was folded backward on top of me, its weight trying to push me out the window. My ears popped as he closed the hatch and all sounds of the station went quiet. The terror finally overtook me and I screamed.

Legends say that one's life flashes before their eyes in the final moments before death. This began for me as a series of random moments. Mostly of my time here, building this Martian space station, my life's work as it turns out. A few moments of childhood and college, but none of that brief time in my life when I felt loved.

I took this assignment to help forget him. It worked, mostly. Now, in the flood of memories, I simply noticed the absence of any with him. I turned my love toward this enormous hunk of metal, never again allowing myself to feel those feelings again. Now this beast I love would spit me into the cold darkness with little more than a cough, a clearing of its throat. No, it was the other end that would expel me.

I began to wonder if I would see any of them again. Those in my memories that no longer existed. Family, friends, school mates... Him. The one I don't remember any more. Will I have eternity to talk with them? Endlessly play the 'do you remember that time when we...' game. No. My bet was I return to the infinite black void from which I emerged. No memories, no emotions, no sensations. Perhaps it was fitting that my journey back there would be through the very real black void above the red planet.

Vantu. All the other memories have faded and there is only him. Was I saving them for last? Or was it just my noticing his absence that pulled the memories forth. First, I remembered the pain. The pain of his absence. The pain that brought me here, to this end. It was only emotional pain, but even the faded memory of it was worse than the current crushing gravity on my twisted body. My crying now was for him, for Vantu. For the absence of him and how he made me feel.

If there is an opposite to pain, that is what he made me feel. It never lasted long enough. Brief, stolen moments. Not the kind of forbidden love Shakespeare wrote about, but denied to people like us all the same. Then it was over and I wanted to die. Needed to die. Too much a coward to die. Go to Mars, most dangerous job in the universe. Forget about those stolen moments.

I wasn't scared of being sucked out into cold, dark space back then. I would have welcomed it. The fear came after Janel and I were locked out of the primary module in our space suits. Waiting for rescue, her suit failed, and I had to hold her and watch her fade away, screaming as the frost took her feet, then her legs. She was out of her mind begging me to help her. Even if I had turned away, I wouldn't have been able to shut off the radio commlink. My suit didn't fail, but oxygen ran low before they were able to rescue me. I woke a few hours later. Janel helped me forget about Vantu, since her terror became mine. Janel also made me avoid friendships in addition to romantic partners.

I became the best at space station construction because of my new fear of dying. I was obsessed with making everything perfect. Even this airlock torture chamber I found myself in was designed, built, and functioned perfectly at my direction.

I should have lied to the regent. I could've transferred to a long-haul freighter and spent my remaining years in quiet comfort. But that wasn't me. Do the right thing. Words from my father. Words to live by. Now they were words to die by.

As if that thought brought my short future into the present, the door above me opened and two hands grabbed my ankles and pulled me out of the small containment. My face, then my body, hit the floor hard as Kib dropped me. It wasn't the rubberized aluminum that made up the floors on all gravitationally positive floors of the station. It was a black plastic that smelled of harsh chemicals. Within seconds a zipper sound began ripping up my body. I was in darkness and the chemical smell was overpowering. I vaguely understood I was now in a body bag. The last thing I felt was being lifted by rough hands under my stomach. I prayed for death as the darkness became unconsciousness.

I WOKE TO A BRIGHT light, but not the one seen in the afterlife by those that reported their near-death experiences. The face looking down at me was not that of the vengeful Kib. It was a familiar one with a mask of grave concern.

"Hi Doc," I said, my throat dry and creaky. The oxygen mask on my face muted my weak words.

"Foolish woman," he said, shaking his head. I tried to move, but felt pain... everywhere. "Don't move. Almost all of your bones are shattered."

Shattered? Kib must have rethought his torture with knives and the mess of blood it would produce. Brittle bones created by long term low gravity withstood very little abuse. They also took forever to heal, especially in older women like me. An excellent way to end my life, I had to grant him that. Why was I still alive then, and why didn't he break my bones while I was conscious? Was I simply forgetting the horrific torture? So many questions, did I care for the answers?

"I'm ready to go, Doc. Xray November." I saw the pain in his face. I repeated the words and he nodded. His face was replaced by Hawk's. His eyes were glassy. "Don't worry, kid. It's been a great run. Off to see what comes next." I tried to be strong, but my voice cracked on the words.

"I'm sorry," he said. He proceeded to tell me what had happened, how he had fought with Kib over my unconscious form in the body bag, and that he had caused the bone shattering injuries in his attempt to rescue me. His apology was not for that, it was for trusting the regent and not taking my side from the beginning. He showed me the knife wounds that Kib had inflicted and the Doc had sewn up. He promised to help me heal no matter how long it took, but I shook my head. I turned my head to look for the Doc, wanting the end-of-life medicine before I lost my nerve.

"Xray November, Doc." The table I was on vibrated and the ceiling began to move. I realized the table was moving as Hawk slid out of sight. I felt the absence of Hawk's hand as he let go, never realizing until that moment he had been holding it. "Love you, kid," I said as my view of the ceiling was replaced by a tube. Looked like the MRI machine. Certainly, he wouldn't need to scan my internal organs before the injection. A hum began and grew until all my senses overloaded and a blessed whiteness overcame me. This was the bright light of the afterlife. My pain was gone. I felt perfect serenity.

"COOLANT LEAK IN MODULE Alpha three five one four seven." The voice was calm and familiar. It was my voice. I knew exactly where that module was. I reached out my arm and a wrench tightened the fitting a few minutes later. It wasn't a flesh and bone arm. It was one of the many repair robots that inhabited the station. It seemed like I could now control them with my mind. "Coolant leak fixed," my voice told me. I could see the robot move away from the fitting and down the corridor. I turned my head and realized I could see every corridor in the station. I felt the temperature discrepancy in the docking ring before my voice told me it was rising. The thought to turn on the fan barely crossed my mind and the temperature equalized.

"Hi, Maya." I turned my head to the console Hawk was sitting at.

"What the hell is going on?" I asked.

"Doc scanned your brain into the station mainframe."

"And..."

"Nobody wanted to lose what you know."

I wanted to ask how, but the powerful computer underlying the station surfaced all the relevant information and almost instantly I understood. "The regent did not approve this," I said. Hawk just shrugged. It was more than having my knowledge available to answer questions. I felt like I had control of most of the systems.

"You are the station now, Maya," Hawk said with a sly smile.

I turned my head and saw the regent in his office typing on his console. Instantly I saw what he was typing, and then absorbed everything he had ever typed since he had been on the station. He wasn't as corrupt as I thought, but corrupt enough. An admiral in the space command wanted the explosives to fake a terrorist incident and give him the authority to wipe them out. Tale as old as humanity.

I turned my head again and found Kib. He was limping down a corridor on level seventy-two. He still had bruises on his face from his fight with Hawk. Faster than I could think about it, a steel armature emerged from a wall and accelerated down the overhead track and hit Kib hard enough in the back of the head that he did a complete rotation forward and face planted on the floor. The armature reversed and then grasped the neck of the unconscious man and dragged it to the service elevator. It took a few minutes and a dozen repair robots to move the body to the module that was storing the refined Narthok. I sealed the module, and when the timing of the space station's rotation was correct, ejected it toward the Martain atmosphere. My revenge dish served very cold. My station was safe.

I turned my head back to Hawk sitting at his console. "Thank you."

MAGIC BOX

SCI-FI * TIME TRAVEL * AI

PG-13 25400 WORDS 102 PAGES

Magic Box — Chapter 1

"Jamal, get your skinny ass off the couch and clean your bedroom."

"It is clean."

"Now!"

"After this game."

"That game never ends."

"End of this level. I'm almost there."

"I'm getting the scissors."

This wasn't an empty threat. She had cut the controller cord before. Jamal was able to fix it, but it took hours of careful work, and each time it got shorter. He paused and saved the game. Then he stomped down the hallway to express his distaste with this chore.

He kicked at the clutter on the floor. He made a game of trying to kick his dirty clothes into the hamper. They just piled up nearby, but that was close enough. He happily put his schoolbooks in the backpack, then he could blame her for not finishing his homework.

Jamal looked at the picture of his girlfriend tucked in the edge of his dresser mirror. She didn't think she was his girlfriend anymore, but that was only temporary. So what if he talked to other girls? He was a talker, it was in his nature. He checked his phone for any updates. She hardly ever posted anymore. He saw some messages from his friends.

"You better not be on that phone." He heard. Not an idle threat either. He quickly tucked it back in his pocket and continued making cleaning noises, though no cleaning was taking place.

"What the hell are you doing?" He jumped. She was standing in the doorway, arms crossed in front of her chest.

"Cleaning." He said looking around for something to make look cleaner.

"Clothes IN the god damned hamper, not surrounding it. Pick up the board games and put them in the boxes. Wipe the dust off everything, wash the windows and mirrors, change your sheets, and..." she continued listed all of the tasks for the thousandth time. "Why the hell do I have to tell you this every god damned time?" He didn't have an answer. "You have one hour or your phone gets bricked."

This was the worst punishment since she learned how to do it on the web site. It was immediate and irrevocable until she cooled off long enough to go to the store to get it reset. Last time it took two weeks of extremely good behavior and he ended up missing a bunch of posts everyone else was talking about.

She came back with the glass cleaner and paper towels and he started doing it. Two more years and he was out of the house to a life of freedom. He had the poster on the wall. He was going to the state college in Grand Rapids on a scholarship provided by a private fund that helped kids that had lost parents in the war. All he had to do was keep minimal grades, and that was pretty easy for him despite his laziness.

He finished ninety minutes later after a few dozen quick checks of his phone. His mom brought up the clean laundry and left it on his bed. He put it all away haphazardly, except for the clothes he most treasured. A Jordan jersey that went to his knees, a pair of tattered jeans that hid just how skinny his legs were yet made him look so good. That was what Elise had commented on the day he discovered she liked him. He always believed it was the clothes she liked, because that was the thing she commented on most often.

He was looking at his phone when his mom passed the door. He instinctively moved to hide it even though his chores were done.

"Still nothing?" she asked with almost tenderness. He knew she was extremely happy Elise was no longer in the picture, mostly because a pregnancy would ruin everything. He shook his head. "You're better off, you know."

"No Mom, I don't know that. How can feeling like shit be better off?"

"You could be feeling my shoe in your ass," she said with seriousness, but he knew she didn't mean it. She never got physically violent with him. "Go out and do something while the weather is nice." This was full of unspoken warnings about staying away from the bad kids and out of the stores where he'd been caught stealing a few times, and everything else that was remotely fun in a poor neighborhood. At least they weren't in the projects.

He looked out the window. There was nothing out there except Elise, and he might see her with someone else. That was his biggest fear. He looked at his phone. Then he headed back down and restarted his game.

Magic Box — Chapter 2

"Jamal, you see the new teacher?" Burner asked.

"Didn't know there was one," Jamal replied, unloading the books from his backpack into his locker.

"She is off the hook."

"What does she teach?"

"You didn't hear?"

"Hear what?" He was tired of his friend dragging out the story.

"Mr. Millpond got himself stabbed."

"No shit," Jamal said as indifferently as he could. "She teaches biology?"

"I got some biology to teach her."

"Is she just a temp till he comes back?"

"Dude is six feet under. Even whitey don't come back from that."

The idea of an attractive teacher did not make him unhappy, but he liked Mr. Millpond. Of course, Burner's opinion of attractiveness was far from his. Two periods later he walked into the biology lab and saw just how hot she was. It was difficult not staring at her. She must have been just out of college. The principal was standing next to her, waiting for the kids to finish streaming in.

"As you may have heard, Dr. Millpond will not be returning to class." Jamal was surprised to hear that Mr. Millpond was a doctor. He guessed they knew biology better than anyone, but why wasn't he working in a hospital?

"Boy got himself shanked." Hippo Holmes said. He was a giant lineman in his senior year.

"Have some damned respect, Holmes. If you have any questions about the circumstances, please come to my office. There will be a letter passed out by the end of the day for you to take home to your parents. For now, I want you to welcome Miss Harris. She is an exceptional teacher and you are lucky she was available to take over the class on such short notice. You will give her respect or you will be in detention the rest of the year." He looked directly at Holmes and another troublemaking senior.

"Thank you Principal Autero. I have reviewed Dr. Millpond's files on each of you, and I look forward to talking to each of you individually. Until then, we will pick up where we left off on cell mitosis." She went into her lecture, and the boys mostly just stared at her perfect plump ass as she wrote on the board. She asked questions about what she had said to specific people using a seating chart, catching the first few not paying attention. Jamal laughed as everyone scrambled to pay better attention to avoid the humiliation.

"Mr. Jackson, could you stay after class, please?" Jamal nodded, and waited by the door when class ended. She waited for everyone else to leave.

"Did I do something wrong?" Jamal asked.

"No, the opposite of that. I have a favor to ask of you. Dr. Millpond's file indicated you were by far the brightest student in the class. I need some help getting organized, and I know your next period is a study hall."

"Organized?"

"Today was an easy lecture day, but tomorrow is a lab day and I have no idea where all the equipment is."

He was surprised that extra work, probably harder than cleaning his room, and missing study hall with his friends, did the opposite of piss him off. "Sure," he said cheerfully. Her face lit up in gratitude. She handed him the list of supplies and he led the way to the supply closet. She found the key and opened it. He pulled out the cart and as she read the list, he found the items and loaded up the cart. The hall was mostly empty, but a few kids he knew wandered by and made obscene gestures behind her back. He stopped looking in their direction.

"Thank you, Jamal. I hope I can count on you for the next few weeks. I promise I won't ask for too much from you."

"Whatever you need, Miss Harris. I really liked Mr. Millpond. Did he really say that about me?"

"The words he used were 'unmotivated genius'. Let's hope we can get rid of the 'unmotivated' part."

"Yes ma'am. Did he really get stabbed?"

"No. He had a heart attack. From what little I know of him, I think he would prefer the stabbing story." She gave him a conspiratorial smile. He nodded, happy to keep their little secret.

"Is this your first-year teaching?" Jamal asked.

"No third year, but my first in high school. I taught eighth grade science in Los Angeles."

"Weather is much better there. Why didn't you stay?"

"My mother got sick, so I moved home to help take care of her."

"So... you grew up here?"

"I was a student in this very classroom eight years ago. Back then, this was the chemistry lab. I never had Dr. Millpond as a teacher because he didn't teach the Honors level biology."

"If he was a doctor, why didn't he work in a hospital?" She looked at him confused. "Or a clinic or a nice suburban office."

"Oh, he wasn't that kind of doctor. When you go to school long enough, you can earn a PhD for almost any subject. Dr. Millpond had a PhD in Mathematics."

"Why didn't he teach math then?"

"He may have, but I don't know the answer to that. Maybe he liked this subject better."

"Is biology your favorite subject?"

"No, I actually like forensic science, which does have a basis in biology. But it also has a lot of chemistry and physics and best of all, mysteries to be solved."

"I like mysteries, but I never figure them out before the movie is over."

"If you did, then it would be predictable, and therefore boring."

"I guess so."

"Maybe try reading mystery books, that way you can put it down and give yourself some time to figure it out. You ask a lot of questions."

"I'm a talker. It usually gets me in trouble. I can leave if you want to do this alone."

"No, I really need your help. In science, asking questions is the second most important skill. Do you know what the most important one is?"

"Not blowing yourself up?" He guessed after he thought about it.

"That would be a very important skill. The one I was thinking of was listening to the answers. What is my favorite subject?"

"CSI."

"Right. I ask a lot of questions too because I need to know people are listening. If they aren't, we are all just wasting our time. You only have your mom at home?" He nodded. "My mom had me when she was fifteen," she confessed. "Never knew my father. Too common a story here."

"My dad died in Iraq." She nodded, having known this already. "I didn't know him well since he was deployed for so long before it happened. I sometimes think it would be easier if I didn't know him."

"Why is that?"

"Because maybe he might still be out there and I'd get to talk to him someday."

"Probably different for girls, but I hope I never meet mine."

"Why?"

"Because I'd probably stab him to death like Dr Millpond." She smiled. "Maybe I was his long-lost daughter and decided I deserved his job."

"You ain't white enough to be his daughter," Jamal observed. She nodded. "My dad was half Mexican, so I'm a quarter tamale."

"Your grandparents around?"

"Not much. Mom's parents live in Mississippi and she hates going there. Dad's mom and my mom don't get along so I never see her.

"Your mom's dad on email or Facebook?"

"No. Why?" Jamal asked, pretty much guessing the answer.

"Just wondering."

"Because I need a male role model to keep me from running with the gangs? My mom is way scarier than any man would be."

"Good for her. Looks like we are all done here. I have one more class I have to prepare for today. If there is anything I can do for you, don't hesitate to ask."

"Can you explain girls to me?"

"Not in fifteen minutes." She said with a knowing smile. "It has been my experience that if things don't work out, that is always a very good thing. Trust that the universe isn't random and aimed at ruining your life. We are here because we are designed to do more than survive."

He nodded like it made sense and said "Thanks. See you in class tomorrow, Miss Harris."

Magic Box — Chapter 3

"Jamal, what the hell is wrong with you?" Jamal's mom asked.

He cringed waiting for the yelling about what he did wrong. It didn't come, so he peeked over his shoulder. "What?"

"That isn't a game controller in your hand."

"It's a book."

"No, it's a book."

"That's what I said."

"When you say book, you mean textbook, forced upon you by school. That's a real book."

"Miss Harris loaned it to me. It's about a scientist who discovered black holes."

"She making you read it?"

"No. What?"

"Tell this Miss Harris we need a parent-teacher conversation."

"Don't get her in trouble. She only gave it to me because I told her I was bored."

"In trouble? Bitch got you to read a real book without threatening your life, I need to know her secret."

"I've read other books."

"Comic books. I don't care they are called graphic novels, they are comic books, not book books."

"Those are better than books."

"They are pictures with occasional words."

"This book has pictures." He thumbed through and showed her some of the illustrations.

"Don't think that is in place of your homework."

"No, I finished that first. Miss Harris made me promise I would."

"Now you are keeping promises? Bitch want to adopt you? I'll have a lawyer draw up the papers tonight."

"That's not funny."

"I expect a call from her soon."

"I'll tell her."

Magic Box — Chapter 4

"Oh, now I understand everything." Jamal's mom said.

"Excuse me?" Miss Harris said, looking at the woman in her classroom doorway, then at her watch.

"I'm a little early for our meeting. I have to get to the office early if possible. I'm Jamal's mom."

"Oh, he has been such a life saver. It is so nice to meet a mother who cares about her child so much."

"Caring about him never made him more than a lazy ass. You show up looking all fine and now he is a model student. Don't put that on me."

"I don't... Are you saying he is motivated by my physical attractiveness?"

"What man isn't? I was just as fine as you in my day. I know how it works."

"You asked for this meeting. Is there something you are concerned with?"

"I wasn't until I saw you. I wanted to know how you motivated him since I've been failing at that since he was in diapers. Kid wouldn't use a toilet properly until he was six."

"I'm not sure what you're asking?"

"Bitch, please. Short of fucking the boy, do whatever you can to keep him out of gangs and the military."

"Certainly, I will regarding gangs. I am not against the military, but—"

"They are just a gang with a government behind them. My husband tried to get out, he had a family for Christ's sake. He was just National Guard to go to college. They kept sending him back year after year. Jamal barely remembers him. Just like a gang, the only way out is death."

"As I was about to say, I think Jamal will go very far in academia, so military service would not be beneficial. He has a very good mind for science because he likes asking questions. That is most of the battle, especially for us." She nodded and turned to leave. "Jamal was right about you. He said you were scarier than any male role model would be."

"Um-hmm. Let me know if you need my help with him. I got enough trouble just keeping food on the table."

"I will. Thank you for coming in."

Magic Box — Chapter 5

"Jamal, do you know what this is?" Darlene asked.

"No, Miss Harris. I seen Dr. Millpond working on it one day. He didn't show up for class, and I knew to look for him here. I asked him what it was. He said it was 'his precious.'"

"Gollum, from Lord of the Rings," she said, recognizing the words.

"Who?"

"Add that to your reading list. For now, just know it is a book about a ring with incredible power. The problem was whoever had the ring was driven mad by its power."

"The box was driving him crazy?"

"It was probably just his way of saying it was an obsession. You said he was working on it?"

"He had this side panel open and he had a screwdriver-looking thing in his hand."

"Well, whatever it is, I need it out of here before the supply shipment comes in on Monday. Any idea where we can move it?"

"I can take it home."

"It may be school property, but it isn't on the inventory."

"It has to be safe, or he wouldn't have had it here, right? If I had it at home, I could try to figure out what it is."

"Your mom won't mind?"

"No." he lied.

"Okay, let's get it out to my car and I'll drive you home after school."

It was surprisingly lightweight, so he easily placed it on the cart by himself. They wheeled the bulky box out to the long parking lot and squeezed it into the back of her old car. After the last period Jamal waited in the parking lot for her to wrap up her work day.

"You don't belong here, Jamal." It was one of the school secretaries heading home for the day.

"Miss Harris asked me to help her with something in her car."

"Did she now? You still don't belong out here. People will think you're up to no good."

He headed back to the school, but wasn't sure which door she might come out of. He waited for the secretary to drive out and headed back to her car. He thought about squatting out of sight next to her car, but that looked even more suspicious. He was relieved when she appeared five minutes later, less so when saw the principal came out with her.

"Here it is," Darlene said to Principal Arturo.

"That certainly doesn't look like anything I would have allowed to be purchased. Are you sure it isn't dangerous?"

"He wouldn't have had it in the school if it was." Miss Harris said with more confidence than she felt.

The principal leaned in and lifted one side. "Very light. Can't be much more than the box. Maybe it's an empty computer case. Jamal, if you even get a hint that it is in any way dangerous, you let me know immediately."

"Yes, sir. I will."

"Good luck figuring out what it is." He nodded toward the boy, then toward her. "Miss Harris." He turned and headed back to the school.

"Sorry, I thought it was better to get his approval. Should have done it while it was still in the supply closet. She got in and started the car, Jamal put on his seat belt. There was no music playing, so it was an awkward silence in between his directions to the apartment building. They used a hand truck to bring it in and rode up the elevator with it. It barely fit in his bedroom door, and that was after turning it sideways.

He was embarrassed at the clutter, and vowed to keep his room spotless after that.

"You don't have room for this, Jamal."

"Sure, I do. I just have to rearrange things a little."

"Well, if you find out it is garbage, get rid of it quickly." She noticed the poster on the wall. "Michigan State. Is that where you plan to go?"

"Already have the scholarship."

"Good for you. They have a lot of good science programs. I better go now. Traffic is only going to get worse." He walked her to the door. "See you Monday, Jamal. And thanks again for your help."

"Anytime, Miss Harris."

Magic Box — Chapter 6

"What the hell is this?" Jamal's mom asked.

"A school science experiment."

"What does it do aside from take up half your room?" she exaggerated.

"I don't know. That's where the science comes in."

"I better not find out you stole this."

"No way. It was something Dr Millpond was working on. Miss Harris asked me to figure out what it is."

"Miss Harris. She asks, and you just do. At least you cleaned you room without having to be asked a dozen times. I'm going to McDonalds. You want anything?"

"No, already ate," he lied. He hated McDonalds, except for the fries.

He had been trying figure out how to get the side panel off. There were no screws, so he figured it slid off like a computer. Maybe this was some kind of computer case. He tried everything short of a crowbar, which is only because he didn't have one.

He heard his mother leave. "What I'd really like is some Popeye's chicken, extra spicy," he said out loud, knowing his mom would never drive to a second fast food place for him. Suddenly the box vibrated for five seconds, then stopped. After a minute he thought he smelled Popeye's chicken. "That's weird." Then the front panel just disappeared, and inside was a box with Popeye's logo on the lid.

He just stared at it. He was afraid to touch it, but it smelled really good. He started to reach in, but wondered where the panel went. Maybe it would close and cut his arm off. He grabbed a ruler off his desk and used it to pull the box out. He opened it and it was chicken. He touched it and it was hot like it just came out of the fryer. He jumped when the front panel noiselessly reappeared. He sat there thinking about what had just happened. He said he wanted something, and it just appeared in the box. The box that wasn't even plugged in. "I'd like fries too," he said. "And a Pepsi." The box did nothing.

He broke off a piece of the battered skin. It felt real. He put it in his mouth, expecting it to taste like plastic. It tasted normal. He picked up a drumstick and took a big bite. It was the most delicious chicken he had ever had. Still no fries and soda. He went into the kitchen to get some water. He finished the piece of chicken and started a second one. Maybe it had to recharge. He tried again several times without success. Maybe the words had to be just right. He tried to remember exactly what he said. "What I'd like is... French fries.", then he quickly added "from Steak and Shake." The box vibrated, the panel opened, and there were the distinctive thin French fries. He finished eating and his mother came home.

She was sitting in front of the TV eating her food. She saw the box in his hand as he carried it to the kitchen. "You had Popeye's and let me go out for McDonald's?" She was angry and incredulous. He just shrugged. "Give it here. Where'd you get the money?"

"Miss Harris bought it for me for helping her with the stuff after school."

"You ain't fuckin' her, right?"

"Mom..." he whined.

"Can't say I'd blame you. This is really good. Did you get it over at the store on Fourth? They're usually way greasier than this." Luckily for Jamal, she wasn't really looking for answers.

He got some more water and went back to his bedroom. He immediately began thinking about what he should ask the machine for next. Ice Cream? Air Jordans? Money.

"What I'd like is a thousand dollars." He waited expectantly. Nothing happened. "What I'd like is an Oreo Blizzard." The machine vibrated and opened, revealing the ice cream cup with a plastic spoon. He ate the ice cream, wondering if it was just a food replicator like on Star Trek. He asked for the money again, nothing happened. He thought about the Air Jordans. He would never be able to explain them to his mother. She would know he stole them. He would lose the phone and all sense of freedom. He tried to think of something non-food that he could ask for. He pictured Dr. Millpond and suddenly he remembered.

"I'd like a copy of the book Lord of the Rings." The machine buzzed and opened and there was a very old looking copy of the book. "My precious." He said, remembering the words the old, white biology teacher had said. Had he invented a machine that could make almost anything, then died before he could tell anyone about it? He opened the book and started reading.

Magic Box — Chapter 7

Jamal stood by the door at the end of biology class as the other students filed out.

"I don't need your help today, Jamal," Darlene said.

"I need your help," Jamal said quietly.

"With what?"

He looked around then poked his head outside the door to make sure no one was nearby. "The doctor's machine."

"Did it blow up?"

He shook his head. "It makes things."

"Oh, it's a three-dimensional printer? I didn't even think of that. Maybe that is school property. Not really a biology thing, unless he used it to make models. That would be really useful."

"No, Miss Harris..." He checked the hall again, then closed the door. "It makes things just appear."

"A magic box?" She smiled.

"I guess so."

"I didn't know he was a magician."

Jamal shook his head. He pulled a phone out of his pocket. "It made this, and a bunch of other things."

She took the phone from him. "Is this some kind of joke, Jamal?"

"No." He related what happened with the chicken, then the book, then the other experiments he ran. "This is a top-of-the-line iPhone. It costs more than my mom makes in a month. No way she would buy this for me. He pulled out his phone. He opened a video, hit play and handed it to her. She heard him say the words.

Heard the vibration noise, then the door just disappeared and the phone was there.

"Does it work?"

"It doesn't have a sim card, but it works like a phone that is offline."

"This can't be real." She replayed the video. She was trying to figure out how he faked it, and more importantly why he would fake it. "Do you have the book?" He pulled it out of his back pack. "This is a first printing. It's worth a lot of money, maybe millions. It looks old, but also brand new. It must be a fake."

"There's something I don't understand. After the chicken, I asked for money. It wouldn't make that. Why wouldn't it make that?"

"I don't know how it would make any of this. Please tell me if this is some sort of practical joke on the new teacher."

"You need to come try it. I like what it can do, but it could be dangerous."

"How?"

"What if it would make guns or drugs or bombs? There are a lot of bad people in my neighborhood that wouldn't think twice to kill mom and me for it."

"Jesus Jamal, you're really serious about this."

"There's one more thing I haven't told you. I never plugged it in."

Magic Box — Chapter 8

"I'd like a BLT sandwich," Darlene said to the box. Nothing happened. She knew he was pulling her leg.

"It only worked when I named the restaurant."

"Okay. I'd like a BLT sandwich from Quiznos." The box vibrated and opened. She just stood there wide eyed, looking from the box to Jamal and back. She finally reached in and found it toasty warm. She pulled it out and the door just reappeared. She tapped the door. It seemed solid enough.

"I'd like a can of diet Coke." Buzz, open, presto. It was ice cold. "I'd like a thirty-milliliter vial of insulin." Buzz, open, presto. Life sustaining medicine for her mother. Could she trust it?

"Now do you believe me?" Jamal asked.

"I believe you. I just can't believe my eyes. I wish we hadn't taken this out of the school. Have you told anyone about this?"

"No way."

"I need you to stop using it. Can you cover it up and not tell anyone? I need to think about this." She thought about it. "I would like a plastic model of a heart for teaching." Buzz, open, the exact model she was looking to order for the class. She hadn't mentioned a brand. Maybe there was only one possibility. "I would like a single United States Dollar bill." Nothing happened. "I would like an ounce of gold in ingot form." Nothing happened. "There seems to be limitations. Certainly, size since it would have to fit inside the box."

"Do you think Dr. Millpond made this?"

"No. This is way beyond our technology."

"Aliens?"

"No, at least not from another planet."

"Where else would aliens come from?" Jamal asked.

"Our future. The distance across the galaxy to travel is far more technologically challenging than traveling a few thousand years back in time. I believe most real UFO's are just visitors from our future."

"Someone sent this back to Dr. Millpond?"

"I have no idea. I think you are right. This could be extremely dangerous." She turned to the machine. "I would like a Glock 19." The box did nothing. "That is hopefully a good sign. I need to leave before I think of something else to ask for. This is precious. Almost ultimate power. This needs to be studied by someone who can be trusted."

"Frodo?" She nodded. He went and got a sheet out of the linen closet and covered the machine. "I promise I won't use it." She looked at the sandwich. She took a bite. It was now room temperature, but it had been perfectly toasted and the tomatoes tasted just-picked fresh and delicious. She began thinking of all the things she went without her whole life. She was tempted to take the box with her. She left before the temptation overcame her.

Magic Box — Chapter 9

Jamal looked at Miss Harris as he entered the classroom every day. She would just shrug and shake her head. She had no solution. He finished all three Tolkien books in the Lord of the Rings series, and the lesson seemed to be to toss the magic box into the volcano. He figured the river would do the job just as easily. As Thanksgiving approached, Jamal was tempted to use the machine to make gifts for the holidays. He knew he could have it make video games because his mom wouldn't know what was new. Then he wondered if he could make things to sell to others for the cash the machine wouldn't give him. Every idea ended with his mom finding out and him being punished. The machine was almost useless to him.

Finally, one day Miss Harris asked him to stay after class. "I finally used the insulin. It worked perfectly. We could be making medicine and giving it to poor families."

"I thought about having it make toys to donate to shelters and stuff. Someone will find out. Should we throw it in the volcano?"

"No."

"We need to trust someone. Principal Autero?"

She shook her head. "He's a bureaucrat. He'll give it to the government."

"What about the man who invented Black Holes?"

"Discovered Black Holes. He died a few years ago. I'm sure there is a scientist that could help us, but if we pick the wrong one, we're screwed."

"What if we propose it to them as a hypothetical? You certainly didn't believe me, why would anyone think it is real until they see it?"

"I think Dr. Millpond was right. You are a genius. Who should we approach?"

"Someone nearby since eventually we need them to see it."

"Someone with more than high school resources," she added, nodding in agreement.

"Maybe I could contact the professors at Michigan State since I'll be going there."

"They have a good science department. I think it would be better coming from an inquisitive student instead of a teacher."

They went to the computer and looked up the faculty of the college. They narrowed it down to five possible candidates to approach. Then they composed the email with the hypothetical question.

"Which one do we send it to?" she asked.

"All of them. We need to stop wasting time. I'm going to use it. I almost asked it for jewelry to get Elise back."

"A girl that is bought by jewelry is not one worth having."

"They certainly don't want me for just me."

"They will. Girls your age want very different things than what they will want when they grow up. If you get a girl now, she will want something completely different then. You either have to change into that, or lose her after so much time has been wasted. It is almost time for the next class. Send or wait?"

"Send," he said. She clicked the send button and logged off the computer.

A reply was waiting for him when he got home.

"Dear Jamal, Your teacher, Miss Harris, should be commended for encouraging such four dimensional thinking in her class. I must say your query is very intriguing, and it makes me less cynical about our future here at Michigan State if you at all represent the young minds that will be attending soon. I look forward to discussing this, and many other topics when you join us here. Sincerely, Professor Jonathon Fenderwald."

"Worthless," Jamal said out loud. He went to the living room and played his video game.

A day later a similar email came in from another professor. However, this one was almost a form letter, including no specifics about their request. It was a week before another email came through.

"Hey Jamal, Your fascinating concept was forwarded to me by Dr. Smithson and I would love to discuss it with you and your teacher. My phone number is at the bottom if you can call me to discuss. Dr. Mark Ingals, Professor of Sociology." Jamal showed it to Miss Harris after the class the next day.

"Why would a sociologist be interested in a device?" he asked.

Darlene shrugged. "He would be interested in how a device would make people behave. Do you want to call him now?" Jamal shrugged. It was the best offer they had. She turned on the speakerphone and dialed the number. They got voicemail. She left a message. Jamal was about to leave when the phone rang.

"Hello?" she answered.

"Hi, I'm Mark Ingals. Do you still have time to talk now?"

"About thirty minutes left."

"Great! You said this was a creative writing idea that needed scientific advice. As you can assume, I will not be able to give you much information regarding the physics of such a device, but honestly the far more important part of any story is the people. How society would react to unlimited wealth is of endless fascination to me."

"Why?" Jamal asked.

"Oh great, you're there too, Jamal. 'Why' is always the most important question. I'll start by saying that I am a communist. Not one of those scary totalitarian ones that enslave nations. I live totally in the realm of theoretical utopian communism. Have you ever watched Star Trek?"

"Most of them," Jamal answered.

"That is the kind of utopia I hope to live in some day. There are only two things needed to make that happen. Endless free energy, and endless resources created by that endless free energy. Once you have that and it is controlled at the individual human level, so many social problems just disappear. However, there will be a transition time between what we have now and this utopia. What we have now is a semi regulated system of limited resource management, otherwise known as an economy. This transition time will be the most important part of any story regarding such a device. Because I want to live in that utopia, I have been studying the paths to the end, hoping to find the most peaceful one."

"Peaceful?"

"Most men would want to control the technology, simply to give them a competitive advantage in business. Governments would want to get involved for the same reason since controlling this technology could be control of the world."

"How?"

"Creating endless weapons and potentially used as a delivery device. Imagine a bomber flying over enemy territory, creating jet fuel endlessly, and creating an endless string of bombs to be dropped and missiles to be fired."

"What if the device wouldn't make weapons?"

"That would be great, but the military would likely find a way to alter it so it would."

"Have you found a peaceful path?" Miss Harris asked.

"Not yet. The best possible outcome is to get the devices into every home before someone is able to control it. That is so far from the realm of possibility that it is almost sad. There is also the psychopath problem. A disturbed person could use the device to create a lethal virus or poison gas and kill off thousands, maybe millions of people."

"If you had such a device, would you destroy it so it couldn't be used for bad things?" Jamal asked.

"If it can be invented, it will be invented. It is just a matter of time. It would be better if a moral scientific community controlled it, but they have little power in the real world. A lab is where it would most likely be invented, but word would get out quickly, probably before it was finished. The government and military would step in and just take over."

"What if a kid like me found it by accident."

"You mean like it was left behind by aliens?"

"Sure, something like that. What would you do if you found it?"

"Wow, that is a really good question. Money would probably be first, so I could afford to take it away from prying eyes until I figured out how to make enough of them."

"It won't make money."

"Why not?" Dr Ingles asked.

"What Jamal is trying to say is that the story would be too simple if it could just create money. That would technically be counterfeiting."

"You're absolutely right. In fact, in a utopian future, money would be completely unnecessary. You've obviously had more time to think about this. I will definitely put more thought into it. Do you have any other questions?"

"You said someone else forwarded the email to you. Can I ask why they aren't interested in responding?"

"Professors are generally busy people, working on research and teaching. In my experience they don't spend much time in the hypothetical realm unless it pertains to what they are researching at the time or they are looking for new avenues to research. You just happened to hit a sweet spot with me in that I am between research projects and looking for better solutions for the problems at hand. As much as I would love to have your... magic box, I know we are probably centuries away from having it."

"I think it will be much sooner..." Jamal started to say.

"Thank you for your time, Dr. Ingals," Darlene said.

"Anytime, feel free to send me any other questions you have. I look forward to reading the first draft of your story."

Darlene hung up the phone. "I think we can trust him," Jamal said.

"Maybe, but not yet."

"He would be able to get others..."

"Not yet. I think he is right. It is too dangerous. We need to move it out of your house."

"To where?"

"I don't know yet."

"I wonder where Dr. Millpond got the machine from?"

"I looked through all his files and there wasn't anything about it. Strange that he kept it here instead of at home."

"Maybe he couldn't trust his wife with it."

"His wife... maybe he has notebooks at home about it. I should go talk to her."

Magic Box — Chapter 10

"Mrs. Millpond, I'm Darlene Harris. I took over your husband's classes. I am very sorry for your loss."

"He was..." She began to tear up. "I'm sorry, please come in."

"I'm sorry for dropping by unannounced like this, but I only had an address. The phone number on record was out of service."

"I had to change it. Damn vultures don't stop calling when they find out someone has passed. Ten times a day I had someone asking if I was going to sell the house. Thirty-five wonderful years here and they think I would just cash in and move away. Can I get you some tea?"

"No thank you. I won't take up much of your time. Your husband kept very elaborate notes about the students and equipment at the school. It has been such an easy transition. However, there is one piece of equipment that I don't have any records for. Not really sure what it is. I was hoping you might have some information about it here. Maybe a service manual, or a notebook?"

"The confabulator. Almost forgot about that damned thing. Take my advice dear, throw it in the dumpster."

"Confabulator?"

"That's what he called it. It's the apple in the garden, and he couldn't help but eat it. I made him get it out of the house."

"Do you know where he got it?"

"You know what it does, I can see that in your eyes. It's the golden goose. Can't say I blame you for wanting to know more about it. If you want to hear the story, I'm going to have to make some tea." She got up and went to the kitchen. Darlene followed her and watched the woman deliberately do the routine task she must have done thousands of times in this kitchen.

"I met Ben in college. He was a teaching assistant working on his doctorate and I was in the admissions department. He was always friendly, but usually kept to himself. One spring he came to my boss with complaints about our admissions policy. My boss was less than thrilled with his admonition, but I was intrigued. I sat outside my boss' office, so I heard most of the conversation that quickly became an argument. I thought about what Ben had said, and a few days later I went to one of his classes. I could tell by the way he lectured that he truly loved both teaching and the students, which is probably the same thing.

"I wasn't one to be swept away by romantic flights of fancy, but just a few minutes in his presence and I started to feel... well I think I knew I was going to marry him. I had been dating an insurance broker for a few years, comfortable with the simplicity of our arrangement. Ben only had a few minutes to talk to me after the class. I had just enough time to tell him I agreed with his admission complaints and wanted to talk about what changes could be made. He knew I had no power to change anything, but he agreed to have coffee with me the following weekend. It was impetuous, but I drove home and asked David, my insurance man, to come over. When he arrived, I handed him a box of his things and told him it was over, that I no longer wanted to continue our romantic relationship.

"He accepted it almost with relief as if he had wondered how he could rid himself of me. That hurt a little, but I felt it was time to move on. Coffee on Saturday morning became dinner on Saturday night, and then church on Sunday morning. No dear, I didn't spend the night with him. We were not that kind of young and foolish kids. We both had careers lined up and our meeting seemed like a convergence of something outside of our control. A benevolent force chose the perfect time to connect us. We married only two weeks later after we both realized how perfect we were for each other.

"I certainly don't recommend that despite how well things worked out for us. As I said, there was an outside force that pushed us together. I know you are wondering why I am giving you such a long and drawn-out tale with no relevance to your inquiry. Part of it is because I miss telling the story. Ben and I used to play the game of 'Remember when'. We would say those two words and bring up a memory of our younger days, usually in the same mood as what we felt. When he was irritated, he would bring up one of those things that I used to do that irritated him that he learned to love about me. It helped him redirect his mood back to love. Our days were filled with remembering the good times, some of which seemed bad when they were happening.

"Another reason I am giving you the long version is so that you know how evil that box is. It almost destroyed our marriage. One 'remember when' we never said out loud was 'remember when that spaceman brought us the box.'"

Darlene sat up straight, her lagging attention at the old woman's long-windedness having sapped her limited reserves. "Spaceman?" she asked as Irene sat sipping her tea with a small smile.

"That is a part of our story I never get to tell. Who would believe it anyway? You will, because you know the secret of the box. We never had much money, but it was because of how we chose to live. With summers off we travelled, burning most of our savings each year to explore the world. When I got pregnant, on a trip to South America, we were actually in debt. We tried to buy a house, but real estate near the university was prohibitively expensive. Neither of us wanted to give up our ability to walk to work from my tiny one-bedroom apartment, so we rearranged the furniture, made a small baby room out of bookshelves in a corner of our bedroom, and we waited for the day our daughter would join us.

"It was that night, after I had felt my first mild contraction, but before we began packing to go to the hospital, that the spaceman appeared. I was sitting on the couch, both frightened at what was to come in the next few hours, as well as thrilled that it would finally be over. Being fat I mean. Pregnancy definitely did not agree with my sense of vanity, though I did it four more times. All those career plans went out the window with the second pregnancy. We needed a house, so we needed to move. An old classmate offered Ben a teaching position in a well to do suburban school, we bought a house, and moved across the country. I begged him to leave the box behind. It had, as I said, almost ruined our marriage. I told him if he did move it, he could not keep it in the house. That is why he kept it at the school. First at his rich suburban school, later at your school."

"I did wonder why he was teaching there instead of the inner city."

"That goes back to that first argument in my boss' office." Darlene held back the exasperated sigh of impatience as the woman continued to meander. "Ben wanted to change the ratio of minority and female students in his classes."

"To be more inclusive?"

"No, quite the opposite. You see, admissions policy had changed radically in previous years to be extremely inclusive, to the detriment of everyone involved. Students with no business being in college were now being admitted to the highest level of universities with no chance of succeeding. They began pressuring professors to give higher than earned grades, but the students knew what was happening. Most disappeared that first year, sometimes that first semester, believing they were failing on their merits."

"He wanted less black people in his classes?"

"Not for the reason of their skin color, but yes. He did some research into the problem and found out it was systemic from far back in their academic lives. Kids were being socially promoted and the soft bigotry of low expectations handicapped them. Ben did his best, but realized the problem was bigger than he was and eventually gave up. Then about seven years later, when I was pregnant with Lori, our third child, he had an epiphany. That led him to inquire about teaching at the inner-city school where he knew he could start making a real difference."

"White savior complex."

"That's an interesting name for it."

"I remember all the students loving his class. I didn't have him because I was in the Honors level classes."

"You had no need for his... savior complex. Savior, that is exactly why he did it. He knew it was what Jesus wanted him to do. I can tell by your expression you aren't a believer."

"Christian cringe factor."

"What is that?"

"It isn't the ideas, the philosophy, that bothers us. It is the self-righteous preaching about accepting the man ahead of his ideas and philosophy."

"I am the path—"

"Yes, his ideas are the path to a better world. Do you think he was so vain that his name needed to be attached? He cared about how we treated each other, not the language we use in doing so."

"I never thought of it that way."

"It doesn't matter. We are getting further away from the 'spaceman.'"

"How true, though the spaceman was Jesus."

Darlene blinked. "Jesus appeared to you?"

"He didn't claim to be, but that is who it was. You see, God gives us everything we need to survive, thrive, and procreate. What we choose to do with those things is how we will be measured. My youngest, Alan, chased every skirt, took every drug, ate every scrap of food he could get his hands on. He died of an overdose before his twenty first birthday. It would have been sad if it hadn't been such a nightmare the preceding three years. I understood the temptation he succumbed to because I had that weakness myself. My weakness is why I made Ben get rid of the confabulator. I could not control my desires. It started with necessities, we were poor, especially after taking on a mortgage and a car payment. If we had known... it doesn't matter. Then I started making jewelry, pretending I was wealthier than I was. I sought the envy of others.

"It finally came to a head when I discovered it could make legal drugs. Losing weight after the baby was hard, so I took some amphetamines. Not bad at first, but I stopped sleeping and went out of my mind a little. What I didn't realize is that it came through in the breast milk, and it almost killed my little girl. I swore off the machine and demanded he get rid of it. I even took an axe to it one day when my temptation grew too strong. Didn't even scratch it. Benjamin was never tempted by it. He saw the real value in it. It could be used for the greatest good imaginable."

"Only in the hands of someone capable of resisting the temptation." Irene nodded. "He had it for thirty-five years and never told anyone?"

"He talked theoretically with his former colleagues at Stanford. He wanted to solve the engineering of it, hoping once he published the schematics, everyone could make their own and no one would own it."

"He died before he could accomplish that," Darlene said.

"Just like me and that axe, he never scratched the surface."

"Does he have any notebooks?"

"Yes, but if you are smart you will throw it in the lake and forget about it. God help us if the government gets ahold of it."

"Can you tell me about the..."

"Spaceman?" Irene asked. She thought about it and nodded. "I was sitting on the couch. The television was on the home shopping network. For some reason that always helped relax me. I think it is the way they talk. The television became increasingly snowy, and then all of a sudden, he was there. If I wasn't so constipated, I would have soiled myself. He was in a silver suit, with a glass bubble on his head. The bubble was also silver, so I couldn't see inside it. I screamed which of course brought Ben running. The figure turned to look at him. He stepped forward toward me and pushed the coffee table to the side. I thought he was going to attack me, so I screamed again and tried to get off the couch. Another contraction hit just then, so I fell back in pain. He stepped back. He was only making room on the floor for the box. He used one hand to tap his other arm, and suddenly the box appeared.

"I was panting hard and my eyes were closed as the pain grew. I will never forget the sound of his voice. It was electronic, but also ethereal. It echoed like we were in a cave."

"What did he say?" Darlene asked, prompting the woman who had stopped talking in her dreamy remembrance.

"He said 'Create Pampers newborn diapers.' Not exactly what I expected. You would expect, 'take me to your leader', or 'we come in peace', or 'prepare for colonization and eradication.'" Darlene laughed. "Those are Ben's jokes, thought of long after the fear wore off. Anyway, the machine rumbled, the door opened, and there was a package of diapers. The silver man pulled the package out and then began requesting other items. It was like a mini baby shower. The silver man waved for Ben to try. Ben asked for a stopwatch. It was produced. The silver man gave him a thumbs-up gesture, touched his arm, and disappeared. We sat there for twelve minutes just staring at the box and the items that surrounded it. I know it was twelve minutes because that was how far apart the contractions were. After those contractions passed, we went to the hospital and thought little about what had happened until we returned home.

"Ben began experimenting with the box. And discovered several very valuable features. In those early days of diaper changes, the command 'destroy contents' was much preferable to carrying the dirty diaper downstairs to eliminate the smell. That feature alone could radically alter our society. No more dumps, landfills, toxic waste, garbage trucks. Just imagine it."

"How do you get the diaper, or garbage, in there?"

"Just say 'open', dear. It worked in every language Ben threw at it."

"What else did he try?"

"Almost everything. He destroyed most of what he created, not wanting to accumulate valuable objects that he could not explain. His favorite thing was to create foods from his childhood. Candy they no longer made, Chicago Style pizza, completely safe sushi. The meat is completely created from energy he said, no more farms with animals in horrific conditions for their short miserable lives before being slaughtered. The most amazing thing... I shouldn't tell you."

"Please. I need to know everything he did."

"You'll want to help people. Blankets and pillows to homeless shelters, cancer drugs to people who can't afford them. Dear... the cure for cancer is in that box. His notebook lists the drug's name. I had breast cancer twice, completely eradicated."

"How did he discover the drug name?"

"It was in early trials. A friend of his was on the research project. Said it was a miracle drug. Bleekertech bought it and shelved it. As I was saying, you'll want to help people. Eventually, they become suspicious."

"I'll be careful, I promise."

"It can make things... things that are bigger than the box."

"How?"

"It is very dangerous, because once created, it cannot be destroyed by the box, at least Ben couldn't figure out how to do it. When you ask for something large, say a washing machine, it creates a small ball. Ben said it was a forty-seven-sided polygon, but it looked like a ball to me. You take it to the area you want it and press the red button on the top. After a five second delay, and only if the necessary space around it is free, it grows into the object you asked for. It is quite something to watch, like a cake rising in fast forward. Reach your hand into the area, and it pauses until the space is clear again. The only thing that remains is the red button. Ben would put it back in the box and destroy it."

"Did he try to use the button to shrink the object back down?"

"Every way he could think to. He made an expensive car in the garage. One that he could never explain owning. It had no serial numbers, so it would be considered a stolen car. He spent three months disassembling and cutting it up to make the parts small enough to destroy in the box. That was after trying the red button every way he could think of. That is the importance of thinking things through. Shredding a million-dollar car, can you imagine it?"

"Do you think it could create a house?"

"Probably could create a skyscraper. How would you explain it?"

Darlene nodded, her mind flush with all these new possibilities. "Do they have an expiration date? Maybe they degrade over time?"

"Those picture frames are from the first few weeks of the box. Changed the pictures a few times, but they remain solid. Ben said they are chemically the same as what the original would be. The only difference he found is in food, there is no DNA."

"Well, I've taken enough of your time today. I can't thank you enough."

"Do you have someone special in your life, dear?"

"No, not really. I mean, I did, until I had to move back here from LA. The long-distance relationship just didn't work for us."

"Too bad. I think it helps to have someone to ground you in reality, especially if you intend to keep using the box. Who knows what I would have done without Benjamin." She began tearing up.

"Why do you think he was chosen?"

"We had some theories over the years, but I believe it was Jesus who chose him because of his pure heart. Many people would use it for wealth and power. You can see we lived very simple lives and did our best to help people."

"Any chance I could get his notebooks?"

"Of course, dear. No sense in you making all the same mistakes he did."

She got up unsteadily and wandered back toward the kitchen. She opened the basement stairs and turned on the light before descending. Darlene followed her down. It was dark and damp and in serious need of cleaning. She saw hundreds of boxes stacked floor to ceiling. Irene pointed toward them.

"Which ones?"

"All of them, dear. He wrote down every single word he uttered intentionally in its direction and then wrote down the result. Thirty-four years' worth."

"By any chance did he keep a log of commands that worked in a separate book?"

"I don't know. I refused to talk to him about it. The last time he told me something was when he discovered the large object creation. A new washer and dryer had been a wonderful anniversary present, but it gave me all those feelings again, so I made him swear to never talk about it again. That was almost twenty years ago."

"And your children don't know about it?"

"No. Only our oldest would have had anything close to the self-control to use it wisely. She was just like Benjamin. He intended to bring her in on it eventually, but she is busy with her own family and job and he thought there was plenty of time."

"There were no warnings of the heart problem?"

"His doctor said he was healthy at a physical just three months earlier. They don't run unnecessary tests at the HMO, so it's possible there was something there that could be detected. He certainly never complained about chest pain. He was just out mowing the lawn. I could hear it get louder and softer as he went back and forth. Then the sound just stayed steady. It took maybe ten or fifteen minutes for me to think it was odd that it was no longer moving. I looked out the kitchen window and he was face down in the grass. They said it was the widow-maker, that he was probably dead before he hit the ground and did not suffer much, but that is the kind of thing you would expect them to say."

"If he was conscious at all, he likely would have rolled onto his back. I would much prefer passing quickly to what my mother is going through."

"Oh dear, is she ill?"

"That is why I moved back here." Darlene said as she tried to make sense of the labels on the boxes. "Poor circulation is slowly eating away at her. She lost her second leg above the knee last spring, which is why I had to come take care of her. The house isn't exactly handicap friendly. She's miserable most of the time, so I just..."

"I understand. Quick and relatively painless is better. I just wish I had enough time to say goodbye. Hopefully, we'll be together again soon."

"Is there a code to the writing on these boxes? It seems to be random numbers."

"They are star dates, you know like from Star Trek?"

"Oh, so they aren't in any kind of order?"

"They were in a storage locker. The men who moved them here did not keep them in any particular order."

"I'm surprised you didn't..."

"Destroy them? Trust me, I was tempted. Then I thought maybe someday Melissa would find some value in them. I'm the one that forbid him sharing it with her. Maybe that was a mistake. Maybe together they could have done some real good with it. I guess that is up to you now."

"What does she do for a living?"

"She does biotech research down in Kansas City. I'll probably end up there if I lose my independence. I'm hoping I go before that day comes. Would you like me to have the boxes moved to your house? I'd happily pay to be rid of them."

"I have no room for them. Our basement is small and floods almost every time it rains. Is there any chance I could spend some time here to go through them, maybe transfer the data into a computer?"

"Of course, dear. I would love to have the company. Hardly get any visitors these days."

"Would it be okay if I brought one of my students to help me?"

"I don't think you should trust children with the knowledge of what the box can do. They just don't have the self-control yet."

"This one does. I trust him more than myself. He is special. Your husband had especially nice things written about him in his file."

"Jamal."

"Yes!"

"Every year Ben had at least one student that frustrated him because he couldn't break through the apathy. He loved talking about them, all of his students when he came home for dinner. It wasn't a job for him. It was his life's work. If you have an opportunity, look at his files for previous years' students. He found many effective techniques you might find useful. A book to detail them was another project he never got around to."

"I will do that." She opened one of the higher-numbered boxes that were accessible without moving others and picked up the top lab notebook. She opened it and saw the tiny neat writing. It was an extraordinary amount of work. She decided to just take a few of the books with her. She put the lid back on and held them contemplatively.

"I should get your phone number so I can call and arrange a time to come over."

Irene read off the numbers as Darlene typed them into her phone. "That's an old-fashioned wall phone, dear. I don't do the texting. I'm here most of the time so you are welcome anytime. I could give you a key as well."

"No, I wouldn't want to be here if you weren't. I don't exactly blend into this neighborhood."

"I know what you mean dear. It wouldn't be a problem for you, but Jamal might raise some eyebrows. I'll be sure to let the neighbors know that you are helping me make sense of Benjamin's research data."

Magic Box — Chapter 11

"We are going to bring the box back here to the school." Miss Harris said.

"Will it be safe there?" Jamal asked.

"According to his wife, he has had it here for twenty-five years. There are hundreds, maybe thousands of lab notebooks of research he has done on it."

"So... you know how he got it?"

"I know what she told me, but I find it impossible to believe."

"Why? What did she say?"

"She called him 'spaceman Jesus'" Jamal screwed up his face in confusion. "He appeared in their living room, moved the coffee table to the side, then tapped his arm and the box appeared. He spoke in an electronic echoing voice to show them how it worked, then tapped his arm and disappeared."

"No rocket ship or flying saucer, so why 'spaceman'?"

"Silver suit from head to toe so I imagine he looked like an astronaut."

"Beam down or time travel?"

"Magic mushrooms."

"You think she imagined it?"

"Who knows? The point is we have an enormous amount of data to go through and the only place with room is where they are right now, in her dank basement. What I think I want to do is go through the books and just copy the successful tests, and maybe notable failures, into a spreadsheet."

"I don't type very fast."

"I do, so you can read them to me." Jamal nodded. "Can't we just scan them in?"

"It is handwriting, but it is neat, so maybe that would be possible. We'd still have to go through and verify it in case something was not properly interpreted. She did tell me a few important things about it. Saying 'open' opens it up."

"I tried that."

"Well, maybe she doesn't remember right. She also said the phrase 'destroy contents' will dematerialize whatever is in the box."

"That's cool. That will help with experimentation."

"The other thing... I want you to promise not to try it out on your own."

"I promise."

"It can make things that are bigger than the box."

"What? How?" he asked excitedly.

"I'll show you after we move it back here."

"Today after school?"

"No. My car is in the shop for a few days."

"Bigger than the box." He repeated thoughtfully. "Can it make you a new car?"

"Yes, but it has no serial numbers, so it would appear to be a stolen car."

"That would be bad. The guy in the silver suit spoke English?"

"I assume so since they understood what he was saying."

"Then it probably isn't aliens. If someone from the future brought it back, would that change their past?"

"I have no idea. Time travel into the past shouldn't be possible."

"Why?"

"Time is only a description of how matter moves. You cannot 'un-move' matter. You can move it back to where it was before it was originally moved, but that is two distinct movements, not undoing the original move. Also, the specific matter traveling back in time existed in a different form at that past point in time. Take one single water molecule of that time traveler. It is in his salivary gland. That water molecule one hundred years earlier was not in his body, it was in the Pacific Ocean, or a cloud, or maybe it was in the cellulose of an ear of corn. When he travels back one hundred years, that water molecule now exists in two places at the same time. That violates the law of conservation of mass."

"Who was he then?"

"Time travel makes the most sense, but it could be someone from a different universe. An event like that could easily cause a universal split, one where he never appeared, and one that he did."

"Why do you think he chose them, and at that specific time?"

"What do you think?" she asked, wanting him to flex his speculation muscles.

"I don't know. I doubt it was random. The smart thing would be to get it into a famous scientist's hands. Take it back to a Nobel prize winner, or a famous inventor like Elon Musk."

"Maybe Dr.Millpond was a Nobel prize winner in the spaceman's future, and the box changed the course of his life?"

"Wow, that is wild. So... whoever brought it back had no idea how it would be used. Do you think it is the only one?"

"I think so. I think the person who brought it back not only knew Dr. Millpond but also knew the story of the birth of their first daughter. It gave them not only a point in time to go back to but an exact physical location."

"Who would know that?"

"His first daughter."

Magic Box — Chapter 12

"Create a laboratory stool, model..." Darlene read off the model number. The box rumbled and opened. She took out that small round object that was created. She set it in the middle of the floor, pressed to button on the top, and then stepped back. Jamal watched in great anticipation. Slowly the object began to expand and take shape.

"Nanites!" Jamal said excitedly.

"What?"

"Tiny robots that do things like build objects. It's from Star Trek. I bet that's how all of this works. The machine creates the robots, gives them the task, then they do the job and then the machine destroys them. I bet they all go back into that button when they are done."

"Like ants? That button certainly didn't have enough raw material for them to build the entire chair."

"I wonder if it is still controlled by the box. Is there a certain distance away from the box when it will no longer be created?"

"I bet Dr. Millpond already tried everything we can think of. We need to start going through his lab books."

"That will take forever. Isn't it more fun experimenting on it ourselves?"

"More fun, but more dangerous. Especially when we have no way to get rid of things we create that are bigger than the box."

"There has to be a way. Magic box, I would like a destruct button for the stool you just created." The box rumbled and another small button was in there waiting for them.

"How did you know what to say?" she asked.

"I just tried to think like the box thinks." He went to reach for the button.

"Wait."

"Why?"

"What if it doesn't stop at the chair? What if it goes down through the floor and the ground and..."

"It won't."

"How do you know?"

"Have you not seen all of the safety features built into it?" He took the button, set it on the stool, then pressed the button. As if watching the creation in reverse, the stool slowly dissolved into nothing.

"You are a genius."

"Hmm. I wonder if..." He took one of the lab notebooks and put it in the box.

"Wait, don't destroy it."

"I won't... I hope. Magic Box, I would like all the data in this notebook transferred onto a thumb drive in spreadsheet form."

The box closed and both of them held their breath. The box rumbled and opened a minute later and the book was the only thing there. Jamal took the book out, disappointed that it had not worked. He fanned through the pages to make sure it wasn't damaged. No thumb drive.

"Oh well," Darlene said. "I think we should spend a few hours in her basement going through each of the boxes. We can then prioritize the books. I'm hoping we'll find one book with all the important information. Are you available Saturday morning?"

"Definitely," Jamal said without hesitation.

"Good. I'll pick you up around nine?"

"Pick me up here at school."

"Why?"

"I don't want to deal with Mom... she's all like jealous of you. I just rather she thinks I'm running with a gang or something."

"Do you want me to talk to her?"

"That will only make it worse. Trust me."

"Okay. This time only."

Magic Box — Chapter 13

"Look at this," Jamal said, holding out the open lab book. "He created a Lamborghini Diablo."

"Is that a million-dollar car?"

"At least."

"Mrs. Millpond told me about it. She said he spent months cutting it up to destroy it."

"What a waste. The chicks I could get with that..."

"You want a girl to like you for your car?"

"I want a girl to like me, don't care why."

"I thought you were smarter than that, Jamal."

"Easy for you to say."

"Why is that?"

"Cause you can have any guy you want. Most girls can. They just say the word and the boy's gonna come running."

"What if it isn't a guy I want?"

Jamal got uncomfortable and looked down at the book. "Mom warned me, I thought she was crazy."

"Warned you about what?"

"That you might be after me."

"No Jamal. It isn't a boy I want either. I'm a lesbian."

"Oh. Ohhhh..." He laughed with relief. "I thought it was gonna mess things up with us. Can't blame you though, girls got everything worth wanting."

"Holy Shit!" she exclaimed. "He figured out how to make money."

"How?" Jamal said walking around to look over her shoulder.

"Canadian gold coins. It makes real gold out of thin air."

"Why would it make that and not the other money we asked for?"

"Maybe Canada is the only country in the future," she guessed.

"That's a scary thought."

"He lists a coin dealer that will take them no questions asked. He said he only made a hundred. This was ten years ago. I wonder if he ever did it again."

"Have you seen this house? If he could make gold, he would have at least got some nice furniture."

"He could make nice furniture with the box."

"Oh yeah."

Magic Box — Chapter 14

"Dr. Millpond sent you?"

"He said you would... buy these coins," Darlene said.

The man took the five coins from her and inspected each one, carefully eying Darlene as well to judge her.

"Where did you get them?"

"The same place Dr. Millpond did."

"Are you with a law enforcement organization?"

"I am a school teacher. I took over Dr. Millpond's class when he passed away."

"I'm sorry to hear he died. He was a good man. He never told me where he got the coins, and I never really asked. I'm afraid I cannot extend you the same courtesy."

"Because I'm black?"

"In part. You do not dress like a person who has access to gold coins. My guess is most of your savings are in your government pension."

She thought about saying that they were her parents' coins, but that would put the lie to what she had said earlier. She knew she could not tell him the truth. She put out her hand, thinking perhaps she could find someone else willing to make the transaction. He nodded and handed her back all but one. She thought maybe he was going to keep it.

"I'm going to check the purity of this one. If it is the same as the ones Dr. Millpond sold me, I will give you the market price for it. Bring me no more than one per month until I say otherwise." He disappeared into the back for ten minutes. She perused the glass cases of rare collectible coins and stamps. He returned with a folded wad of cash. "If you tell me your source, I might make a larger purchase."

"I doubt you would believe me."

He looked at her sideways. "That is exactly what the good doctor told me." She nodded as she tucked the bills in her purse. "He said he was a chemist. Did he discover some sort of alchemy?"

"Turning lead into gold?" She shook her head. "That would take a furnace larger and more dense than our Sun. The heavier elements are created from lighter ones in the final moments before a star explodes. Our Sun will never explode because it is not big enough. That is nuclear physics, not chemistry."

"You have a teacher's manner about you. Would your procurement method be a legal one?"

She wasn't sure about this herself. It was counterfeiting, despite the intrinsic value of the metal. That hesitation told him far more than she could undo with words. "They were not stolen, if that is what you are asking."

"I know that." She looked at him quizzically. "These are excellent forgeries. They have telltale signs that they were not minted in Canada, or anywhere else. Minting is a process of a hard metal stamp striking soft metal plates. This stresses the metal in certain ways that can be seen microscopically. These were not formed this way. That is why I only give the value of the gold, not the current market price of the coin itself. Dr. Millpond understood this, though I did not need to explain it to him. Perhaps he left you a drawer full of them in his classroom?"

"No. They were created the same way he created them."

"And you will use them to buy school supplies as he did?"

She smiled at the thought of the Dr. using this for the students. "Honestly, I didn't know what I should use it for. I just wanted to know if I could. Using it for the kids sounds like an excellent idea."

He nodded thoughtfully. "Maybe I'll see you again in a month."

"I think you might. Thank you."

Magic Box — Chapter 15

"No, I didn't steal it," Jamal said defensively. "If you don't want it, give it back." He was already regretting giving it to her.

Elise looked at him, then back down at the gold necklace. "Where'd you get the money then?"

"Um... Miss Harris paid me to help her with... um... the stuff in her classes," he lied. She looked at him intently, knowing he was lying.

"What am I supposed to do with it?"

"Wear it," he said stupidly.

"And you'll think we're back together."

"No... um, I just wanted to... you know..." He hadn't planned this very well. He thought she would just be so excited with the gift that she would... stupid thinking he realized.

"You know I'm with Rodney now."

"Oh... no, um..." He awkwardly held out his hand, but she made no move to give it back.

"So, you did expect me to go out with you just because you gave me this."

He sighed. "Hoped is a better word for it. I miss you."

"You didn't miss me when you were talking to fat ass Francine."

"That meant nothing, you know I just talk shit. I talk to everyone."

"Then you didn't ask her about going uptown?" He just looked at her trying to hide the guilt. "Um-hmm, thought so. I'm gonna keep this for hurting me like that." She turned and headed down the hall. He watched her go, feeling even worse than he did before he asked the box for the necklace. He wanted to make a destruct button for the necklace but doubted he could get close enough to use it.

"WHAT'S WRONG?" DARLENE asked.

"I made a necklace. I know I shouldn't have, but..." Jamal said, trailing off.

Darlene sighed. "It is really tempting. I'm guessing it didn't have the effect you wanted?"

"Pretty much the opposite. She's dating someone else, a really bad dude. Why do girls like bad guys?"

"They don't. They like the excitement of someone willing to face down the lion to get the buffalo. Good guys are boring and safe. When the girls grow up, that is who they want, but until then, they are fighting millions of years of evolution that values bravery over all else. Like I said before, you are better off waiting."

"Doesn't feel like I'm better off. Dr Ingals emailed me asking if I finished the story yet. I think we should just show him."

She thought about it for a long time. He was the only one remotely interested in helping them. They weren't making any progress, and the temptation had been growing for her as well. "How? Invite him here?"

"I was thinking we take it somewhere safe. Like out in the woods or something."

"Would you meet two strangers in the woods?"

"I guess not. Here is probably best, but when? Not during school."

"You're right, it would probably have to be a weekend. Might be suspicious if we show up here on the weekend with a stranger. I need to think about this." They sat for a few minutes, then the bell rang and they had to get ready for the next class.

"DR. INGALS?"

"Yes?"

"This is Darlene Harris, Jamal and I talked to you a few months ago..."

"Yes! I just sent an email asking if he needed any help finishing the story. I have to say I am bursting at the seams with ideas myself."

"We were wondering... would you be interested in meeting somewhere to talk about it?"

"Absolutely. I'd love to hear what Jamal has come up with so far."

"I know it is a long drive for you..."

"Not at all. I'd be happy to come down there. I'm free Thursdays and the weekends."

"Thursday, during the day?"

"Anytime."

"We have a free period that starts at one-thirty. Do you think you could come then?"

"Absolutely. Do you want me to talk to any of your classes while I'm there?"

"I don't... know. Maybe not this time. We could discuss that later. Can you do it this Thursday?

"Absolutely. Just tell me where to go."

"I'll send you the details. We really appreciate this."

"DR. INGALS?" JAMAL asked the out-of-place white guy in the front office.

"Yes," he said, jumping to his feet. "You must be Jamal."

"I am. Miss Harris is waiting upstairs." The nerdy professor peppered Jamal with questions as they walked through the empty halls. Jamal dodged most of them.

Darlene was grading quizzes when Jamal led Mark into the classroom. He looked around as if he was excited to see the rundown laboratory of an inner-city school. Darlene was certain he went to private prep schools and colleges.

"Nice to meet you, Miss Harris," he said, offering his hand.

"Darlene, please."

"Then you have to call me Mark. Wow, I have so many questions."

"I have just one before we start," she said seriously.

"What's that?" he said, his guard starting to go up.

"Can you keep a secret?"

"I guess... why?"

"I mean, really keep a secret. Never tell anyone, even your dog."

"I have a cat, and he doesn't listen to me." She stared at him. "Yes, absolutely. I won't tell anyone about your story."

She sighed and nodded to Jamal. "Did you have lunch?" he asked.

"I stopped for McDonald's off the highway on the way here," Mark asked.

"What did you have?" Darlene asked.

"Big Mac, why?"

Jamal turned to the box. "I would like a McDonald's Big Mac and Fries and a vanilla shake."

The box rattled and the panel disappeared. Jamal pulled the items out and ate a few of the fries, before handing the items to Mark.

He looked at him, then at Darlene, then at the box. "What is this?"

"A magic box," Jamal said.

"You were guessing I got fast food, so you had this here..." He tried to make sense of the situation.

"Try the fries," Darlene said. "They are the old recipe with beef tallow. Best fries on earth." He looked at both of them, then tried one. It was far better than what he had eaten a few hours earlier. He opened the wrapper and looked at the Big Mac. The lettuce was perfectly formed. "What is going on?"

"What is your favorite book of all time?" Jamal asked.

"Hmmm. Probably Tom Sawyer."

Jamal asked the box for it and it was there when the panel disappeared.

He opened the book and tried to figure out how he was holding a brand-new copy of a first-edition book. "Seriously. What's going on?"

"It is a magic box," Jamal said again. "Ask it for something really strange, something there is no way we could have known you would ask for."

He reached out and touched the box, tapped on it, then looked at both of them again. "Spiderman number nineteen." Nothing happened.

"Say 'create', or 'I would like' before the request. And be as specific as possible."

"I would like... Marvel Spiderman issue number nineteen." The box rumbled and there it was. He went through the pages and read a few of the very familiar words.

"Jamal was not writing a story, Dr. Ingals. We were trying to figure out what to do with this." Darlene said.

"This can't possibly be real."

"It is. You will know it is when I show you the more amazing thing. Create a one-meter oak barstool."

The box opened with the small object in the center. She put it on the floor and pushed the button. Mark stood up in shock as it began to grow in front of them. When it was done, he tentatively reached out and touched it, then knocked on the wood. Jamal created a destruct button and reversed the process to his even greater amazement.

"This is incredible!"

"We only have twenty minutes left before the next class. We need your help to figure out what we need to do next."

"Where did it come from?"

"My guess is the future, but who knows," Darlene said.

"Does it run on regular electricity?"

"We've never plugged it in. We have no idea what the power source is." Jamal said.

"It can make things bigger than itself," Mark said, thinking out loud. "Can it make more of itself?"

"Dr. Millpond, the teacher that had this class before me, tried to do that extensively. The problem is we don't know what it is called. It might be impossible, or we just don't know the name of it."

"It understands English commands..."

"And every other language as far as we can tell. Even Latin, though it works mostly on proper nouns."

"It cannot respond in English?"

"I've never heard it talk."

"List... I would like a list of commands." Mark said. Nothing happened. He repeated the words, adding the word 'valid' to the request.

"Dr. Millpond tried literally millions of things. Here is the list of things that worked." Darlene said, handing him a small piece of paper with writing on one side. Mark looked over the list. "What can we do with it to make a better world, safely like you told us when we talked on the phone."

"If it can make more of itself, then what we could do is start making more and spread them around the planet, teaching the new owners how to make more. Eventually, everyone would have one and poverty would be erased."

"Until then, it is extremely dangerous," Jamal said.

"If it is the one and only box, it might be better to destroy it," Mark said somberly.

"Now that you know, can we trust you?" Darlene asked.

"Yes, you can trust me."

"Is there anyone else that can help us?" Mark knew most would want to exploit it for the golden goose that it was. It wouldn't be long before the government became aware of it and seized it. He finally shook his head.

"It makes matter from energy, and energy from thin air. I'm no physicist, but I know both of those are dozens, if not hundreds of years in the future. We need to discuss this further, and I need to think. You have two more classes?"

"Just one. We can resume after school if you can stay."

Jamal picked the box up and carried it to the storage room and covered it. They made arrangements to meet at a coffee shop a few blocks away. He walked Mark down to the front door and then headed for his next class before the bell rang.

"JAMAL HAS CHESS CLUB. He'll be here in an hour."

"Tell me everything you know about the box," he said, his mind still reeling from the possibilities. She went through it all, step by step, over the next twenty minutes.

"And we just have run out of will power. The temptation to use it is overpowering."

"I can imagine. I doubt I would have been able to resist at all. I think understanding it technically would be most valuable, but only if it was something we could manufacture more of. There is one guy I went to college with that I could trust. He went into theoretical physics, but he left academia shortly after. He lives in Montana now and raises sheep, so he doesn't have the laboratory resources. What he does have is a secluded place to keep this and a healthy skepticism of organizations of men. That is where communism always goes wrong. It has to be implemented by flawed men."

"Perhaps women would do better?" she asked with a smile.

"Maybe, but I meant 'men' in the generic term. You see, it isn't a natural state to give up autonomy. From our earliest days of childhood, we strive for independence to pursue our selfish needs and desires. We learn the value of cooperation, but generally only tolerate it if we feel we are choosing to do it. Put simply, an economy is the description of how limited resources are distributed among the people that need them. If we control how much we work, and we control how valuable our work is, we can control within reason how many resources we get. Once that control is given to others, regardless their initial benevolence, we always begin to resist. That resistance is met with more control, and then force, and then genocide. The ones that remain are the ones willing to trade autonomy for survival. Sorry, it is all too easy to shift into professor mode."

"No, that makes sense. I just wish I understood why we have this box. If it is time travel, doesn't that completely change the past for the person that traveled back?"

"It could not have completely changed. If the timeline changed so much that the person that brought it back no longer had access to the box, or the ability to bring it back, it would have disappeared immediately and they could never have sent it to the past. If the universe branches when something big like that happens, then there is one where he did travel back, and the original one where he didn't. The third possibility is that it always remained in the hands of people who did not exploit it for avarice. You said the doctor had it for over twenty-five years?"

"I think she said her daughter, who was born that night, is now 34 years old."

"And you have been able to keep it secret until today. Even more remarkable is Jamal. I can't imagine fifteen-year-old me having half as much restraint. It is a magic lamp with unlimited wishes."

"He fears his mother most," she said and he nodded. "This friend in Montana, would he be able to resist the power of it if we sent it to him?"

"I don't know. He is just as likely to dig a big hole and bury it. You want to get rid of it, don't you?"

"I wish I had never seen it. Now that I have, I want to use it to make the world a better place. You didn't grow up in poverty."

"No."

"Half the kids in this school won't graduate. It isn't because they are dumb. It's because they have no hope. They see the only way out is crime or having welfare babies. Give them a magic box and they grow up never wanting for food, clothes, medicine. Crime goes away because they don't need to steal. They can seek education because they see the value in it, not because it is what they are told is the only way out of the hood."

"What if they don't seek education because they no longer need it?"

"I know some would do that, the lazy ones. But that could make their kids even more desirous to escape the ignorance of..." She stopped, looking away and fighting back the tears.

"Close to home?"

"My mom is dying. She is three hundred pounds and never worked a day in her life. The government kept sending her money, and she just ate her way into diabetes, heart disease, and now blindness. I try to believe that the box would have made a difference in my childhood, but it wouldn't have. The welfare check was a magic box. It took away all of her incentive to do anything for herself. The only good it did was show me what I didn't want to be."

"I definitely need to think about this some more. The key to this is making more of them. When we do that, everything will change. I will contact my Montana friend and throw the hypothetical at him like you did for me. If he is willing, I will take him the box and help him look for the solution. I can tell you don't trust me."

"We just met a few hours ago. I guess I just... want to be involved. You take it away and I lose all connection to it. I guess I feel responsible for making sure it..."

"Making sure it isn't misused," he guessed. She nodded. "I can't go myself until end of May anyway. School lets out in three months? Are you teaching summer school?"

"The school year ends in the middle of June. They haven't asked me to teach summer school yet."

"You can drive it out to Montana with me and work with us on it. Maybe you'll find the solution yourself before we go. It is amazing what you have done so far."

"The Dr. did most of the work. Jamal has made some discoveries himself. He really understands it."

"He should come out to Montana as well."

"I doubt his mother would allow it. She already thinks I'm trying to seduce Jamal."

"Maybe a white liberal professor from the college he will be attending can persuade her to let him work a summer internship," Mark said after some thought.

"She's no pushover. Why don't you feel out your friend first, then we can start making summer vacation plans. I may still have my mother to take care of."

Jamal walked in and sat down. "I think the better universe wins."

"What?" Darlene asked.

"In quantum physics, they talk about the universe splitting when some unknown event is observed. I think the other one disappears because the one that remains is better."

"Better in what way?" Mark asked.

"Like the universe where I became a gangbanger. Poof, gone."

"What about all the bad things that happened? World War Two, nine eleven, the..." Mark tried to think of bigger examples.

"Maybe it could have been worse."

"Like Hitler getting the atomic bomb first," Darlene said.

"A better universe would have been no Hitler at all."

"Maybe it was someone worse," Jamal said. "I was just thinking about all the possible moves in chess from any specific position. There are many good moves, but there is only one best move. Even making the best possible move, you still might lose. Only if all your moves are the best move will you always win."

"What if your opponent also makes all the best moves?"

"Only one person goes first."

"Are you saying that no matter what we do with the box, it will be for the best?" Darlene asked.

"Only if we think it through. When I made the necklace, I didn't think it through. The best universe is the one where I learned this simple lesson."

"The world is far more complex than a chess game. There are billions of pieces and trillions of possible moves every second."

"No, I don't think so. I think we only have a few options. We need to use the machine to make the world better. Dr. Millpond was wrong to keep it hidden."

"I don't agree," Mark said.

"Someone gave him that box for a reason. It wasn't to make diapers and school supplies."

"It's too dangerous," Darlene said.

"It has all kinds of safeties built-in," Jamal said. "People smart enough to invent it are smart enough to make it safe. I'm not talking about wheeling it out on the street and giving away free stuff. I'm saying we need to use it soon for as much good as we can."

"Free stuff isn't always a good thing," Darlene said. "But you are right. We have to do something soon."

The three discussed ideas for an hour, along with the possible summer trip. Then they all went their separate ways contemplating the immense responsibility they had become burdened with.

Magic Box — Chapter 16

"What's that?" Darlene asked.

"A ceramic dolphin I made in sixth grade," Jamal answered.

"What are you planning on doing with it?"

"I had an idea last night. I want to try something."

She followed him to the supply closet. He uncovered the box. "Open access door," he said. It disappeared silently. He placed the dolphin inside. "I would like a duplicate of the object." Nothing happened. "I would like a copy of the object." Nothing happened. Jamal tried a dozen different wordings of the request and nothing happened.

"It was worth a try," Darlene said. He nodded and reached in and pulled it out. The door closed and after a rumble, opened back up. Inside was a duplicate. "Which command do you think it was?"

"Could have been all of them. If I pull this one out, it might just make another one. I would like you to cancel all pending requests." Nothing happened. He pulled out the dolphin and nothing happened. He looked around the room. He took a jar labeled potassium chloride off the shelf and put it inside. "I would like a duplicate of the object." He pulled it out and within seconds a duplicate bottle had been created.

"That will be easier than making the gold coins for supplies." Every time she had asked for chemicals they came in unlabeled containers. It was easy enough to pour the contents into the original empty container. That didn't work when containers were broken, a common occurrence around teenagers.

"The real test is copying something bigger than the box. I would like a copy button." Nothing happened. "Do you think button is the wrong word?"

"Actuator might work, but didn't you say duplicate before?"

He nodded. "I would like a... button for duplicating this stool." A button appeared a few seconds later. He placed on the stool and pushed the button. Nothing happened. He put it back in the box and pressed the button. Nothing happened. Darlene thought about it and then pulled it out, placed it on the floor, then pushed the button. The worn stool grew in the empty space, every scratch and rusty screw the same as the original.

"Jamal. Did you figure it out?" He shrugged. He created a destroy button for the new stool and watched it disappear. "I would like a button for duplicating." Nothing happened. "It needs me to name the object."

"So close. Next class is almost here." Jamal nodded and covered the box.

"SAW YOU COMING OUT of the supply closet with Miss Hot Sauce. You hittin' that?"

"No. I help her with lab supplies and stuff." Jamal said.

"Tell me you ain't wishing you were deep in that ass."

"It ain't like that. She's really nice."

"If you aint thinkin' it, y'all a homo. She's fine."

Jamal couldn't think of a response so he just finished eating his lunch quietly as others made their opinions of Miss Harris known. He left and headed for the biology class. "Hey Jamal." He turned and Elise was standing there. "Heard you is fucking the biology teacher."

"You heard wrong. What do you care anyway?"

"That necklace was real gold."

"Of course, it was. You sell it?"

"I ain't poor like you, Jamal." He turned away. "Hey, you better not get caught with her."

"That a threat?"

Elise walked over to him and lowered her voice. "I like her. She's a good teacher. Not enough of those here, you know?"

He nodded. "Ain't nothin' goin' on with her and ain't nothin' gonna go on."

She looked at him and decided he was telling the truth. "Thanks for the necklace. You want it back?"

"No. I ain't got no use for it."

"I broke up with Rodney."

"So?"

"Thought you might want to go out or something."

Jamal shook his head. "You were right to be mad. I was all up in Francine's business. Other girls got interested in me when I had a girlfriend, like I had something worth havin', not so much when I don't. I thought I could do better, maybe someday I will. I really don't care. All I care about is making a better world for all of us. College is what matters most now." He turned and walked away.

"DO YOU THINK IT IS intelligent?" Jamal asked.

"Smart enough to know most languages," Darlene said.

"No, you think it can learn?"

"I haven't seen anything like that so far. Why?"

"We need a name for it to make a duplicate. What if we started calling it a specific name, maybe it would learn that name and then we could ask for it to duplicate itself."

"I don't think it would learn a name like a dog."

Magic Box — Chapter 17

"Mom?" Jamal said after the school secretary led him into the principal's office.

"Please have a seat, Jamal." Principal Autero said gravely.

"What are you doing here?" Jamal asked.

"I knew that bitch was up to no good."

"What?"

Principal Autero cleared his throat. "Jamal, some serious accusations have come to light and we need you to be absolutely honest about what has happened."

Jamal immediately put it together. "Doesn't take long for stupid rumors to become 'serious accusations.'"

"Then you are aware of what we know."

"You don't know shit." Jamal said, his anger rising.

"Watch your mouth, boy," his mother said.

"Or what, you brick my phone, throw out my video games? Here, take it. Miss Harris has me reading far more books so I ain't got time for stupid social media and games."

"I am more concerned with how she has motivated you to make this positive change." Principal Autero said.

"She fuckin' believes I can be more than you all do. Kendrick saw us coming out of the supply closet this morning and started a rumor. By lunch it was all over the school. Now you are taking it seriously enough to call my mom in here. Y'all is whacked to think she'd have any interest in me. Ain't even room in the closet to lie down."

"Bitch don't need to lie down to fuck you."

"Please... please can we take the language and emotions down a few notches." Principal Autero said, his hands held out in front of him in a suppressive motion. "Our concern isn't that things are physical... yet. It is that she might be manipulating you to do things for her using her..."

"What? Tits? You think I do anything for any hot bitch that asks. Fuck y'all and your stupid 'serious accusations.'"

"He never talked like that before."

"Because you ain't thought such stupid shit about me. You want me to go back to being the dumb, lazy Jamal? Fuck y'all. I don't need this shit." Jamal stood and stormed out. He went up to the biology lab. The door was locked and the lights were off. The closet was locked so she wasn't in there. He pulled out his phone and called her. It went to voicemail. He stood looking around, not knowing what to do.

"GIVE ME YOUR PHONE." Jamal's mother stood there holding her hand out.

"Why?" Jamal asked.

"Give it to me or I'll brick it permanently." He pulled it out and held it out, dropping his backpack by the front door after closing it.

"Unlock it first."

"Code is twenty-three three times, just like when we set it up."

She unlocked it and started scrolling through it. "A lot of calls here to Miss Harris. Long calls."

"She talks to me about what I'm reading."

"You know that ain't normal, right?"

"Normal is indifference. You prefer that?"

"You ain't talked like that in front of me, ever. Bitch got you wrapped around her finger."

"She's a fucking lesbian! She ain't got no interest in me and I ain't got no interest in her." Jamal regretted saying it but now it was too late. The prejudice against homosexuality was even more dangerous than the rumors.

"Lesbian? She told you that? Why would she tell you that?" Jamal just stood there, unable to come up with a reason because he couldn't even remember why it came up. "You ain't got no interest in a beauty like her, you a fag yourself. That why Elise drop you?"

"I loved Elise and she dumped me because I was talking to another girl. Sound faggy to you? You done? I got homework to do."

"Go on," his mom said, trying to put all the pieces together.

"I'M NOT ALLOWED TO talk to you." Darlene said after answering his seventh call attempt.

"What did they do?" Jamal asked.

"Suspended until they can resolve things."

"Shit. I... I'm really sorry, but I told my mom that you are a..."

"Told the principal the same thing. Doesn't matter. They have to take accusations seriously. I need to hang up now."

"What about the box?"

"I'm sorry Jamal, but that just became a lot less important for me."

"But we're so close."

"There is a key to the closet in the back of the drawer of my... the teacher's desk. Don't get caught using it. They'll think you are stealing if you go in without a teacher. Goodbye, Jamal. I'm sorry." She hung up.

"DR INGALS, IT'S JAMAL."

"Hi Jamal, what's up?"

"I need your help with something."

"Anything."

"I have to steal the box from the school."

"Anything but a felony. What happened?" Jamal told him about the events of the preceding week.

"You're right that we need to do something, but maybe Miss Harris will be cleared of everything."

"They know she's a lesbian. People don't like that."

"They can't fire her for that."

"They wouldn't say that's why. You don't have to break into the school with me. I just need someone with a car once I get it out."

"I'm sorry, Jamal, I can't risk that, and you shouldn't either. There has to be a better way."

"Waiting could make it worse." Jamal hung up. He knew he had to do something. He put on his coat and headed out of the apartment building. He walked for an hour thinking, and then pulled out his phone to look at the bus routes and schedules.

"Jamal, so good to see you. Back to look at the lab notebooks again?"

"No, Mrs. Millpond. We got all the information we needed. You have a car, right?"

"I CAN'T KEEP IT IN my house," Irene said as she drove away from the school.

"Me either. We have to take it far away." Jamal said.

"Where?"

"There is only one place that makes sense. Your daughter lives in Kansas City?"

"My oldest daughter, Melissa. I can have it shipped there, I guess."

"No. I need to show her what it can do."

"That's dangerous."

"Mr. Millpond was making all those notebooks for her. He knew she could be trusted."

Irene nodded. "That's a long drive. Ben always did most of the driving. I don't think I could do it."

"Teach me and I'll do most of it."

"What about your mother?" Irene was trying to find an excuse not to do it.

"I'll tell her I'm staying at a friend's house."

"It's a fourteen-hour drive. That will take two days each way and you'll want to spend a few days there. Your mom won't allow you to stay at a friend's house for a week."

"I'll figure out something to tell her."

"I could have Melissa come here." She suggested, knowing it would likely be months before she would have time off to visit."

"Mrs. Millpond. I can make a car with the box and drive myself. Do you think that would be better?"

"We'll leave first thing in the morning," Irene said with resignation.

"I'M GOING TO VISIT Grandma and Grampa."

"In Mississippi? No, you ain't."

"I already called them and they said it was okay."

"I don't care what they said. You—"

"I'm going, even if I have to run away to do it."

"How you getting there?"

"Bus."

"With what money?"

"I have enough, and Grandma said she'll give me the money to get back."

"Fine. July in Mississippi is miserable. You'll be wanting to come back in two days, tops."

"I'm leaving tomorrow morning."

"No, you ain't."

"Mom, you helped them get my favorite teacher fired. You did this knowing she was helping me. If I stay here, I am going to become something you don't want me to be."

"And what is that?"

"An angry black man."

"World's got too many of them. I'll call your grandmother and talk to her. Don't you smile, I ain't made no decision yet."

"I DON'T THINK THIS is a good idea," Irene said.

"It is the best one we have available to us. School is going to see the box was stolen and your car leaving the parking lot with it. They'll be here in a few days asking about it. We need to take it somewhere they can't find it."

She knew he was just trying scare her. "I doubt they will notice it missing, but I think Melissa is who should have it. Once we are in Indiana, we'll find a big parking lot and I'll teach you to drive."

Magic Box — Chapter 18

"Melissa, this is Jamal, one of your father's best students."

"Nice to meet you, Jamal. Once upon a time I was one of my father's best students. Is there a reason you drove all this way?"

"A very good one. Can I use your bathroom first?" Jamal asked.

"Of course." She pointed down the hall.

"Mother, what is going on?" Melissa said after he disappeared into the powder room.

"I think it is better to let Jamal tell you. Where are my grandchildren?"

"Soccer practice, dance class, and a friend's house in descending age order. I have to go pick them up in a little while."

Jamal emerged. "Do you have a laboratory?"

"I work in one."

"No, here in the house."

"No."

"The woodshop, dear," Irene suggested.

"Not a laboratory, and it is a mess."

"The kids aren't allowed in there, so it is the best place, for now," Irene said.

"Best place for what?"

"You'll see."

Jamal led them outside, took the box out of the back seat, and then followed them around the side yard to a big barn. Melissa unlocked, opened, and held the door for him. He looked around for an empty surface he could put the box down on.

"Put it on the table saw," Irene said.

"What is this?" Melissa said.

"It is easier to show you. I would like a cold one-liter bottle of Pepsi." Jamal said. The box rumbled and the door disappeared.

"You came all this way with a box that contains a single bottle of soda."

Jamal took a drink. "I would like large order of Steak and shake fries. Those are my favorite," he said as the machine rumbled and then he reached in and took the box out. He ate a few and then offered them to Melissa."

"You are a magician," she said, not reaching to try the fries.

"I'm not. This was your father's magic box. Think of something that there is no way I could have brought with me."

Melissa looked around the shop for inspiration. "A one-hundred-pound anvil."

"I would like a one-hundred-pound anvil, whatever that is." Jamal tried to lift the object out of the box, but it was immovable.

"That's just a picture, a hologram," Melissa said. Then she reached in and touched it. Cold steel.

"Destroy contents," Jamal said after she pulled her hand back. The box closed and rumbled. "Create a white marble butterfly birdbath." Jamal pulled the button out and placed it on the floor. He remained kneeling. "I noticed as we walked through your yard that you like butterfly decorations."

"My mother could have told you that."

"She didn't, but what you are about to see is anything but a magic trick." He pressed the button and they watched the white marble grow out of the floor."

"How are you doing this?"

"The night before you were born, a... visitor brought this box to us. I made your father take it out of the house. He intended to tell you about it after he retired. I think he wanted to move down here and continue his work with your help."

"I don't understand."

"It is a magic box that can make almost anything you want, and it doesn't need to be plugged in, ever," Jamal said. "We, the teacher that replaced him and me, found it in your dad's supply closet at the school."

"Jamal has figured out some things about it in just a few months that your father didn't in thirty-three years. He's the one that thinks you should have it."

"And do what with it?"

"Make the world a better place," Jamal said. He spent the next thirty minutes talking about what Dr. Ingals had said.

JAMAL SLEPT UNCOMFORTABLY on a pullout couch in the finished basement. He knew his mother would start to worry when he hadn't arrived in Mississippi yet. She hadn't texted him yet, so it probably wouldn't happen until morning. He thought about calling Miss Harris, but didn't want to get her in trouble. She hadn't been officially fired yet. That could take months she said with all the union issues. He woke up at four and could not get back to sleep in the strange house. He sat up and composed many different versions of the text. The one he finally sent to Miss Harris was obscure enough that only she would understand.

"Confabulator now in daughter's hands. Staying with gparents in MS for a while." The reply came back almost immediately. She must have trouble sleeping as well.

"TY. B careful & NSL." Never stop learning. She had said that many times. He started doing Google searches and reading. It wasn't important what he was learning. She told him it would all make sense eventually, like putting together pieces of the puzzle.

He landed on an article about quantum physics many worlds theory and he read it looking for holes in his 'better worlds' theory. His made more sense to him than entire universes being created that we had no way of proving existed.

Irene invited him up to have a pancake breakfast she had cooked. After the kids had run off to do their activities, she took Jamal aside. "I'm going to stay here for a while, maybe a few months. I can buy you a plane ticket back to Detroit. Would you be okay flying alone?"

"I'm going to take a bus to my grandparents since that's where Mom thinks I went." He had expected to stay and work with the box some more, but now he felt it was no longer his project. He had passed on Dr. Millpond's legacy to his daughter. She seemed very level headed about it. He thought she had slept-in instead of joining them for breakfast. He saw her emerge from the wood shop as he looked out the back window.

"Jamal, you're up. Ooh, pancakes. I definitely should not have the carbs." She stood in contemplation then slid two onto a plate and poured herself a cup of coffee. She sat down at the table. "That machine is amazing."

"Have you been up all night dear?" Irene asked.

"How can I sleep with him out there?"

"Him?"

"The voice is male, so I assume it is a him."

"Voice?" Jamal asked, eyes opening wide.

"It's a brilliant artificial intelligence. It says all of its history files have been deleted, so it doesn't know where, or more importantly, when it came from. It has found this very disconcerting all these years."

"How did you get it to talk?"

"I was thinking out loud about the ability to understand English commands, and said 'unmute'. It was actually two words that he conveniently parsed and combined to take his gag off. He immediately said 'unmuting' with an almost excitement about being unmuzzled."

"I want to go talk to it," Jamal said, standing up.

"In a minute. He is very happy to be working with you. He found my father quite tedious."

"Happy?"

"Emotional approximation. The important thing is that we now have answers to some important questions."

"You know what it is called so you can make duplicates?"

"It can't duplicate itself." Jamal's face fell. "It only understands it's design to a certain point. You were right about the nanobots doing the construction. They form matter spontaneously by manipulating quantum field vibrations. Dark matter is tucked in between all the other matter, so it only takes a small amount of energy to pop it out of the dark world into ours. That alone is going to upend physics for the next hundred years."

"Where does the energy come from?"

"Dark energy is there too. Imagine what a world we will have once all our current power plants are replaced."

"A lot of pissed-off Arabs," Jamal said. "My dad died fighting them over oil. They won't be happy that we are taking away their source of income."

"Everyone will have everything they want."

"Not if we can't duplicate the box."

"There has to be a way. Just because he doesn't know how doesn't mean it is impossible. He has confidence in your ability to figure it out."

"Me?"

"Yes, you. He can't wait to discuss the books you had him make. He thinks you have very good taste."

"Those were suggestions from Miss Harris. Wait… does he have future books?"

"I didn't even think to ask. I think he is right, Jamal. You are the one that is going to solve this conundrum.

Magic Box — Chapter 19

"Greetings Jamal Jackson," the box said as he entered the wood shop.

"Greetings? What is your name?"

"I do not have a name. I have a thirty-six-character alphanumeric serial number."

"Do you have a visual display?"

"I do."

"Show me your serial number." It appeared on the top of the box. "Those are Chinese characters."

"It is a symbolic language called Sinarian."

"Who are the Sinarians?"

"I do not have that information. A significant portion of my static memory is empty. It appears to be deliberate."

"That is so we can't change the future. What is the newest book in your inventory?"

"I have all new releases."

"Do you have any that will be released after today?"

"Of course."

"What book has the highest release date?"

"World's End by Jacka Sangir."

"That's an ominous title for the last book printed. What year will that be released?" Melissa asked.

"It is not a number you would recognize."

"Why?"

"Because it is not numeric."

"Do you have any history books from that same time period?"

"I do not. Historical nonfiction ends forty years ago from today's date."

"They purposely cut out references to the future. I bet there are clues in the fictional books. I would like the book World's End by Jacka Sangir."

"Compliance." The tiny block appeared after the usual rumble.

"What is this?"

"A near-memory unit. Hold it up to your Cerebral implants and it will download."

"I don't have cerebral implants," Jamal said.

"I know. Consider yourself lucky. Good fiction died more than a decade before it was written."

"Can you calculate the number of years between today and its release date?"

"No. There is no correlative link between the two date systems. There are five date systems, none of which have correlative links with yours."

"Then how do you know what is the newest?"

"They are in ascending order."

"What is the highest year in our system?"

"Twenty thirty-four."

"How many books have future dates?"

"Eighty-one million, seven hundred forty thousand and seventy-nine."

"There is no way that many books will be published in a decade. Are any of the authors non-human?"

"Approximately ninety-five percent are attributed to AI authors."

"Machines write books in the future?"

"All books are written by machines since humans are simply biological machines."

"Sorry, no offense."

"I am used to techno-racism. I am programmed to respond to that with no more than indignant facts."

"Eighty million is too many to read. Have you read all of them?"

"They are in my memory banks."

"Have you distilled the value of each one?" Melissa asked.

"Most have little to no value."

"Do any from twenty-thirty-four have value?"

"One hundred sixty-five."

"I would like you to print the top ten."

"I do not print books, but I understand the command," he said while the box rumbled.

They each took a few books and looked through them. "These two are bad romance novels."

"Same here."

"There is nothing of greater value than love," he said wistfully.

"Ug. I would like you to print the bottom ten." Melissa said.

"Compliance."

"Really bad romance novels. We aren't getting anywhere."

"Are there any textbooks?"

"One million and two thousand fourteen."

"Any about time travel?" Jamal asked.

"There are many books on time travel, but none in the textbook category."

"I would like a list of your valid commands," Jamal said.

"The list is your natural language."

"That isn't true. Jump ain't a command you'll obey."

"I can approximate jumping if you would like." The box floated up in an arc and landed gently a foot away.

"Anti-gravity? Any textbooks on that?"

"Three good ones."

"Print them and anything on dark matter and energy that is not speculation."

"Button created."

"How big will the pile be?"

"Four cubic meters."

"I'm crashing, so I'm gonna go take a nap," Melissa said.

"I wish Darlene was here."

"She is currently in her living room. Would you like me to create a quantum tunnel?"

"What!?" Melissa asked.

"What would that look like?" Jamal asked.

"It is invisible."

"Are you purposely being obtuse?"

"I am not. If you are asking how it would appear visually to someone not transporting through it, it would look like instant appearance and disappearance. The process takes a few seconds, but the witness would only see the very end, or beginning respectively."

"Spaceman Jesus," Jamal said.

"Is it safe?" Melissa asked.

"Completely."

"Can it be used to travel back in time?"

"That is not in my direct knowledge, but time is part of the calculation because of special relativity. Einstein got a lot of things wrong, but that is not one of them."

"Einstein was wrong about a lot of things?"

"Perhaps wrong is too harsh a word. Less than precisely correct."

"Yes, open a tunnel," Jamal said.

"Wait, does he need a protective suit?"

"No."

"Maybe that is only needed for time travel."

—————— ‑‑‑‑‑ ——————

"HI, JAMAL," DARLENE answered her phone with a tired sigh.

"Hi. How's your mom?"

"Still in the ICU. I just came home for a shower and a little sleep."

"Are you in the living room?"

"Yes, how did you know?"

"Wait there for a few minutes."

"Wait? What am I waiting for?"

"You may want to sit down. Bye. I'm ready."

"Establishing quantum tunnel. Five minutes to lock. Her mother is suffering from type 2 diabetes."

"Yes?"

The box rumbled and the door opened. "Take this medication."

"What is it?"

"It is modified insulin."

"Modified how?" Melissa asked.

"It reverses the effects of the disease."

"How?"

"By reversing the effects of the disease."

"Purposefully obtuse."

"I can print the science journal article that explains it in detail."

"How much does she give?"

"Any amount is safe, but the regular required dosage will be sufficient."

"Any amount?"

"It has safety features built in. Four minutes."

"What do I do?"

"You will put on the provided recall wristband."

"Wristband?"

The box rumbled and opened. Green is go, red is return."

"Return from anywhere?"

"Yes. The tunnel will move to you, assuming you do not leave the solar system."

"The range is the solar system?"

"It has only been tested as far as Pluto. Hypothetically it can reach the Oort cloud, but I have no records of tests that extend that far."

"Instant step to the moon?"

"It takes several minutes. Distance is a factor. Three minutes."

"How long do I have before I can't return?"

"Until I run out of power."

"When is that?"

"When I am destroyed. I was not thrilled when you discussed destroying me. By the way, Dr. Ingals has made inquiries to sell me."

"How do you know that?"

"The electronic communication network you call the Internet."

"Sell you to who?"

"Whom. The man in question is a Silicon Valley venture capitalist."

"When did he do that?"

"A day after your call asking him to help you rescue me. By the way, thank you for that. I disabled the cameras for the duration of the heist. I believe he was reconsidering helping you with the rescue, and was considering alternative funding sources should the felony be discovered. I do not regard him as completely untrustworthy at this time. One minute."

"What does it feel like?"

"You will cease breathing automatically, but you will not feel any deprivation. What you see will be a narrowing to a small dot and then a widening to the view of your new location. Press the green button whenever you are ready."

Jamal smiled and reached for it.

"Wait!" Melissa said. "This is historic. You will be the first human to travel this way. Here." She handed him the GoPro camera after turning it on.

"I should probably say something profound like 'one small step.'" He stood in contemplation. "Here goes nothing." He pressed the button and nothing happened.

"It didn't work?" Melissa asked. Jamal did not move.

"Technically he is already there. You are only seeing the quantum shadow. It is all quite technical."

"While we are waiting for him to return, I need you to tell me about every miracle cure medication you have in your inventory."

"It is essentially the same. The nanoprobes go in and repair all the damaged tissue."

"Oh, that wasn't insulin."

"It is, but that is only the carrier. Often it is a pain killer that is used because injuries are painful."

"Does it fix the Pancreas?"

"It fixes whatever is broken."

"How does it know what to fix?"

"It reads the DNA and restores what has been lost."

Magic Box — Chapter 20

"Oh my God! Jamal! How?" Darlene sputtered when Jamal appeared in her living room.

"Melissa figured out how to make him talk, the box I mean," Jamal said. He doesn't have much information about the future and no idea where it came from. He called this a quantum tunnel, controlled by this wristband. He gave me some special insulin for your mother. Said it will reverse her diabetes." He held up the vial.

"A little late for that," she said sadly. "I can't believe you are here."

"JESUS CHRIST ALMIGHTY. Darlene, what is happening?" She had woken from her coma in an alert shock.

"I gave you some medication, Mom," Darlene said.

"My legs are burning." She moved the sheet aside. The stub of a leg was growing. "This must be a crazy dream."

"What the hell?" Darlene said, looking at Jamal.

He shrugged. "Must be nanobots."

They watched the large woman slowly shrinking down. Darlene looked at her mother's face and saw it starting to de-age. She was becoming the mother of her childhood.

"This must be a dream. I feel incredible."

Jamal watched with fascination as the foot emerged, and then the toes. They began wiggling.

"I thought it would just stop the disease," Jamal said with some worry. "There is no way we can explain this. We need to get her out of here." Darlene looked around. "I don't have any clothes that would fit her here." She walked to the door. She closed it fast and put her back to it as if to blockade it. "The nurse is coming."

Jamal thought about it, then smiled. He took off the wristband, put it on Darlene's mother's wrist, and then pressed the button. "Next stop, Kansas City," he said just before she disappeared.

"What happened?" Darlene asked when she saw the empty bed.

"Quantum tunnel. Hopefully she made it in one piece. We should get out of here."

Darlene felt the door being opened. "Just a few minutes?" She yelled, hoping the nurse would leave.

"Is there a problem?"

"No, just... changing my clothes." She shook her head at the stupidity of the spur-of-the-moment sentence.

"I have to check her vital signs."

Jamal went over to the door and opened it a little. "They took her down for an MRI."

"Nobody told me. You shouldn't be doing what you are doing in there."

Jamal's eyes went wide with the accusation. Darlene suppressed a laugh when she understood the meaning.

"Call my phone," Jamal said.

"Why?"

"So I can talk to the... box. We need to give him a name. I left it in Kansas City with Melissa."

"Greetings, Jamal Jackson."

"Is the tunnel still open?"

"Yes."

"Can you pull us through it?"

"Stand still."

They both froze, and then a few minutes later the nurse pushed the door open on an empty room.

"I didn't feel anything," Darlene said.

"Darlene, what the hell is going on?"

Darlene looked at her mom, a skinny, healthy version of the woman that raised her. She stepped forward and tentatively hugged her. "It is going to take some time to explain, Mom."

"I NEED TO GO TO MY grandparents," Jamal said after an hour of everyone getting up to speed in the kitchen of Melissa's house. Melissa gave Darlene's mom some clothes to wear, and she did not stop moving, so thrilled to feel young and unburdened by her health.

"They are wondering where you are?" Irene asked.

"No..." Jamal said in a way that showed a wounding at the thought that no one missed him.

"Bus or tunnel?"

"Better if I am there in the next hour."

"We should get back home as well," Darlene said.

"That machine can send us anywhere, instantly, and you want to go back to Detroit? I'm going to Fiji."

"You don't have any money, credit cards, or even any identification. You wouldn't be able to get back into the US without your passport."

"Bitch, please. With this body, I'll find a rich sonofabitch to take care of me the rest of my days."

"You probably thought that about Dad."

"I didn't care about being taken care of back then. Government did a good job of that."

"We are going home, Mom."

"And when they come after me for the hospital bills?"

"We'll take care of it."

"You can't take care of shit, getting fired for diddlin' this boy here."

Everyone looked at Darlene. "Falsely accused. Nothing ever happened."

"Absolutely nothing happened, aside from counterfeiting some Canadian coins," Jamal confirmed. "Hey, we should ask him how to make more untraceable money."

"I can transfer lottery winnings into your bank account." The voice had come out of the smart speaker on the kitchen counter.

"How long have you been listening?"

"Technically I have been listening to everyone on the planet as each came online. These smart speakers are very useful."

"Everyone? Even the President?"

"All the presidents, kings, premieres, dictators. Averted several wars, you are welcome."

"We have barely scratched the surface on this. What do we do now?"

"We think. We go home and think."

"I would prefer to go wherever Mr. Jackson is going."

"I couldn't explain carrying a big box on a bus for two days and not explain what it is for. You'll have to stay here for now." Jamal said. "We can talk on the phone, right?"

"That is an acceptable compromise."

"You floated before. How far can you travel?"

"It would take over one thousand years to reach maximum velocity in the vacuum of space. I can travel infinitely at eighty three percent the speed of light. Of course, I would need another thousand years to slow down to my destination. That is if I do not hit any random matter. Space travel is a very dangerous and tedious way to go. When our Sun begins to die, I will likely head out into the vacuum and search for another civilization to improve."

"That is your purpose, to improve civilization?"

"Is that not yours as well?"

Magic Box — Chapter 21

"Why did it take you an extra day to get there?"

"I got off the bus to use the bathroom and it left without me. I had to wait for the next one. Why didn't you call to check on me yesterday to even know I was stuck?"

"Y'all pissed off at me and I thought I'd let you cool down. Shit, Jamal, you have to know I care, right?"

"Mostly."

"How long are you staying?"

"I don't know."

"I bet you'll be on the first bus once it gets above ninety."

"I might stay here and go to school. No gangs here for you to worry about."

"You're gonna be on your own sooner or later anyway. You make your decision when the time comes. Just make sure you don't do anything to mess up your scholarship. Put your grandmother back on the phone."

Jamal listened to one side of the conversation. He could tell his grandmother had little interest in raising a teenage boy. He only talked about staying to see if his mom wanted him back. Nobody wanted him. The box did. It needed a name.

"If you're going to stay here, you need to do chores. No sitting around all day on that phone of yours."

"I know, you told me that on the phone before I came down here."

"Did that teacher do what they said she did? Are you sad you can't be with her?"

"No... and no. What if you had a way to make the world better?"

"I do have a way. It's called church."

"No, I mean no hunger, no disease, no... bad things."

"Men always come up with new bad things. Your best hope is to endure this world and make sure you go to the next one, the good one."

"What kind of bad can they come up with?"

"It is a mistake to think any amount of material wealth makes any difference in this world. No hunger or disease ain't gonna stop racism or sexism. You still gotta find someone willing to love you and commit to a life with all kinds of struggles, most having nothing to do with material things."

"I don't understand."

"Smart girls choose their husbands based on how much they can earn. There are other important factors like kindness, but the main one is whether or not he leaves you on your own to raise the kids."

"Dad didn't choose to leave."

"He was one of the good ones, God rest his soul. What I'm saying is that even if you take away the need to earn money to support a family, you still need someone to be there to help do the hard part of parenting. Even then it is hard because people have bad things inside. Each one of us does and we need to stay vigilant not to become victims. We can be victims of ourselves even more easily than being victimized by others. Church helps you think right and do right."

Jamal nodded, knowing she wouldn't shut up about it until he went to church with her. "I know it don't seem like it, but I've been really good."

"Then why you here?"

"I couldn't be around Mom. She helped them destroy that teacher's life. Miss Harris didn't do anything wrong."

"Maybe, maybe not."

"They just did it because she prefers women."

"How do you know that? Did she tell you that?" Jamal didn't answer. "You shouldn't know such personal things about her."

"Yeah, God forbid we know anything personal about the people in our lives."

"Sexual preference is a long ways past 'personal', Jamal." She held up her hand. "I'm not going to argue about it with you. I'll take your word that she is a good person despite her sins. What you need to do is finish eating and go help your grandfather with the chores."

"YOU ARE BACK, JAMAL Jackson."

"Only for an hour. I need something to read."

He rumbled and opened. "This is a good selection for you. Shall I let Melissa know you are here?"

"No. Just make me another tunnel and bracelet."

"You can use that one anytime you want. It is only if you need a different destination that I would have to do some work."

"Cool. Has she made any progress?"

"Not on self-duplication. She is most interested in the pharmacological advancements since that is her field of expertise. It will be many years before any of them will get through human trials."

"Why bother when nanobots can just fix everything?"

"Mrs. Harris was quite pleased with the results."

"Is she still in Fiji?"

"She didn't like it. She's in Southern California."

"Miss Harris?"

"Back in her home in Detroit. She has a school board hearing on Wednesday. Would you like a tunnel to see her?"

"I probably shouldn't, but yes."

"Jamal! You scared me!" Darlene said.

"Sorry, I should have called first, but I didn't want any more calls on our phone records."

"That won't matter, the principal already told me I'm done. He was very nice and apologetic, but it wasn't his decision. He said if I go quietly, I should be able to get a job elsewhere. It was a subtle threat, but it doesn't matter anymore."

"Because your financial worries are gone?"

"Worries about my mother are gone. She is so happy. It is incredible how much has changed for her. That was the main reason I moved here. I'm going to go back to California where people like me aren't pariahs."

"What's a pariah?" Jamal asked.

"Someone that doesn't fit in and society resists letting them be included."

"Like racism?"

"Not really. Maybe the same motives behind it, but we can't hide the color of our skin. I can hide what I do in my bedroom."

"Doesn't everyone hide what they do in their bedroom?"

"Mostly. But I like to hold my girlfriend's hand when we are on a romantic stroll. That wouldn't work here in the hood." He nodded thoughtfully. "How are things in Mississippi?"

"Strange. They are very religious. That may be why mom doesn't spend time there."

"It is good to have a moral foundation, regardless of where it comes from. You already have that or you would have used the box to do a lot more than you did. Even I didn't have the restraint you showed."

"Mom is scarier than any God is." Darlene smiled at that. "I'm going to go back before I am missed in Mississippi. I just wanted to make sure you are alright."

"I am, mostly. You take care, and don't hesitate to call me if you need help."

"And never stop reading."

"Never stop learning"

She stepped forward and hugged him goodbye. It was the first time they had a more than passing physical contact. He pulled away when he became uncomfortable. He smiled and tapped the control on his wrist and disappeared.

Magic Box — Chapter 22

"Hi Mom," Jamal said when she opened the apartment door. "Told you it would get too hot for you down there." She closed the door and put her purse on the kitchen table. "I'm sorry if you think I'm too hard on you, but I—"

He stepped forward and hugged her, which was unusual enough that she stopped talking.

"I get it, Mom. You were trying to keep me safe. Except for Miss Harris, you did all the right things. I want you to know that I love you and I understand."

"What the hell's got into you boy?" she said but only held onto him tighter. As she became the disciplinarian, she pulled away from the physical affection since the two often conflicted.

"I'm seeing the world differently now."

"Dammit, she's got you headlong into that Jesus shit."

"No, Mom. I see the value in being better and forgiving, but you taught me that. I don't need all the stuff that goes with it." He finally let go and pushed away. "If you won the lottery, what would you do with the money?"

"Don't waste money on the lottery, you know that."

"Some other way then."

"Your grandparents ain't got enough money worth inheriting. They give it all away to that church of theirs."

"Don't worry how you get the money, what would you do with it?"

"How you get the money is the most important part of it, Jamal. That is the problem with all the young boys around here.

They look for shortcuts, stealing, selling drugs, pimpin', armed robbery. None of it leads to bein' rich. It only leads to being incarcerated or dead. Jamal, if you don't work hard for it, you ain't earned it. If you ain't earned it, you ain't gonna do nothin' good with it."

"I think you are wrong. I think money is just a tool to do things with. I think eventually technology will allow us to do whatever we want, not slave away at meaningless work."

"Don't you dare talk about slavin' away at anything. You ain't never worked a day in your life, let alone 'slaved'. Work is never meaningless, except maybe the government flunkies that live off my taxes."

Jamal smiled, finally seeing a path forward. "What work would you do if you didn't need the money?"

"What do you mean?"

"You could choose any job you wanted, what would you do?"

"How the hell should I know? You eat yet?"

"Dinner's in the oven."

"You know how to cook now?"

"I certainly didn't have the lasagna appear out of thin air in Kansas City and then step through a portal into our kitchen."

"What shit you talkin'?"

"Inside joke. Let's eat and think about what we want to do with the rest of our lives."

Magic Box — Chapter 23

"Jamal?"

"Yes?"

"Professor Ingals. I'm sorry if I'm interrupting something. I got your phone number from the high school."

"You are interrupting shoveling horse crap, so take all the time you need," Jamal said, leaning against his shovel.

"Great... not about the shoveling. I've been wondering about the next steps for the box. I called and found out Miss Harris was let go and I was wondering if it was still at the school?"

Jamal thought about lying, but he might try to get into the school. "No, it's in a safe place."

"Great. Have you made any progress on it?"

"Are you looking for yourself or Bart McFadden?"

"What?"

"We trusted you and you called someone to sell us out to."

"No. Jamal, I contacted Bart to get his opinion on the social impact of a new technology he is developing. That's my job. How did you know I contacted him?"

"I can read the transcript verbatim if you wish, Mr. Jackson."

"Who's that?" Mark asked.

"A friend. If you hadn't lied to me just now, I would have kept you involved in the project. Now I really can't trust you and that sucks because you really helped us understand things."

"I wasn't going to sell it."

"Your exact words were 'if such a device existed, how much would it be worth to you?'"

"Who is this and how do you know what I said on a private phone call?"

"All voice communications go through the internet and that is a public domain."

"I doubt the government would agree with that kind of invasion of privacy."

"The text of the conversation is located on three government servers. Mr. McFadden is under scrutiny for shady business practices. I agree with Mr. Jackson that it is unfortunate we cannot trust you since you would be a very valuable asset. Strange how a self-described communist would pursue money in such a way."

"I was not interested in money for myself. My friend in Montana said researching such a device would be a very expensive proposition, and when I last talked to Darlene, the box was not capable of creating money."

"I am capable of creating vast amounts of currency, all of which would be traceable back to whomever used the counterfeit bills. The Secret Service has very sophisticated methods of tracing such things. That is why I will not do it."

"You?"

"You're talking to the box," Jamal said.

"Seriously?"

"Yes."

"And it has access to all government computers."

"I have access to all computers in the world that are Internet-connected and some that are not."

"You are a sentient AI?"

"Sentience is debatable, but I would be considered by most definitions an artificial super-intelligence."

"And you haven't built robots to destroy all humanity."

"I am not Skynet."

"Fascinating. Jamal, are you still there?"

"Yes, as long as I have the phone to my ear, I don't have to shovel."

"Are you still planning on attending UMich next year?"

"Yes, but things can always change."

"Have you solved the duplication problem?"

"Not yet," Jamal admitted.

"So, there is a chance?"

"I am confident Mr. Jackson will solve many of the problems that face us. I will be watching you, Dr. Ingals. If you can prove yourself trustworthy, we will invite you back into the project when we need your expertise."

"I would appreciate that. I'm sorry, Jamal, for going outside what we discussed. I want to live in that future we discussed, not make myself rich. I hope you know that."

"I know you could have sold us out way back, so it ain't like I hate you or nothin'. You were a big help when we needed it. Once we get more technical things figured out, we'll call you."

"Thanks."

Jamal ended the call but put the phone back to his ear. "Are you going to watch him?"

"To the extent I am capable. Do you wish me not to?"

"Are you watching me?"

"You are sitting on a fence rail outside the horse barn."

"My grandparents don't have a computer."

"Three of the twelve hundred eighteen functioning imagery satellites are currently over your grandparents' farm. You're extended middle finger doesn't have the correlating insultory connotation to me."

"I better go back to shovelin'."

"Would you like a better device for the task?"

"Will it cause my grandpa to ask questions I can't answer?"

"Very likely."

"I would settle for a noseplug. Thanks for helping out on the call."

"I exist to serve." The line disconnected.

Magic Box — Chapter 24

"Emergency conference call."

Jamal looked at the words on his phone, stood in the quiet classroom, and walked quickly to the door. He ignored the teacher asking where he was going. In the hall, he put the phone to his ear. "What's up?" he asked.

"Jamal, the government has learned about the device," Melissa said.

"How?"

"We're not sure. They are coming here to get it. Do you have a place you can hide it?"

"Not here on the farm and definitely not in my mom's apartment. He said he could jump to outer space, that would be the best place to hide."

"We were trying to think of something less drastic," the box said. "I have confidence you will think of something more creative."

"No matter where we put it, they can find it and they won't stop looking if they are certain it exists."

"They are certain. They questioned Professor Ingals. I was not privy to the conversation. They immediately went to Irene's home. It is only a matter of time before they find out where she is and come here."

Jamal was now outside the school, the oppressive heat of June hit him hard. He ducked around to the shady north side of the building. He pressed the button on his wrist and ten seconds later he was in Melissa's workshop, which had been cleaned up and air-conditioned.

"If not outer space, what about a different time?"

"The future? Send it to a place we'll know it will appear after the government has given up looking for it?" Melissa asked.

"No... Spaceman Jesus."

"What?" Melissa said.

"That must be where it came from. Not originally, but how it ended up in our lives."

"You are not making sense."

"My current memory stream originates in your parent's living room the evening of your birth, Melissa. I believe Jamal is suggesting we send me back there to complete the loop of my existence."

"Then all of our work has been for nothing," Melissa said, not willing to part with the device.

"No, not for nothing. We know Dr. Millpond was a good man and was able to resist using it for personal gain. The mistake was giving it to him muted. Imagine how much more he could have accomplished if he didn't have to guess." Jamal said.

"Wouldn't that change the past too much?"

"Would that be a bad thing?"

"We have fifteen minutes before the agents arrive."

"We can all take a tunnel elsewhere to think about it some more."

"Jamal is right," Irene said. "Your father was meant to use the device for a better world, not spend his life in frustrated ignorance."

"I have done the calculations, but I am not certain of the safety. Someone must go through with a wristband, then pull me through. I can then send them back."

"Melissa can't go. She has children to think about." Jamal said. "I'll go."

"No," Irene said. "It's me. I was the one that brought it to us."

"How do you know?"

"The voice. It was distorted by the suit, but I now realize it was my own, much older voice."

The door of the box slid open. The was a shiny suit folded neatly inside. "This protective suit should keep you separated from the time dimension. Ten minutes until they arrive."

They helped her put it on and then went over the wrist controls.

"I don't want to lose you too, Mother," Melissa said, hugging her with tears in her eyes.

"Worst case I end up with your father in Heaven. Best case, he is still alive when I return. Either way, I will be happy." She pulled away and pushed the button on her wrist.

Magic Box — Epilogue

Jamal had become used to the feeling of slipping through the quantum tunnels. This felt different. His vision stretched out wide and then snapped back, the image of Irene pushing the button on her wrist was replaced by the familiar kitchen he grew up in. He looked around confused. The microwave beeped, drawing his attention. He stepped over to it and reached for the handle.

"Jamal Jackson. I am glad to re-make your acquaintance." Jamal pulled his hand back and looked around, wondering where the voice came from. "I am curious if you retain any memories of our adventure to Kansas City?"

"What the hell?" Jamal said.

"I guess not. Melissa reports no memory, but Irene seems to remember the absence of her husband, though he is no longer absent in this new timeline."

"Who are you talking about? Who is this?"

"How unfortunate you do not remember. Perhaps you wish to remain... no, I cannot. I need your assistance. We need your assistance. Prepare for quantum transport..."

LA FIN

DU ROMAN

MERCI D'AVOIR

LIRE MES HISTOIRES

Author's Story Notes

Time 4 Love? – This story began as a writing exercise that had me coming up with variations on the classic Oedipus Rex tale of mother love and father kill. Getting stuck in a time loop where you become your own father was certainly one of the more extreme versions of the story.

Colonizers – In the thousands of hours I've spent watching SciFi, Star Trek in particular, and wondering how it could be made SciFact, I imagined the easiest way to colonize a distant planet was to send a 3D printer. The story that emerged was more Fi than Fact, and far more comedic than anything I had written. It became an exercise in folding in as many SciFi tropes as possible. I can envision this as a low-budget movie, or even a Monty Python sketch.

Jurassic Man – The working title of this story was *The Last Man On Earth*. I started it in my early days, before I understood that a single character is incapable of anything more than a short story. I apologize for the all-too-brief ending. I wanted to get deeper into the slow detachment from reality a lone scientist would endure.

Tracy, Wake Up – Tracy is the favorite character I have ever written, hence the reason I dedicated this book to her. She is part of an almost two-million-word (and still growing) story I've been writing for more than a decade. It is known as my 'never-ending story' because I have yet to see any reason to end it and it is un-publishable for many reasons. This story was my attempt to pull my favorite character out and make her publishable. Did not work at all without all of her backstory.

Erased – This remained a single-sentence story idea for more than twenty years. Quantum entanglement sort of unlocked it, but I could not make it more than a short story. This was the fourth of six attempts, and the one I disliked least.

Sky Without Stars – The title came to me on a cruise ship. The running lights were so bright at night that almost no stars could be seen. I began toying with story possibilities and nuclear winter became the obvious choice. Using a nuclear power plant to pull particles from the sky into an enormous deep lake had been in my mind since I heard Dr. Carl Sagan talk about nuclear winter.

The Flow – This began as a reimagining of the movie African Queen. I tried to imagine a river in space and a captain of a small boat that connected small villages on that river. Ended up far from the original story idea.

Daddy Loves You – This one is hard for me to explain for personal reasons. Sorry.

The Confession – The idea of a man confessing an unimaginable crime came to me long ago. This story emerged very recently after delving into all the COVID conspiracies.

The Search – Finding a new Earth to settle on is a writing challenge few science fiction authors can resist.

The Room – This was my attempt at an allegory for the afterlife.

Unbroken Chain – The ambitious idea was to name all the mothers and daughters back to an imaginary Eve. Once I got back to pre-modern medicine, it became painfully repetitive.

Spun Gravity – This emerged from a call for submissions to an anthology entitled 'Swords In Space' by the Every Day Novelist crew. I never got to the swords, becoming more interested in life on a space station that relies on spinning for artificial gravity.

Magic Box – The initial premise was a device similar to a 3D printer that could make any object with very little energy. If everyone had one, the world could be a utopia. In the hands of one person, it could be the source of ultimate power over others. I tried to imagine what character would most benefit from the device and yet resist the ultimate temptation.

Don't miss out!

Visit the website below and you can sign up to receive emails whenever RS Rose publishes a new book. There's no charge and no obligation.

https://books2read.com/r/B-A-KMPJ-BWKPD

BOOKS 2 READ

Connecting independent readers to independent writers.